Y0-AES-065

This book is printed on 100% recycled paper.

SAUNDERS GOLDEN SERIES IN ENVIRONMENTAL STUDIES

ECOLOGY, POLLUTION, ENVIRONMENT by Amos Turk, Jonathan Turk, and Janet T. Wittes

ENVIRONMENTAL SCIENCE by Amos Turk, Jonathan Turk, Janet T. Wittes and Robert Wittes

ECOSYSTEMS, ENERGY, POPULATION by Jonathan Turk, Janet Wittes, Robert Wittes and Amos Turk

SCIENCE, TECHNOLOGY AND THE ENVIRONMENT by John T. Hardy

IMPACT: SCIENCE ON SOCIETY edited by Robert L. Wolke

SCIENCE, MAN AND SOCIETY (*Second Edition*) by Robert B. Fischer

OUR GEOLOGICAL ENVIRONMENT by Joel S. Watkins, Michael L. Bottino and Marie Morisawa

JOEL S. WATKINS
University of Texas Medical Branch

MICHAEL L. BOTTINO
Brooklyn College, CUNY

MARIE MORISAWA
SUNY at Binghamton

SAUNDERS GOLDEN SERIES

Our Geological Environment

1975

W. B. SAUNDERS COMPANY / Philadelphia / London / Toronto

W. B. Saunders Company: West Washington Square
Philadelphia, PA 19105

12 Dyott Street
London, WC1A 1DB

833 Oxford Street
Toronto, Ontario M8Z 5T9, Canada

Library of Congress Cataloging in Publication Data

Watkins, Joel S 1932–

Our geological environment.

(Saunders golden series in environmental studies)

Bibliography: p.

Includes index.

1. Geology. 2. Environmental protection. I. Bottino, Michael L., joint author. II. Morisawa, Marie, joint author. III. Title.

QE33.W37 550 74–21018

ISBN 0–7216–9133–1

Our Geological Environment ISBN 0-7216-9133-1

Last digit is the print number: 9 8 7 6 5 4 3 2 1

PREFACE

Our Geological Environment is a synthesis of environmental geology courses developed and taught by the authors at the University of North Carolina, Brooklyn College and the State University of New York at Binghamton. We, the authors, have specialized teaching and research interests. We have merged our knowledge derived from these interests in order to present to the beginning student of environmental geology a broad and balanced view. A balanced view is important in any topic, but it is especially important in environmental geology, which encompasses such disparate geologic subdisciplines as geomorphology, geochemistry, geophysics and others. Individually, we could competently discuss only small pieces of the expanse of environmental geology; collectively, we believe we have covered in a competent and scholarly manner the major points desirable in an introductory course.

Our Geological Environment was prepared primarily for use as a text in courses for students with no prior geologic instruction. Terms are explained as they occur in the book, and outside references will be required only by those students wishing to pursue topics beyond the scope of this text. The introductory format of this book should make it suitable as supplemental reading in courses in introductory geology, environmental engineering, environmental biology and other environmentally related courses.

We suggest that those who use this book as a primary text supplement it with readings from other environmentally related disciplines. The human environment is best studied in its totality; to examine geologic, biologic, engineering, or other individual pieces of the environment may lead to serious misconceptions.

The book contains three major sections. The first section discusses geologic processes and their environmental implications. Students having completed courses in physical geology will find considerable repetition in this part of the text. These students should avoid the temptation to skim, however, because they may miss environmental implications of physical processes taking place inside and outside the solid earth.

The second section is devoted to resources. Human history, development of civilization and contemporary division of wealth and power among nations result largely from irregular distribution of natural resources. In our human vanity we often believe that the people of one race or one country are "better" than the people of another race or country. Examination of the data, however, shows distribution of natural resources to be a far more likely explanation of such diversity. An important lesson to be learned from the data presented in the second section is that the rise and fall of civilizations are closely related to technologic discovery and depletion of natural resources, respectively. The United States today faces serious problems of resource allocation. The future of our country and probably of western civilization depends on how we and our global neighbors react to the challenge of growing demand and declining resources.

The third section of the book is perhaps the most important. In this section we show how lack of awareness of geologic constraints imposed on our culture can result in disaster, how positive action can result in improved quality of life and how geologic factors can be considered in planning and management. The material in this section is also the most subject to change because inclusion of geologic considerations in planning and management is a new and rapidly evolving concept. We suggest to our readers that they use this section as a springboard for discussion and that they freely implement the material presented with new material from contemporary magazines and journals.

Finally, we would like to acknowledge the assistance of the many people who have contributed to the preparation of the material in the book. Our geologic colleagues have given generously of their ideas during discussions of environmental problems. They have also provided many of the illustrations that appear in the text. In addition, we wish to thank Gene C. Ulmer, Temple University; David A. Stephenson, University of Wisconsin; John A. Ciciarelli, Pennsylvania State University; and John B. Droste, Indiana University, for their constructive comments. The wife of the senior author, Ms. Billie Watkins, typed much of the manuscript and assisted with editing and revision. Ms. Jane Anepohl assisted with the compilation and editing of illustrations. Most of the drafting was done by Linda Swickheimer. Mr. S. A. Fair assisted with the photography in Chapter 14. The publishing staff of W. B. Saunders Company have been extremely helpful. Ms. Amy Shapiro's assistance was invaluable in the final editing and collation of manuscript material from three geographically separate co-authors.

Last but not least, we are indebted to our students, whose enthusiasm challenged us to undertake this work, whose criticism impelled us to discard simplistic models and search for more realistic answers, and on whose shoulders rests much of the burden of finding solutions to the many environmental problems facing us today. We dedicate this book to our students and their contemporaries as they search for ways in which humankind can live in harmony with the constraints imposed by Our Geological Environment.

J. S. W.

Galveston, Texas
January 8, 1975

CONTENTS

1

INTRODUCTION

The ancient city of Jericho owes its location to springs along the edge of the Dead Sea valley. The modern city is in the center of the picture; archaeological excavations are in the foregound.

Prior to 1973, few Americans had considered the possibility of an immediate shortage of petroleum and petroleum products. The storm clouds which had gathered almost unnoticed on the horizon broke suddenly in the summer of 1973, when gasoline stations began to ration purchases, operate shorter hours and, in some cases, close permanently because of a shortage of gasoline. Gas wars, giveaways and other gimmicks to encourage gas sales stopped. Prices increased suddenly. Commercials appeared in the news media encouraging consumers to limit use of energy in all forms. Consumer groups suspected that the sudden shortage was a result of collusion among large oil companies for purposes of self-enrichment. Government agencies planned immediate hearings. The oil companies claimed that the real reasons for the shortages were excessive taxation, small profit margins and lack of incentives for exploration and drilling. Everyone was unhappy. No one seemed to have a solution.

What happened?

Who was right?

The entire story may take years to unfold, but several facts are clear.

The shortage was not as unexpected as it seemed during the summer of 1973. For several years, data on projected oil supply and consumption were widely available in geologic and government planning circles and showed not only that consumption would overtake supply but that the demand-supply "squeeze" would gradually worsen during the next few decades.

The projections omitted several important factors, however. They did not take into account that much of the remaining supply of petroleum and natural gas lies deep beneath the earth or in rocks of the continental shelves and continental rises. Drilling for oil and gas from these regions is much more expensive than drilling in near-surface continental rocks. The projections were based on total world reserves, but much of the remaining reserves lies outside the political jurisdiction or sphere of influence of the United States. The biggest reserves are in the Near East, an area currently in the throes of political unrest and nationalism. The availability of these supplies is therefore uncertain at best. The recent movement to reduce air pollution resulted in restrictions on high-sulfur petroleum and created shortages in low-sulfur petroleum. Electric companies are now saying that without relaxation of low-sulfur controls, their areas face increasing electric power shortages. The convergence of these factors was largely overlooked by a nation whose attention was focused on an unpopular war abroad and political scandal at home. It was overlooked, at least, until the day when there was no more gas at the pump.

Few aspects of environmental geology have results as dramatic as those of the present energy crisis. Most geologic processes operate so slowly that they go almost unnoticed until they cause some sudden change in our mode of living, which focuses attention on geologic processes or problems.

In this book, we will consider the aspects of geology that are most important to our daily lives. Although this is an introductory text, the treatment of some of these aspects goes beyond the introductory level. We introduce geologic concepts normally found in advanced undergraduate and graduate courses because these processes affect our lives, our children's lives and in some cases the future course of our civilization.

These advanced concepts are explained in a manner which we hope will be understandable to beginning students. We must understand these facets of geology and their interaction with man's activities in order to make intelligent decisions regarding our collective future.

We selected the topics in this book on the basis of our experience and an evaluation of what topics are most relevant to environmental considerations. The selection process was not easy. Much of the difficulty derives from the relative youth of environmental geology as a subdiscipline and the absence of established guidelines. The remainder of this introductory chapter will be devoted to a discussion of the major divisions of the text, our rationale for the divisions and a brief discussion of some aspects of the organization of the book. Our objective is to reveal the framework of the book and to explain to the reader how the topics discussed in individual chapters relate to man.

All geology is environmental. Geology is the study of the earth, and the earth is a major component of our environment. Our cultures, our societies and our civilizations are based on mineral resources taken from the earth, energy obtained from either the earth or the sun and water from the earth and atmosphere. All aspects of geology impinge in some way on our environment. What, then, is "environmental geology"?

We define *environmental geology* as those geologic processes which *interact with man's activities in a significant and observable manner or vice versa.* The best way to clarify this definition is to give several examples.

Natural disasters such as earthquakes and volcanic eruptions in populated areas clearly interact with man's activities. Lives are lost, people are maimed, the economy of the region is disrupted and the life-style of the inhabitants is changed at least temporarily, sometimes causing major cultural upheavals.

In a more subtle manner, slow changes of climate interact with man's activities. As a result of climatic changes, it becomes possible to grow crops in regions formerly too hot or too cold; in other areas drought or cold drives people from their homelands. During the last advance of continental glaciers, sea level dropped in many areas as more and more water was transferred from the seas to the glaciers. This lowering of sea level exposed thousands of square miles of continental shelf to grazing animals and prehistoric hunters, a fact demonstrated by the bones and tools that are occasionally dredged from formerly exposed portions of the continental shelves.

Earthquakes, and volcanic eruptions are for the most part *unidirectional* interactions; that is, they influence man's activities, but the reverse is not true. Man's activities have little effect on these phenomena. In other cases, man's impact on geologic processes has been significant. Man is a geologic agent just as wind and rain are geologic agents. Among the geologic processes that man has affected are weathering of rocks, erosion, mineral distribution and composition of the atmosphere and terrestrial waters. There are many well documented cases of intensified weathering that is the result of increased acid pumped into the atmosphere by man. Each time a new housing development is built, land is stripped of its protective blanket of grass and trees. This stripping results in erosion of thousands of tons of topsoil. In some instances streams are choked with sediment, causing fish to die. In other cases, formerly navigable waters are filled with obstructions. As each development is completed, concrete and asphalt surfaces cover the soil. Rainwater that formerly percolated into the soil now rushes into nearby streams, increasing the hazard of flooding.

Man removes thousands of tons of minerals from the earth each day. He then redistributes these minerals and their derivatives over a wide area. Geologists living a few million years hence will be able to identify the twentieth century easily. An iron-enriched layer of soil will be found throughout the world as a result of the widespread use of iron for containers, construction and implements of all kinds.

The natural redistribution of minerals generally has little impact on long-term geologic process. The most important exception to this rule may be man the geologic agent. We like to consider ourselves a long-term geologic agent, but without mineral resource management our species may become like a shooting star, burning brightly for a short time, then disappearing.

It should be clear from the preceding examples that man is both a passive recipient of the effects of certain geologic processes and an active participant in others. The reason for studying environmental geology is to obtain information that can be used to counteract destructive geologic activity and optimize utilization of resources. Man must use his resources in such a way that he will not harm himself by undue pollution nor will he prematurely exhaust essential resources and deprive succeeding generations of their use.

Optimal interaction of man and his geologic environment requires a knowledge of geologic processes—what causes them, how they work, their various stages of development, their impact on man and his impact on them. The next few paragraphs are devoted to discussion of fundamental geologic concepts, concepts which underlie much of our knowledge of the how, what and when of geology. An appreciation of these ideas will help the reader to understand the more detailed descriptions of particular processes.

Two major concepts are pervasive in geologic thought. These concepts (or theories or laws) were formulated years apart. Each resulted in a revolution in geologic thought and in a far-reaching re-examination of contemporary geologic data. These concepts are *uniformitarianism* and *plate tectonics*.

In principle, one could study environmental geology, which is more concerned with current developments than past processes, without reference to these two concepts, because both uniformitarianism and plate tectonics are long-ranging processes. However, to omit them would deprive the student of a description of the most important foundation blocks of geologic thought. Knowledge of these two laws is essential to an understanding of the interrelationship of important environmental processes, and, most important, such knowledge increases the student's ability to understand predictive aspects of geology, and to answer the question: what results can we expect from continued action of geologic processes? Because these concepts are so essential and pervasive, we will define them at this point.

The first of the two great concepts, *uniformitarianism,* was formulated in 1795 by the pioneer geologist James Hutton. Briefly stated, the law of uniformitarianism says that ancient rocks were formed by the same processes that are acting today. Before the discovery of uniformitarianism, geologists had proposed a number of theories to account for the diversity of rocks outcropping on the surface of the land. These theories ranged from a purely Biblical explanation to the theory that all rocks had been deposited on the floor of the sea and subsequently were raised to their present elevation. To explain the differences between *igneous, metamorphic* and *sedimentary* rocks, the latter theory suggested the igneous rocks were laid down first and were therefore the oldest, metamorphic rocks were deposited on top of the igneous rocks and sedimentary rocks were deposited last. Unfortunately for this theory, several localities were discovered where this sequence was clearly not in effect.

Uniformitarianism was a major breakthrough in geology. Its impact was comparable to the impact of Newton's theory of gravitational attraction on astronomy and physics. As geologists gradually accepted uniformitarianism, they found that they could classify rocks according to their age and environment of formation. It also had far-reaching effects in other areas of science. Darwin's theory of evolution was an outgrowth of uniformitarianism. He reasoned that changes perceptible in contemporary organisms could account for their evolution over a long period of time, just as geologists reasoned that present-day processes could account for past formation of rocks.

The definition of this important concept is often paraphrased: "The present is the key to the past." During the 1960's geologists recognized that uniformitarianism is a two-way street. Not only is the present the key to the past, but the past is the key to the present. Observation of past geologic processes can be useful in predicting the behavior of present-day processes. The earth is a vast library. Its rocks record thousands upon thousands of cycles of erosion, deposition, climatic changes, diversification of some species, extinction of others, earthquakes, volcanic eruptions and formation of rocks and mineral deposits. Data stored in this library can guide us to a better understanding of our earth, its capabilities, its limitations and ways in which we can best *live in harmony with our earth*. The records in this library tell us that if we fail to live in harmony with the earth, our species is destined to a reduced role on earth, or perhaps to extinction, the ultimate fate of species that lose their ability to adapt to the changing demands of their environment. The credo of environmental geology is "*the past is the key to the present—and the future.*" It is only through study and understanding of the past that we will be able to maintain the rich diversity of culture, art and other worthwhile aspects of modern civilization.

The second important geologic theory is *plate tectonics*, which was discovered only recently. Its discovery derives from studies of continental drift, a theory advocated most actively in the early part of the twentieth century by the pioneer German geophysicist Alfred Wegener. In its modern form, the theory of plate tectonics dates from the early 1960's, when an American geologist, Harry Hess, proposed that new sea floor was being formed along the axes of the mid-ocean ridges as the continents moved apart. This aspect of plate tectonics is usually referred to as *sea floor spreading*. Because the total surface area of the earth remains fairly constant, for every newly formed segment of sea floor an equal amount of crust must disappear somewhere. Most of the crust disappears into and beneath deep ocean trenches. This process is called *crustal convergence*.

It was discovered that the earth is composed of large plates bounded by mid-ocean ridges, zones of convergence and long, linear *faults* (tears or rips in the earth's crust). Subsequent investigation revealed that plate tectonics is important in many ways other than continental drift. These investigations have shown that plate movement controls locations of most large earthquakes, most volcanic activity and emplacement of many mineral deposits. Plate tectonics indirectly affects virtually every geologic process. It determines the location of mountain ranges, thereby affecting climate, rainfall, erosion, sedimentation and even distribution of faults. The great petroleum accumulations of the Persian Gulf are forming in a trough created by the crustal convergence of Arabia on the west and Iran on the east. Repeated movements of African and European blocks have periodically restricted inflow of water from the Atlantic Ocean to the Mediterranean Sea, causing salt precipitation. A similar process probably contributed to the formation of the great salt province that rims much of the Gulf of Mexico.

Plate tectonics is a great unifying concept in modern geology. An appreciation of the mechanics of plate tectonics will assist the reader in understanding the basic mechanisms of many geologic processes important to man. For this reason, this theory and its implications are discussed at some length in following chapters.

Let us now discuss the organization of this book and the objectives of a program of study of environmental geology.

This book is divided into three major sections. The first deals with geologic processes, the second with resources and the third with policy and planning.

Students with previous exposure to geology will find much familiar material in the first section, Earth Processes. In that section we discuss earthquakes, plate tectonics, seismology, volcanoes, igneous rocks, the atmosphere, the ocean, glaciation, weathering, erosion, deposition and sedimentary rocks. These topics are generally included in elementary geology texts. However, several of the discussions in this book go considerably beyond the scope of elementary texts. For example, volcanism and seismology are examined in depth, because predictions of earthquakes and volcanic eruptions are important to our environment.

The second section is devoted to Earth Resources. This section is especially relevant because of the growing shortage of energy and potential shortages of other resources. We include a discussion of water in this section because water is probably the single most important of man's natural resources, although its importance is seldom recognized. Water is in short supply in the United States. Only in the southeast and northeast are water supplies less than fully utilized. A nationwide water shortage may soon follow the petroleum shortage. With adequate foresight and planning, the impact of the impending shortage can be minimized.

The third section is devoted to a discussion of Policy and Planning. This is the most important section in the book; in it, we attempt to show what can be done to fully utilize geology in our everyday lives. A description of what happened when geologic, economic and social considerations were not integrated in rural West Virginia emphasizes the importance of federal, state and local governments working in a unified manner to achieve geologically sound goals. Finally we discuss the responsibilities and the role of the geologist in environmental planning. This latter aspect is especially important, because most people are unaware of the capabilities—and limitations—of the environmental geologist.

We based the preparation of each chapter in the first two sections on the philosophy that we should explain as clearly as possible the present state of knowledge about geologic processes. We realized, however, that this objective alone is not sufficient in a course as broad as environmental geology. Therefore, we have included within each chapter a discussion of the relationship of geological factors to contemporary culture, society and economy.

These discussions are subject to the unavoidable limitation of every study of current events: the examples cited become out of date from the day the manuscript leaves the author's hands. The discussion sections also suffer because much more is known of the geologic aspect of most problems than of the human side of the problem; man's interaction with geologic processes has only recently begun to receive the attention it deserves. We urge the reader to use the discussion sections in conjunction with reports in contemporary news media. Hardly an issue of a major news magazine appears without an article related in some way to a geologic-social-cultural-economic problem. In this way, the reader can overcome the limitations imposed by preparation and publication delays and can derive an appreciation of current problems.

Taken as a whole, the three sections of the book will provide the reader with a balanced view of the mutual interaction of man and geology. We hope that the information presented in this text, when viewed against a background of contemporary events, will provide the reader with a useful guide for the evaluation of different strategies in man's attempt to live in harmony with his environment.

Unit I

EARTH PROCESSES

2

EARTHQUAKES AND TECTONISM

Geologist examining a recent fault scarp. Shifting of large blocks of the earth's crust caused the fault.

Snow was falling lightly in southern Alaska late in the afternoon of March 27, 1964. Schools were closed because it was Good Friday. By 5:30 PM, businesses were closing for the day. Many people were at home as the evening darkness approached.

Suddenly, at 5:36 PM, an immense section of the earth lurched northward, sliding a few tens of feet into the ground beneath southern Alaska. Fast-traveling waves of vibration radiated outward from the zone of slippage, violently rocking and jarring half of Alaska. The most violent earthquake to occur in North America in the twentieth century had just taken place.

The point of initial movement, or the *focus* of the earthquake, was deep within the earth beneath the Chugach Mountains (Fig. 2–1). Fortunately, the nearest large city, Anchorage, was 90 miles away. Fortunately also, much of the area surrounding the focus was sparsely settled. The death toll of 114 would have been far greater had the earthquake occurred in a more densely populated region. Property damage was esti-

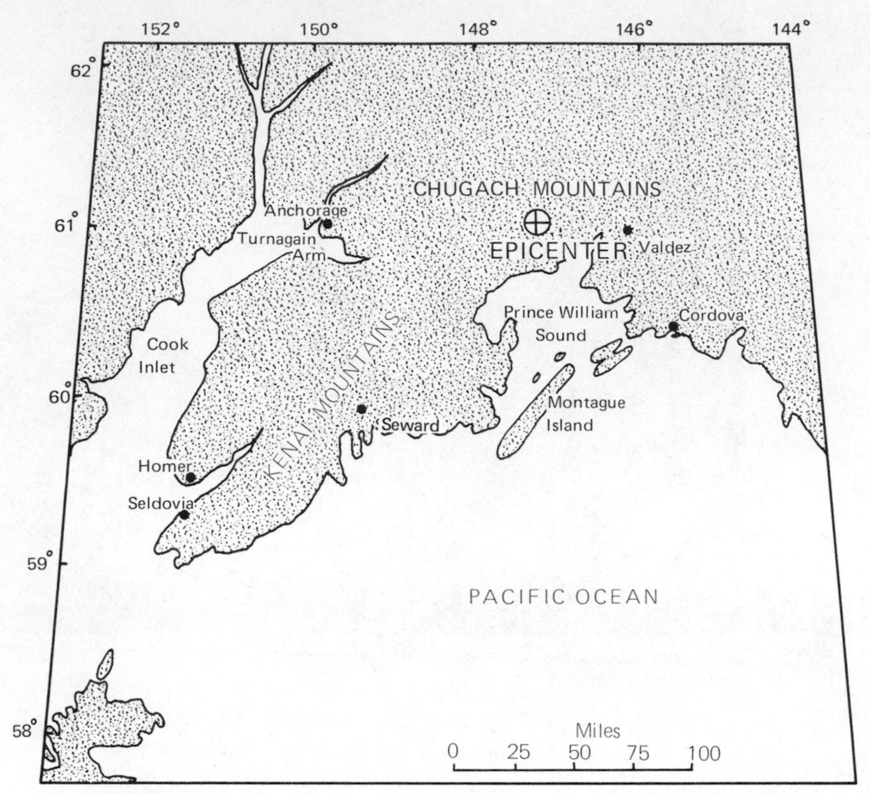

Figure 2–1. Area of the Good Friday Earthquake, Alaska, 1964. (Adapted from Hansen, W. R., et al.: U.S. Geological Survey Prof. Paper 541, 1966.)

mated to be in excess of $300 million. An earthquake of equal magnitude occurring in a densely populated area of California, for example, would have resulted in thousands of deaths and billions of dollars of property damage.

As a cause of deaths, earthquakes rank high among natural disasters. An estimated 830,000 people were killed as a result of the Shenshi, China, earthquake in 1556; approximately 143,000 died in the Tokyo earthquake in 1923; and about 160,000 died in 1908 due to the earthquake in Messina, Italy. The largest earthquake ever to occur near a major United States population center was the 1906 San Francisco earthquake, in which approximately 500 people were killed.

In this chapter we will examine earthquakes from four points of view. We will review what is known about the *origin of earthquakes,* we will discuss the *physics of earthquake waves,* we will study *four individual earthquakes* and we will discuss progress in man's attempts to *live with earthquakes.*

1 ORIGIN OF EARTHQUAKES

Sea Floor Spreading

Almost everyone who has looked at a world map will have noted that the east coast of South America bulges opposite a recess in the west coast of Africa, and that the west coast of Africa bulges opposite a recess in the east coasts of North and South America. Seen on a globe, the "fit" of the coastlines on opposite sides of the Atlantic is even more remarkable. A still closer correlation exists between edges of continental shelves (see Fig. 2–2).

In the late nineteenth century, the Austrian geologist Edward Suess "pushed" all of the continents together to form a single land mass, which he called Gondwanaland (after Gondwana, a geologic province in India). In the early twentieth century, a German geophysicist named Alfred Wegener, after collecting all the evidence available to him, put forth the comprehensive theory of continental drift. Wegener's theory generated great excitement and controversy among geologists and *geophysicists* (geologists and/or physicists who study the physics of the earth, e.g., magnetism, gravity, earthquakes and so forth). Unfortunately for Wegener, the mechanism that he proposed to account for the drifting of continents was proven inadequate, and geologists were able to demonstrate that some of the evidence he presented in favor of continental drift could logically result from other mechanisms.

Geologists in the southern hemisphere, and especially in South Africa, where they were led by A.L. Du Toit, were among the first to accept the theory of continental drift. Much of southern Africa and parts of South America, Australia and India are covered with glacial deposits formed during an ice age which occurred about 200 million years ago. Du Toit and his colleagues realized that the distribution of the ancient glacial deposits could be explained if at that time Africa, South Africa, India and Australia had been contiguous parts of a great southern continent.

Although African geologists presented abundant evidence of drift, general acceptance of the theory was delayed for many years due to lack of a satisfactory explanatory mechanism. It was not until 1960 that real progress was made in the understanding of the mech-

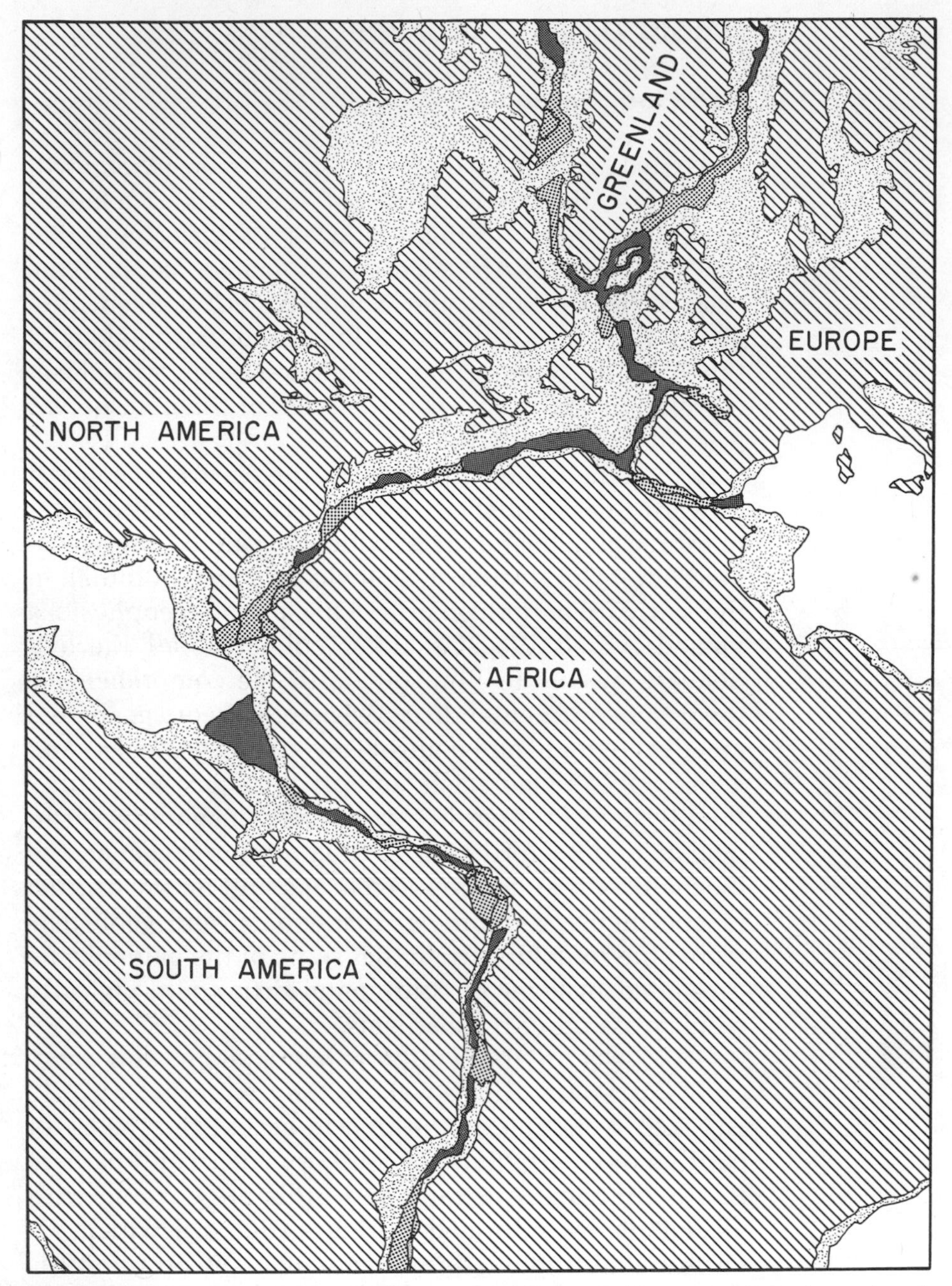

Figure 2–2. Possible predrift arrangement of continents. Subsequent deformation and sedimentation obscure the original configuration of the rift, resulting in small gaps and overlaps. (Adapted from Bullard, E. C., et al.: Phil. Trans. R. Soc. London, VA 258, 1965, p. 41.)

anism of continental drift. In that year, the American geologist H.H. Hess suggested that a kind of convective movement might cause material from deep within the earth to well up beneath axes of the mid-ocean ridges and form new sea floor. The sea floor would spread laterally away from the axes of the ridges and eventually disappear into trenches along continental margins. Robert Dietz coined the phrase "*sea floor spreading*" to describe the proposed mechanism.

At about the same time, scientists from Scripps Institution of Oceanography discovered that the earth's magnetic field off the west coast of North America was "striped"; that is, it consisted of elongated, more or less parallel zones of alternating high and low intensity. (Actually, a more accurate analogy than stripes might be waves on a body of deep water, because the magnitude of the stripes is much smaller than the magnitude of the main field. Nevertheless, the "striped" or wavy character is quite pronounced if the magnitude of the main field is neglected.)

At Cambridge University in England, F.J. Vine and D.H. Matthews had an idea about how magnetic stripes and continental drift might be related. They thought that magnetic stripes might result from changes in the direction of the magnetic field during sea floor spreading, and that new basaltic crust forming and cooling today along the axis of the Mid-Atlantic Ridge would be normally magnetized. As a result, the earth's total magnetic field over the axis of the ridge would be slightly greater than normal because the magnetization of the rocks of the ridge would be added to that of the main field (which is generated deep inside the earth) and increase the total field.

Geologic investigations of magnetized rocks suggested that rocks formed along the ridge between 700,000 and 1.5 million years ago should have been magnetized in the opposite direction from rocks currently being formed. The magnetization of these ancient rocks would decrease the total field slightly because their polarity is opposite to polarity of the present field (like polarities are additive; unlike polarities are subtractive). According to Hess's Sea Floor Spreading theory, rocks formed 700,000 to 1.5 million years ago have now moved away from the ridge. The magnetic field in the vicinity of the ridge consists of a strip of relatively higher intensity over the ridge and for some distance to either side. The stripe of high intensity coincides with the belt of sea floor formed during the past 700,000 years. On either side of the high stripe should be low stripes coinciding with belts of sea floor formed 700,000 to 1.5 million years ago. According to theory, this

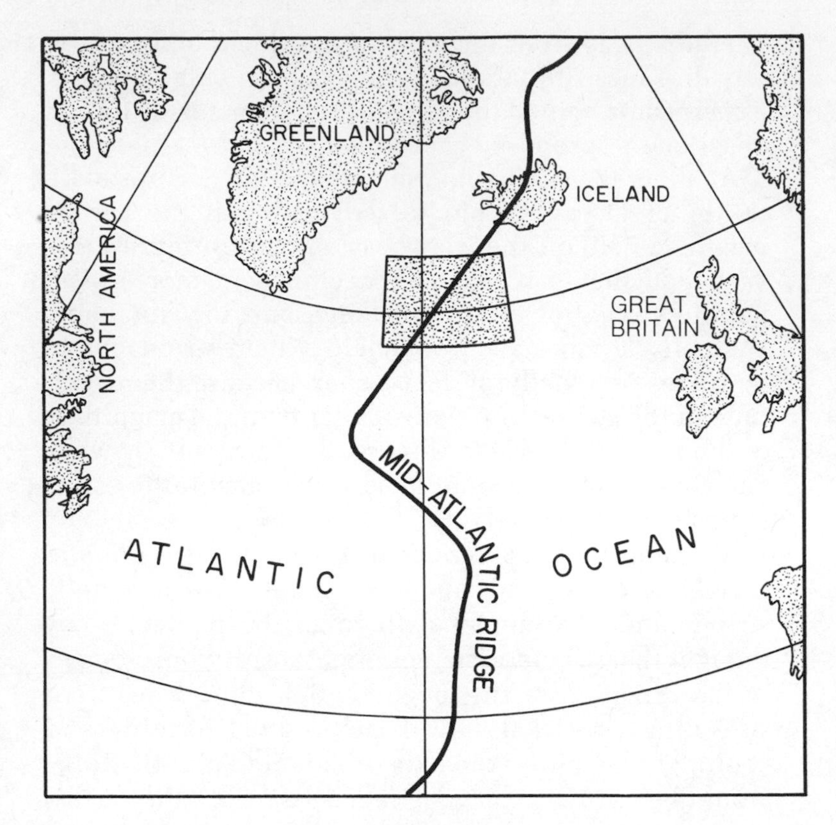

Figure 2–3. Area of the Mid-Atlantic Ridge Survey. (Adapted from Cox, A., et al.: Reversals of the earth's magnetic field. Sci. Am. *216*:44–61, 1967.)

striping effect should continue outward. The width of the stripes would depend on the rate of spreading and changes in the earth's magnetic field intensity and direction.

In 1963, Vine and Matthews were able to test their hypothesis. A United States Navy airplane flew to Iceland, which is located on the Mid-Atlantic Ridge. From there, it flew a grid pattern of an area south of Iceland. The airplane flew one direction for a time, then, at a predetermined point, it turned 90 degrees, flew a few miles, turned 90 degrees again, and flew back in the opposite direction. The airplane continued flying back and forth until it had systematically covered the area shown in Figure 2–3.

As the airplane crossed and recrossed the area, instruments on board recorded variations in the intensity of the earth's magnetic field. Results of the magnetic survey and other tests of the Sea Floor Spreading theory were soon to bring about a revolution in geologic thinking and lead to a greater understanding of the distribution of earthquakes and of the basic geologic forces causing them.

Many subdisciplines of geology contributed to the development and acceptance of the theory of sea floor spreading. However, contributions of two areas of research, *paleomagnetism* (variations in the earth's magnetic field during the geologic past) and *seismology* (study of earthquakes and related phenomena), were especially important. Data collected by the Navy airplane and information about the earth's past magnetic field that had been stored in rocks for thousands of years furnished two important pieces of evidence for the continental drift theory. These two pieces not only fit together perfectly but also solved an important part of the overall puzzle. With this new information to guide them, scientists quickly uncovered other vital data and in a short time filled in much of the remainder of the puzzle.

Paleomagnetism

In the 1950's and early 1960's, scientists studying *paleomagnetism* (paleo- = Greek *ancient*; thus paleomagnetists study ancient variations in the earth's magnetic field) made two discoveries that provided the key to solve the enigma of continental drift. They discovered that the earth's magnetic pole had drifted from place to place with respect to the continents, and that the poles had frequently (speaking geologically, every few hundred thousand years) reversed positions; that is, the North Pole became the South Pole and vice versa.

It had been known for some time that most rocks contain small but significant amounts of magnetic minerals. Such minerals act as if they contain small magnetized needles. When rocks containing magnetic minerals are heated above a certain temperature, known as the *Curie temperature* of the minerals, the minerals (and the rocks) lose their magnetism. When the rocks cool below the Curie temperatures of their magnetic minerals, the minerals reacquire their magnetism. The "needles" align themselves parallel to the earth's magnetic field, and remain frozen in that orientation at lower temperatures. The previous magnetic orientation is destroyed, and the strength of the new magnetization is proportional to the strength of the earth's field at the time of cooling.

Each year, somewhere on earth, magnetic minerals cool through their Curie temperatures and record for posterity the direction and intensity of the present-day magnetic field. These and other rocks which have cooled in earlier geologic ages collectively make up a sort of paleomagnetic tape. Geophysicists search out pieces of the tape scattered throughout the world, date the rocks and reconstruct the earth's ancient magnetic (paleomagnetic) field.

Let us examine the ways in which paleomagnetists applied these data to the theory of continental drift. If you suspend a balanced, magnetized needle on a thin thread, the north-seeking pole of the needle will do two things. It will swing around and point toward the north magnetic pole *and,* if you are in the northern hemisphere, it will dip or *point downward*. The needle will point *vertically* downward when located over the *north magnetic* pole. At the magnetic equator, the needle will be horizontal. At points between pole and equator, *inclination* (amount of dip) of the needle systematically increases as the needle approaches the north magnetic pole. In the southern magnetic hemisphere, inclination of the south-seeking pole of the needle increases from equator to pole.

Paleomagnetists use variations of inclination recorded in ancient rocks to discover locations of ancient poles. From inclination data, they first discovered that the magnetic poles had drifted about slowly during geologic time. They found that 600 million years ago, that relative to North America the North Pole was located where the central Pacific Ocean is now. The pole drifted north by northwest into the area now occupied by eastern Siberia, then turned northeast and drifted along a big looping path to its present position. The most exciting discovery, however, was that North American and Eurasian paleopole positions differed systematically, and that these differences were exactly what would be expected if North America had broken from Europe 200 million years ago and drifted slowly westward since the breakup.

The lock in the door was beginning to turn.

The paleopole data, however, were sparse and scattered. Rocks comprising the paleomagnetic tape recorder had been tilted, folded and broken, making it difficult to measure many inclinations precisely. Some

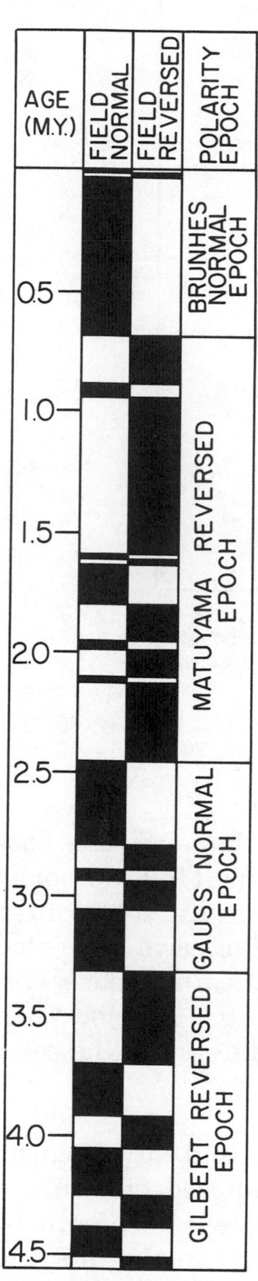

Figure 2–4. Time scale of reversals of the earth's magnetic field. (Adapted from Cox, A.: Geomagnetic reversals. Science *163*:237, 1969.)

rocks had been remagnetized and gave pole positions of the time of remagnetization. Rocks that had been struck by lightning yielded widely improbable pole positions. So most of the geoscience community, although watching intently, withheld final acceptance of continental drift.

The second discovery, the one that would finally lead to an understanding and much wider spread acceptance of continental drift, occurred soon after the first one. This discovery was that paleopole positions determined from approximately half the rocks less than five million years old were oriented "normally"; that is, their north poles pointed in a northerly direction, and their south poles pointed in a southerly direction. Magnetic poles of the other half were "reversed"; the north-seeking poles of the needles pointed south and the south-seeking poles pointed north. When directions of polarity were plotted as a function of age (Fig. 2–4), almost all rocks formed within the past 700,000 years were found to be polarized normally. Most rocks between 700,000 and about 1.5 million years old were reversely polarized. Data from rocks older than 1.5 million years suggested alternating periods of normal and reversed polarization. It was as if the earth's magnetic north and south poles were exchanging positions from time to time.

The enigma was almost solved. It was at this point that Vine and Matthews conducted their Mid-Atlantic Ridge magnetic survey, the results of which are shown in Figure 2–5. The magnetic anomalies closely resemble the predicted pattern.

Discoveries following Vine and Matthews's experiment have shown that the outer shell of the earth consists of six major plates and several small plates. These plates, shown in Figure 2–6, are commonly bounded by 1) *mid-ocean ridges;* 2) *subduction zones* (zones where crust is being carried downward into the earth's interior); or 3) *transcurrent faults* (fractures separating large crustal blocks sliding by one another). *Plate boundaries are especially important to the study of earthquakes because most of the world's earthquakes occur as a result of plates sliding by one another.*

The theory that encompasses all aspects of plate motion, including associated earthquakes and sea floor spreading, is called *plate tectonics.*

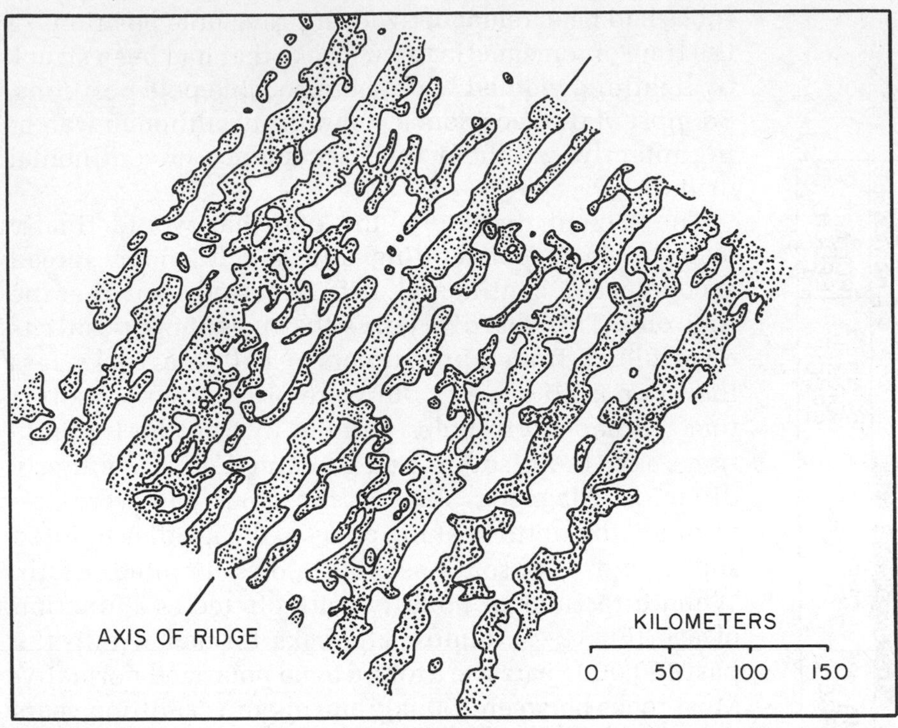

Figure 2–5. Magnetic anomaly pattern found near the Mid-Atlantic Ridge. Dark belts are zones of positive anomalies produced by normally polarized rocks; white bands are negative anomalies produced by zones of reversely polarized rocks. (Adapted from Vine, F. J. Science *154*:1407, 1966.)

Earthquake Data

In order to see how data from earthquakes have contributed to plate tectonic theory, we will follow a somewhat devious route. We will first review some of the major facts about the interior of the earth, provided mainly by study of the behavior of earthquake waves which have traveled through the earth's interior. Then we will see how earthquakes, the interior of the earth and plate tectonics are related.

The Lithosphere, Asthenosphere and Low Velocity Zone. The earth's interior is composed of a core, a mantle and a crust (Fig. 2–7). The core is the innermost zone. Observations of seismic waves crossing the core have shown that it consists of two parts: a solid inner core and a liquid outer core. (Differential movements of the liquid outer core, the solid inner core and the surrounding mantle are probably responsible for the earth's magnetic field.) The mantle comprises roughly the outer half of the earth. Mantle rocks are relatively

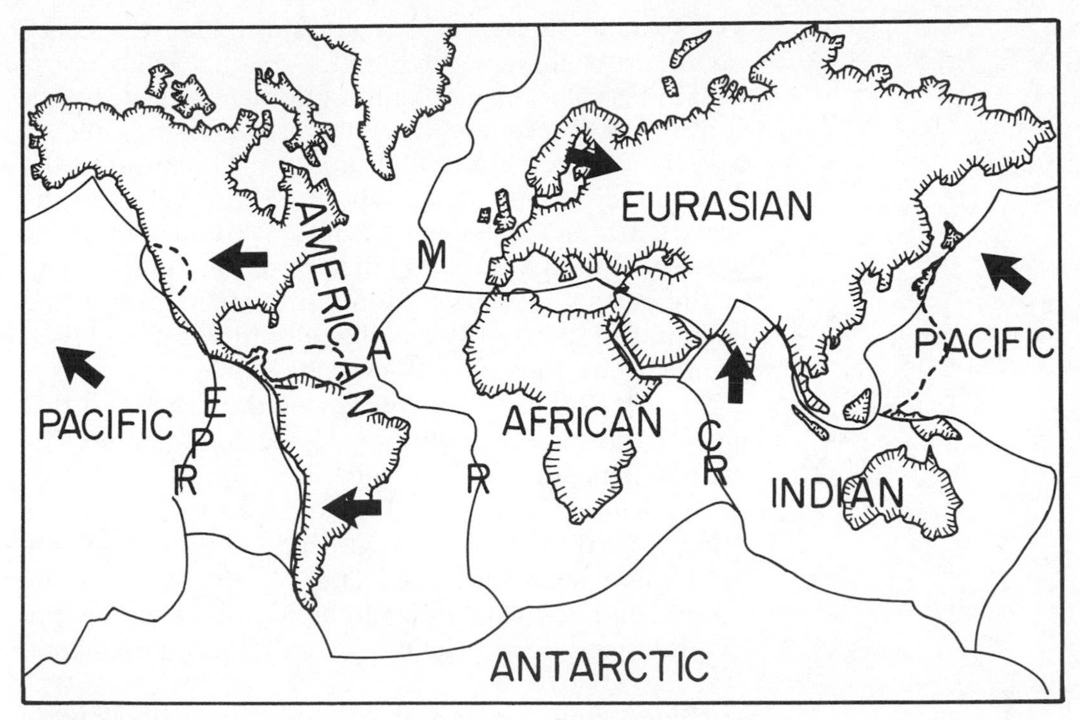

Figure 2–6. The six major plates of the earth's crust and their directions of drift. (EPR, MAR and CR indicate East Pacific Ridge, Mid-Atlantic Ridge and Carlsberg Ridge, respectively.)

rich in iron and magnesium and poor in silica in comparison with surface rocks.

Seismic wave propagation characteristics indicate that the upper part of the mantle includes a plastic zone, which usually lies between 100 and 200 kilometers beneath the earth's surface. In the plastic zone, mantle rocks are not hard, as they appear to be above and below it. The plastic zone is also called the *low velocity zone* because seismic waves travel through it more slowly than through rocks above and below it.

The crust is a relatively thin layer of rock covering the earth. There are two distinct types of crust. Beneath the sea floors, the *oceanic crust* is composed of

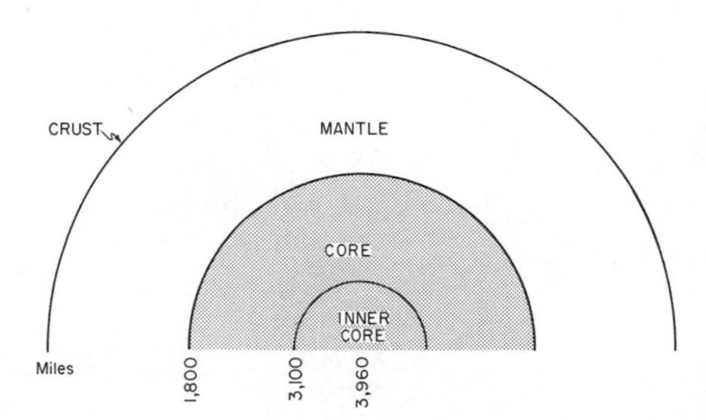

Figure 2–7. Interior of the earth. The crust appears as a line at this scale.

rocks richer in iron and magnesium than the *continental crust* but less rich than the mantle. Thickness of oceanic crust averages about 6 kilometers. Beneath the continents, rocks are more heterogeneous, and are usually rich in silica and aluminum. Thickness of the continental crust averages about 45 kilometers in the eastern United States and about 35 kilometers in the western United States. The history of the development of the earth's crust is a fascinating story, one that is beyond the scope of this book but which can be found in most introductory geology texts.

The *lithosphere* is composed of the crust and that part of the mantle above the low velocity zone. The mantle beneath the low velocity zone is called the *asthenosphere*. The entire lithosphere appears to move during sea floor spreading, gliding over the relatively motionless asthenosphere. The plastic rocks of the intervening low velocity zone yield and seem to serve as a lubricant during the movement of the lithosphere.

Ridges and Spreading Axes. The lithospheric plate system includes three major mid-ocean ridge systems: the Mid-Atlantic Ridge, Carlsberg Ridge and East Pacific Ridge (Fig. 2–6). New lithosphere upwells continuously along the axes of these spreading zones. As near-surface rocks break apart to accommodate upwelling liquid rock, earthquakes occur. Most of these earthquakes take place far from inhabited areas. Although they are seldom felt, they are recorded on *seismographs* (sensitive intruments which record ground tremors).

Mid-ocean ridge systems are generally located far from land. In certain regions, however, the ridges rise to the surface, as in Iceland, where the ridge is well exposed. In Iceland, as elsewhere, the crest of the ridge is marked by a well-defined trough, where spreading occurs as lava wells up into it.

The East Pacific Ridge trends northward through the eastern Pacific, intersecting North America in a complex manner. The Gulf of California is probably a surface expression of the spreading axis of the rise. Some geologists have suggested that at some time in the past, North America overran the East Pacific Ridge, and that the spreading axis now extends northward through Utah. In support of this theory, these geologists point to a linear north-south trending zone of the earthquakes in Utah and to the mountains of much of the southwestern United States. The geologic structure of many southwestern mountains suggests that they were formed by spreading of the crust.

Spreading axes are not limited to mid-ocean ridges. The rift valleys of east Africa are a well-known example of spreading axes crossing a continent. It is possible that east Africa may have started to break up millions of years ago. The present spreading rate, however, is negligible in comparison with spreading rates of Atlantic and Pacific spreading axes.

Trenches and Subduction Zones. To paraphrase an old cliche, what comes up must go down. This is true of the earth's crust. As new lithosphere is formed along axes of mid-ocean ridges, somewhere on earth, old lithosphere must return to the depths. This process is called *subduction,* and the zone containing the sinking lithosphere is called a *subduction zone.* For the most part, subduction takes places beneath deep ocean trenches.

Most lithospheric subduction occurs around the margins of the Pacific Ocean. A deep trench, the Peru Trench, lies off the west coast of South America, where the South American continent is overriding lithosphere of the Pacific plate; another trench parallels southern Alaska and the Aleutian Islands; and a series of trenches extend southward along the seaward margins of Siberia, Japan, the Philippines and other areas of the Pacific. The major trench of the Indian Ocean lies along the southern margin of Indonesia. Lithosphere of the Indian plate is being subducted beneath Indonesia. The Puerto Rican trench is seaward of the Antilles, and a small trench occurs east of the Falkland Islands in the South Atlantic. These Atlantic trenches are small, and subduction is slow or inactive.

Not all subduction consists of oceanic plates subducting beneath continental margins. The Indian subcontinent, originally attached to Africa and Australia, drifted northward and collided with Asia. Before India and Asia collided, there was probably a trench along the southern margin of Asia. The oceanic crust and lithosphere between India and the Asian mainland disappeared into the maw of the trench, but when the lithosphere containing the Indian continental crust began to go down the subduction zone, the situation changed. Continental crust is much lighter than oceanic crust; it tends to "float." As the Indian continental crust slipped under the Asian crust, the buoyant Indian crust pushed upward on the overlying Asian crust, raising the Himalayas to their great heights. At the same time, the force of the Indian crust pressing against the Asian crust folded and faulted the rocks of both margins.

High levels of vibration are caused by the large amounts of slippage associated with subduction. The pressure of the overriding block often prevents slippage until *strain* (deformation) is quite pronounced. When the break finally occurs, it can extend for long distances and generate large earthquake waves. The Alaskan Good Friday earthquake is an example of a large earthquake in a subduction zone.

Transcurrent Faults. In some areas, plates do not spread or subduct but glide past one another. Two prominent examples are depicted in Figure 2–6:

1) the northwestern edge of the Indian plate where the Indian plate adjoins the Asian plate, and
2) the coast of California where the Pacific plate (including a slice of California) is sliding northward relative to the remainder of the American plate.

The San Andreas fault is the principal fracture zone separating Pacific and American plates. Movement along the San Andreas and associated faults has been responsible for many severe earthquakes, including the destructive San Francisco earthquake of 1906.

Taken as a whole, earthquake intensity associated with transcurrent faults is greater than the average intensity along spreading axes but less than the average intensity along active subduction zones. This is probably due to lower levels of pressure pushing the crustal blocks together.

A Few Additional Remarks. We have seen how earthquake wave data revealed the existence of the low velocity zone. Direction of fault motion indicated by earthquake waves confirmed that blocks behind trenches are overriding the blocks sliding beneath the trenches. Earthquake foci beneath deep ocean trenches lie in a narrow zone called a Benioff zone (named after their discoverer, Hugo Benioff), which marks the plate boundary. All of these discoveries were made over a period of several decades. Plate tectonic theory provides a connecting framework for them and for a number of facts that previously appeared unrelated. It explains the distribution of earthquakes, and while it is clear that plate motion is the driving force that produces most earthquakes, many details remain poorly understood. It seems likely that as we fit these pieces into our jigsaw puzzle, we will further improve our understanding of the mechanism of earthquakes.

Other Causes of Earthquakes

Some earthquakes are indirectly related to plate tectonics; others are not related at all. Let us examine the different causes of earthquakes.

Intraplate and Interplate Earthquakes

A significant number of earthquakes are caused by the shifting of large masses of rock within plates. These *intraplate* (within plate boundaries) earthquakes tend to be smaller than *interplate* (along plate boundaries) earthquakes, but they can on occasion be as large and destructive as interplate earthquakes as, for example, the New Madrid, Missouri, earthquake of 1812. The basic causes of intraplate earthquakes are not well understood. In some instances, erosion may have removed varying amounts of rock from the earth's surface. Buoyant continental crust rises more in areas of greater erosion, and the resulting strain may be sufficient to cause earthquakes when these areas of deep erosion are close to areas of little erosion.

Other earthquakes may be caused by poorly understood variations in temperature within the mantle. Many earthquakes in the eastern United States seem to be "fossil" earthquakes, which are due to reduced pressure on rocks that were deformed during the formation of the Appalachian Mountains hundreds of millions of years ago. Although the rocks were deformed at the time of formation of the mountain chain, the pressure of the overlying rocks was sufficient to prevent motion. Erosion removed part of the overlying rock, which reduced the pressure and allowed the rocks to move. Several old fault zones in the Appalachian Mountains may have been reactivated in this manner. Earthquakes associated with these older faults are seldom destructive, however.

Volcanic Eruptions

Volcanic eruptions occur when *magma* (molten rock below the ground; magma becomes *lava* when it flows onto the surface of the ground) forces its way upward through cracks in the mantle and crust until it reaches the surface. As the magma moves, rocks shift underground and generate earthquakes. Volcanic earthquakes interest geologists because it is sometimes possible to trace underground movement of the molten rock by observing changes in location of earthquake activity.

Violently exposive eruptions may generate damaging earthquakes. The ninteenth-century eruption of the Indonesian volcano, Krakatoa, created a seismic sea wave that caused much damage and was responsible for many deaths from drowning.

Slumps and Slides

Occasionally, pieces of hillsides break away and slide downward with force sufficient to generate small earthquakes. Although the slides themselves may be destructive, earthquakes caused by the slides usually are not damaging. Large submarine landslides may cause earthquakes large enough to startle persons on the nearby shore.

Man

Man's activities probably have not been a direct cause of any earthquakes, but they have triggered earthquakes in several areas. Construction of large dams and the subsequent loading of the crust by large quantities of water impounded by the dam has caused small earthquakes. Only a few of these have been damaging, however. Injection of fluids under high pressure into oil wells routinely causes mini-earthquakes. This process, called hydrofracting, is used to crack rocks containing oil and gas. The cracks permit oil to flow more easily from the rock into the well, and may convert an unprofitable well into a profitable one. Small earthquakes accompany the cracking process. Injection of fluids by the Army into a well near Denver inadvertently triggered a series of small earthquakes that caused minor damage in Denver and vicinity (see later in this chapter).

2 SEISMOLOGY

As rocks along plate margins or in plate interiors slowly bend, energy is stored in the same way as it is stored in a stretched rubber band. When the rubber band breaks or the earth shifts, energy is released. Movement of the rock dissipates most of the energy; only a small percentage is released in the form of vibrations, which we call seismic waves. The seismic waves radiate from the focus or point of origin of the earthquake in somewhat the same way that waves radiate when a rock is dropped into water. An earthquake is defined as the shaking that takes place when seismic waves travel through an area.

Seismology is the study of earthquakes and earthquake-related phenomena, including the way in which *stress* (the forces or pressures acting on rocks) deforms or *strains* rocks until they break; the velocity and character of seismic waves (knowledge of which contributes to an understanding of the source and nature of the rocks traversed by the waves); duration and intensity of earthquakes; damage resulting directly from the earth itself (damage due to vibration, for example); and damage resulting directly from the earthquake, such as damage due to landslides. In this section we will review highlights of all these aspects of seismology.

Seismic Waves

Earlier, we compared seismic waves with waves radiating away from the point of impact of a rock dropped into a body of water such as a pond. Actually, this comparison is overly simplistic. In a pond, a single *wave train* forms when the rock is dropped; but three, four or more wave trains radiate from an earthquake focus, each with a different velocity and different vibration characteristics. Each type of wave train plays a specific role in the story of earthquake damage, so let us briefly review the major wave types and their characteristics.

P- and S-Waves

When early seismologists examined recordings of vibrations of distant earthquakes, they frequently found that the amplitude of the *first-arriving* wave train was relatively low. They called these first-arriving waves "primary," and marked the onset of the primary waves with a "P" on the seismogram (Fig. 2–8). A short time after the onset of the P-waves, a sudden burst of larger amplitude waves appeared. These waves were called "secondary," and their onset is indicated by an "S" on the seismogram. P- and S-waves, which rise and fall rapidly, were usually followed by a long train of high amplitude waves. This last group of waves, which rose and fell more slowly than P- and S-waves, were called "long" waves and marked with an "L."

S-waves cause much of the earthquake-related damage. There are two reasons for this. First, S-waves have higher amplitudes than P-waves, and second, they have a *horizontal component of vibration*. Let us more closely examine the phenomenon of wave motion and try to understand this important factor in earthquake damage.

The P-wave is a compressional-dilational wave. This means that the wave pattern consists of alternating zones in which the host media are compressed and dilated. P-wave motion is sometimes called "push-pull" motion, because particles within the compressed zone are being pushed closer together while particles in the dilational zone are being pulled apart. A simple experiment using a coiled spring such as a "Slinky" toy illustrates compressional-dilational wave motion.

Hold an end of the coiled spring in one hand, allowing the spring to hang vertically (Fig. 2–9*A*). The free end should be allowed to touch the floor (a spring 5 to 7 feet in length generally works best). With the free hand, strike downward on the hand holding the

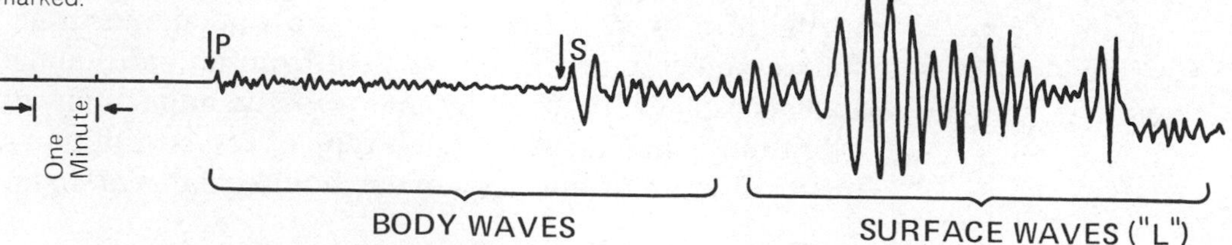

Figure 2–8. A typical seismogram. Time runs from left to right. P, S and L trains are marked.

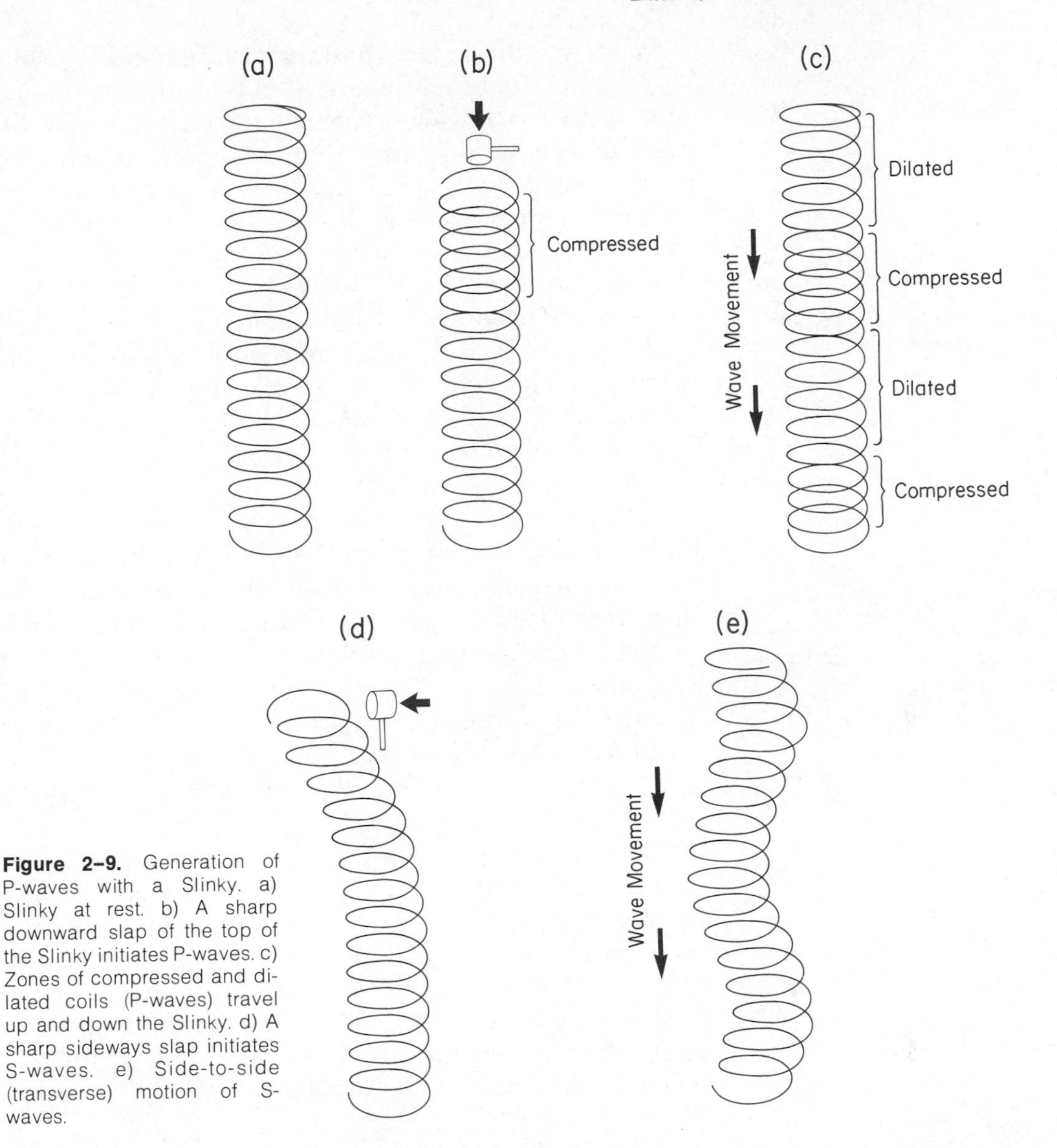

Figure 2–9. Generation of P-waves with a Slinky. a) Slinky at rest. b) A sharp downward slap of the top of the Slinky initiates P-waves. c) Zones of compressed and dilated coils (P-waves) travel up and down the Slinky. d) A sharp sideways slap initiates S-waves. e) Side-to-side (transverse) motion of S-waves.

upper end of the coil (Fig. 2–9*B*). Viewing the coil from a point near the upper hand, you should be able to see one or more disturbances travel down the coil. When the disturbance reaches the floor, it will reverse direction and travel back up the coil. Careful examination of the initial disturbance will show that it consists of a zone in which the coils are closer together or *compressed*. Immediately following the compressed zone should be a zone in which the coils are further apart or dilated (Fig. 2–9C). An earthquake P-wave train usually includes many such alternating zones of compression and dilation.

Another "Slinky" experiment illustrates S-wave movement. Hold the upper end of the coil in the same way as in the P-wave experiment, but this time, instead of striking downward with the free hand, strike laterally or sideways (Fig. 2–9*D*). The wave generated will travel downward as before, but the coils in the train will move from side to side (Fig. 2–9*E*). This side-to-side motion will always be horizontal; that is, all S-waves have a horizontal component of motion. This horizontal side-to-side motion is especially damaging to man-made structures because buildings are usually weak in the horizontal direction.

Sound waves are actually P-waves traveling through the air, which accounts for the low rumbling sounds sometimes heard during an earthquake. S-waves do not travel through air, water or other liquids, a fact utilized by seismologists to detect liquid rock within the earth. Discovery of the liquid core of the earth and the low velocity zone were important consequences of this property of S-waves. Near the surface of the earth, P-waves travel through granite at an average velocity of about 3.5 miles per second; S-waves travel at about 2 miles per second. At greater depths, both P- and S-wave velocities increase.

Observers often report two distinct earthquake shocks, separated by an interval of several seconds. Onsets of P- and S-waves are usually responsible for the two shocks. An observer twenty miles from the center of the earthquake would feel two shocks about four seconds apart. The *P-S interval* measured on a seismogram is an indication of distance from the observing station to the earthquake.

Surface Waves

L-waves consist primarily of *surface waves.* As the name suggests, surface waves travel through near-surface rocks. Two main types of surface waves are Rayleigh waves and Love waves, named after prominent physicists who pioneered wave motion research. Rayleigh and Love wave motions are somewhat analogous to compressional and shear wave motions, respectively.

Surface waves reach a distant point later than P- and S-waves, in part simply because they are slower. P- and S-waves have a further time advantage because they go through the earth's interior, whereas surface waves take the longer circumferential path. Surface waves are more persistent, however, and may circle the earth several times after a large earthquake, whereas P- and S-waves die out at shorter distances. Surface waves several inches high traversed the United States following the 1964 Alaskan earthquake. Because of their slow rise and fall, however, they were not perceptible to most people.

Foreshocks and Aftershocks

The sudden movement of a large segment of the earth's crust and mantle associated with an earthquake is never a single isolated event. Rocks deformed to near their breaking point do not suddenly "snap" to a new position and remain there until the next "snap." Instead, rocks that have been slowly yielding to increased pressure begin to yield faster; at first a little faster, then much faster and finally yielding very rapidly (i.e., breaking) for a fraction of a second.

The period of faster yielding may last several hours or several days. During this period, minor slippage within the rock may result in small earthquakes called *foreshocks*. Foreshocks precede almost all large earthquakes. Unfortunately, they are indistinguishable from the many other small earthquakes recorded in earthquake-prone areas, and therefore cannot be used to warn residents of an imminent major earthquake. It is impossible to examine a single earthquake record and identify a genuine foreshock—until it is too late.

Numerous smaller shocks, known as *aftershocks*, usually follow the main shock. The frequency distribution of aftershock magnitudes is predictable. Within 24 hours after the 1964 Alaskan earthquake, 28 major aftershocks were felt. Altogether about 12,000 aftershocks large enough to be felt occurred during the 69-day period immediately following the main shock.

Aftershock activity dies down slowly following an earthquake as both frequency and magnitude diminish with time. In Alaska, aftershocks continued to be recorded for a year or more. Aftershocks following smaller earthquakes die out more quickly. Figure 2–10 shows the frequency of aftershocks following the 1971 San Fernando earthquake. The pattern is probably typical of many aftershock sequences.

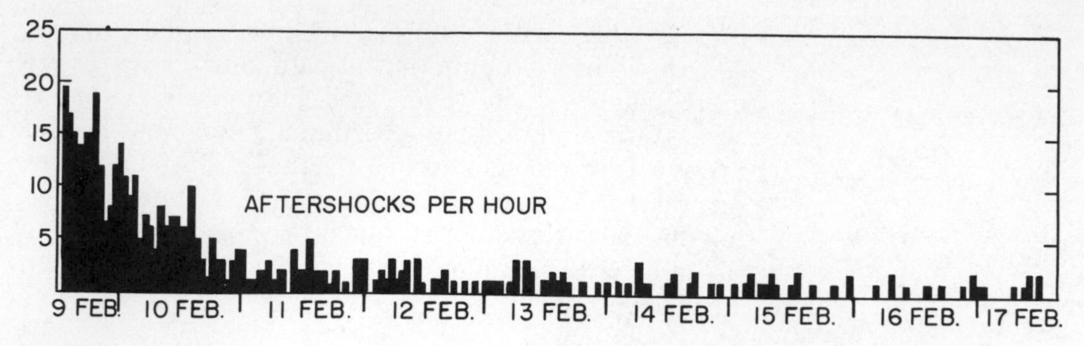

Figure 2–10. Frequency of aftershocks following the February 9, 1971, San Fernando earthquake. Decay of aftershocks with time is typical of most earthquakes. (After Espinosa, A. F., et al.: The San Fernando, California earthquake of February 9, 1971. U.S. Geological Survey Prof. Paper 733, 1971, p. 141.)

Duration

Duration of strong shaking varies widely, ranging from a few seconds to several minutes. An unusually long period of shaking, which observers reported to be from 1½ to more than 7 minutes, intensified damage during the 1964 Alaskan earthquake. Duration of strong shaking during the great 1906 San Francisco earthquake was about one minute, a time interval typical for earthquakes of comparable magnitude.

Recording

Seismographs. The instrument used to record seismic waves is called a *seismograph* (Fig. 2–11).

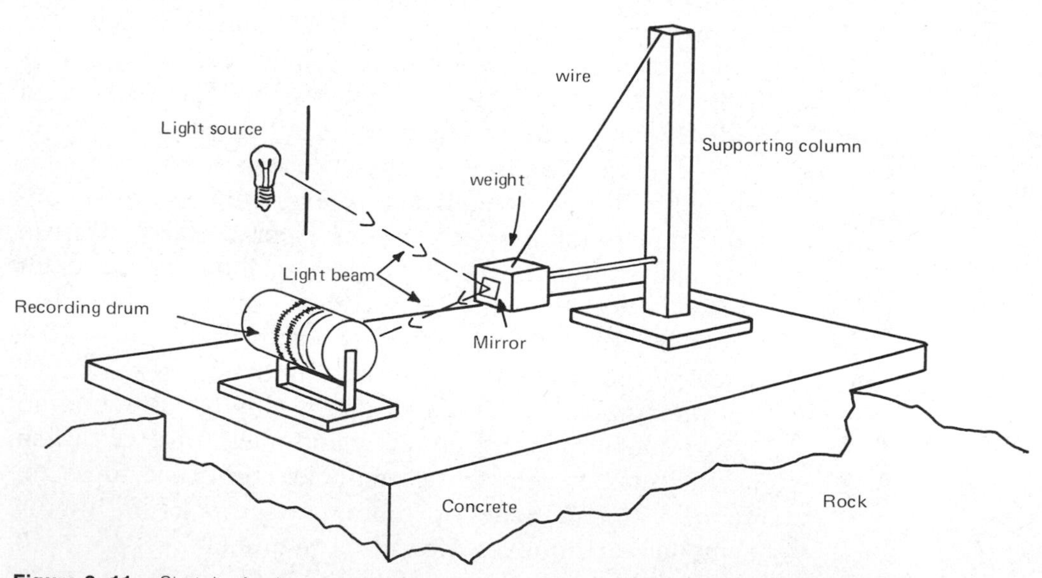

Figure 2–11. Sketch of seismograph recording horizontal earth motion. A light spot reflected from a mirror on the boom moves across photographic paper as the boom oscillates. Springs are omitted from the sketch. Replacing the tie wire with a spring, changing the pivot direction and reorienting the drum convert the seismograph into an instrument that records vertical motion.

Seismographs vary in design and construction, but most consist of a weight attached to a frame by a spring-boom arrangement. The seismograph in Figure 2–11 is typical in design. During an earthquake, the earth may vibrate with up-down-up, right-left-right, forward-back-forward motion or a combination of motions. Seismograph stations usually consist of at least three seismographs, each recording motion in a different direction.

The seismograph is able to record earthquake motion because the weight tends to remain steady as the ground moves beneath it. However, because the weight must be connected in some manner to the ground, it records ground motion imperfectly. It is possible to reconstruct all aspects of ground motion only if several seismometers are used to record each direction of motion. One seismograph will record rapid vibrations quite well but record very slow vibrations poorly. Another seismograph will be insensitive to rapid vibrations but will accurately record very, very slow vibrations. To complicate matters further, earthquake waves generated in Japan and recorded in America may be less than .001 inch high. Waves generated by a nearby earthquake may be more than a foot high. A single instrument cannot accurately record such a large range of wave amplitudes. A major seismograph station may require many instruments to adequately record all aspects of earthquakes.

Magnitude. Seismologists use the *Richter magnitude scale* to compare earthquakes. The Richter magnitude number is proportional to the logarithm of the height of the wave as measured on a seismogram written by a standard seismograph located 62 miles (100 kilometers) from the *epicenter* (point on the surface of the ground directly above the *focus*). The physical characteristics of a standard seismograph and the mathematics of the calculation are not important for our purposes, but it is important to emphasize that the measurement is based on the logarithm of the trace height. Thus, an increase of one unit of Richter magnitude is equivalent to a tenfold increase in height of earthquake waves, an increase of 2 is equivalent to a hundredfold increase and so on. The logarithmic base of the Richter scale can easily mislead the uninformed. The San Fernando earthquake, for example, had a Richter magnitude M = 6.6. The great San Francisco earthquake of 1906 had a magnitude M = 8.3, but the amplitude of the seismic waves near their source was about 60 times greater in the San Francisco earthquake. Earthquakes with magnitudes less than M = 4.5 are not seriously damaging.

The Richter scale has no lower limit. Modern instrumentation can record earthquakes of M = −1 or M = −2 if the foci are close enough to the seismograph station.

Damage and Other Earthquake Effects

When two immense blocks of the earth's crust and upper mantle suddenly shift, there are two apparent immediate and direct effects. First, the surface of the ground is usually offset somewhere, possibly disrupting rivers and streams, raising or lowering land (and possibly submerging it if it is along a sea coast) and breaking structures located over the fault. Secondly, earthquake waves fan out, shaking the surface of the ground and structures built on or beneath the surface.

The vibration associated with seismic waves causes far more damage than the actual fault motion, simply because it covers a much wider area. In addition to direct effects, seismic waves trigger a number of important secondary effects such as slumps, slides, fissures, water spouts, water waves, compaction and settling of ground. Let us now review some of these earthquake-caused effects.

Intensity

The Richter magnitude scale is an effective means for comparing absolute "sizes" of earthquakes, but it tells us very little about the amount and extent of damage caused by an earthquake. Seismologists therefore use a second scale, the *Modified Mercalli Intensity Scale* (MMI) (Table 2–1) to describe *damage* and human *perception* of earthquakes.

The MMI scale is actually the older of the two scales. It had its beginnings in the nineteenth century, when seismologists lacked the accurate and standardized seismographs of today. Reports from human observers were the principal means available for description and

TABLE 2-1. MODIFIED MERCALLI SCALE OF EARTHQUAKE INTENSITY*

I. Not felt except by a very few under especially favorable circumstances.

II. Felt only by a few persons at rest, especially on upper floors of buildings. Delicately suspended objects may swing.

III. Felt quite noticeably indoors, especially on upper floors, but many people do not recognize it as an earthquake. Standing motor cars may rock slightly. Vibration like passing truck.

IV. During the day felt indoors by many, outdoors by few. At night some awakened. Dishes, windows, doors disturbed; walls make creaking sound. Sensation like heavy truck striking building. Standing motor cars rocked noticeably.

V. Felt by nearly everyone; many awakened. Some dishes, windows, etc., broken; a few instances of cracked plaster; unstable objects overturned. Disturbances of trees, poles, and other tall objects sometimes noticed. Pendulum clocks may stop.

VI. Felt by all, many frightened and run outdoors. Some heavy furniture moved; a few instances of fallen plaster or damaged chimneys. Damage slight.

VII. Everybody runs outdoors. Damage negligible in buildings of good design and construction; slight to moderate in well-built ordinary structures; considerable in poorly built or badly designed structures; some chimneys broken. Noticed by persons driving motor cars.

VIII. Damage slight in specially designed structures; considerable in ordinary substantial buildings, with partial collapse; great in poorly built structures. Panel walls thrown out of frame structures. Fall of chimneys, factory stacks, columns, monuments, walls. Heavy furniture overturned. Sand and mud ejected in small amounts. Changes in well-water levels. Disturbs persons driving motor cars.

IX. Damage considerable in specially designed structures; well-designed frame structures thrown out of plumb; great in substantial buildings, with partial collapse. Buildings shifted off foundations. Ground cracked conspicuously. Underground pipes broken.

X. Some well-built wooden structures destroyed; most masonry and frame structures destroyed with foundations; ground badly cracked. Rails bent. Landslides considerable from river banks and steep slopes. Shifted sand and mud. Water splashed over banks.

XI. Few, if any, masonry structures remain standing. Bridges destroyed. Broad fissures in ground. Underground pipelines completely out of service. Earth slumps and land slips in soft ground. Rails bent greatly.

XII. Damage total. Waves seen on ground surfaces. Lines of sight and level distorted. Objects thrown upward into the air.

*Modified from Richter, C. F.: Elementary Seismology. San Francisco and London, W. H. Freeman and Co., 1958.

comparison of earthquakes. Two early seismologists, De Rossi in Italy and Forel in Switzerland, developed a scale with Roman numeral values from I to X, which could be used to describe and compare earthquakes throughout the world. Twenty years after the introduction of the Rossi-Forel Scale, the Italian seismologist Mercalli created a new scale using Roman numerals ranging from I to XII. The Mercalli scale permitted a more refined analysis of major earthquakes. Two Americans modified the Mercalli scale in 1931 to take into account effects on tall buildings and other features of modern construction. The Modified Mercalli Scale is still in use today.

An earthquake always has a single Richter magnitude number associated with it, but there are a range of associated MM intensities. The zone of greatest intensity usually surrounds the epicenter. This zone is surrounded by belts of lesser intensity. Occasionally the pattern takes the form of an island of greater or lesser intensity in a sea of varying intensities. Intensity maps called *isoseismal* maps are widely used in earthquake research (Fig. 2–12).

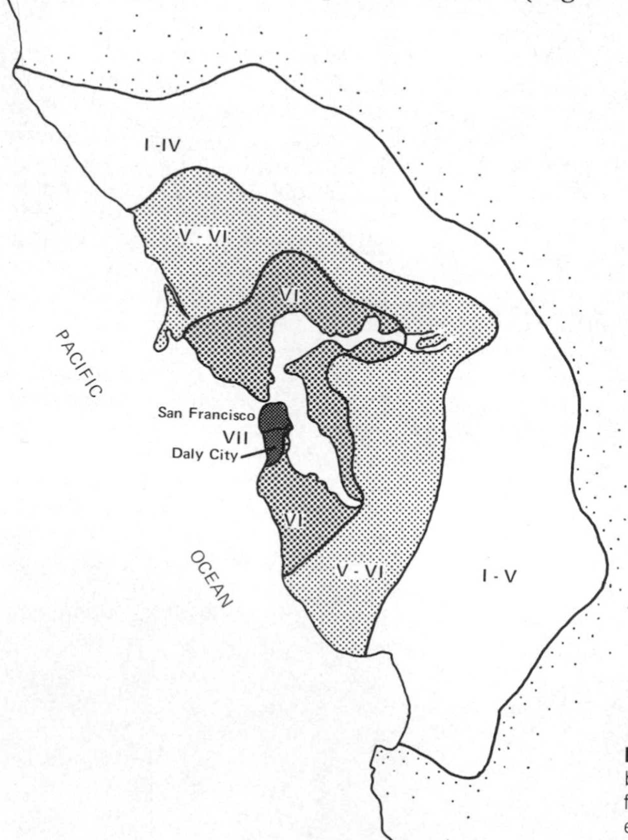

Figure 2–12. Isoseismal map prepared by the U.S. Coast and Geodetic Survey following the 1957 Daly City, California, earthquake. A maximum intensity of VII was observed.

Figure 2–13. The Hanning Bay fault scarp, Montague Island, Alaska, formed during the 1964 earthquake. Vertical offset in the picture is 12 feet; 15-foot vertical displacement was measured near trees in the background. (U.S. Geological Survey photo.)

Tectonics

Faulting. The actual break along which the earth shifts is called a fault. Much fault motion occurs deep within the earth and can be studied only by means of analysis of seismic waves. Surface displacement of a fault occurs during very large earthquakes and occasionally during smaller earthquakes. The longest surface breaks are associated with transcurrent faults such as the San Andreas of California, where, in 1906, surface rocks were broken for a distance of almost 200 miles along the fault. Vertical offsets across the 1906 fault were only 2 or 3 feet, and horizontal offsets of 15 to 20 feet were recorded. Before the 1906 earthquake it was assumed that all fault motion was primarily vertical. The large horizontal offsets observed along the San Andreas fault led to recognition of the role of transcurrent (horizontal) motion in faulting.

Faulting associated with subduction sometimes results in large vertical offsets. In 1897, a piece of the Indian subcontinent slipped northward into the subduction zone beneath the Asian plate. The area of serious damage was one of the largest ever recorded. A vertical displacement of 35 feet was observed along one fault. The lateral extent of the fault was 12 miles, significantly less than the lateral extent of surface fracturing along the San Andreas. (Slightly larger vertical offsets may have occurred on the sea floor south of Alaska during the 1964 earthquake.) Figure 2–13 shows a fault scarp on Montague Island. The scarp resulted from 15 feet of vertical offset associated with the 1964 earthquake.

Uplift and Subsidence. Vast areas may be uplifted or depressed during an earthquake. During the 1964 Alaskan earthquake, the southern edge of the American plate buckled (Fig. 2–14) as subducting Pacific lithosphere moved downward. The elevation of the sea floor of the Pacific plate very probably changed also, but measurements of sufficient accuracy were impossible in the deep waters of the Aleutian Trench. The zone of land level changes extended southwest through the epicenter for a distance of more than 500 miles and in some areas was as much as 200 miles wide. Subsidence exceeded 6 feet in places; uplift locally exceeded 30 feet on land and was even greater on the sea floor. Sudden uplift of the sea floor was responsible for the seismic sea wave (*tsunami*) that spread across the Pacific after the earthquake.

Uplift of large areas of seashore (Fig. 2–15) and marshlands accompanied the Alaskan earthquake of

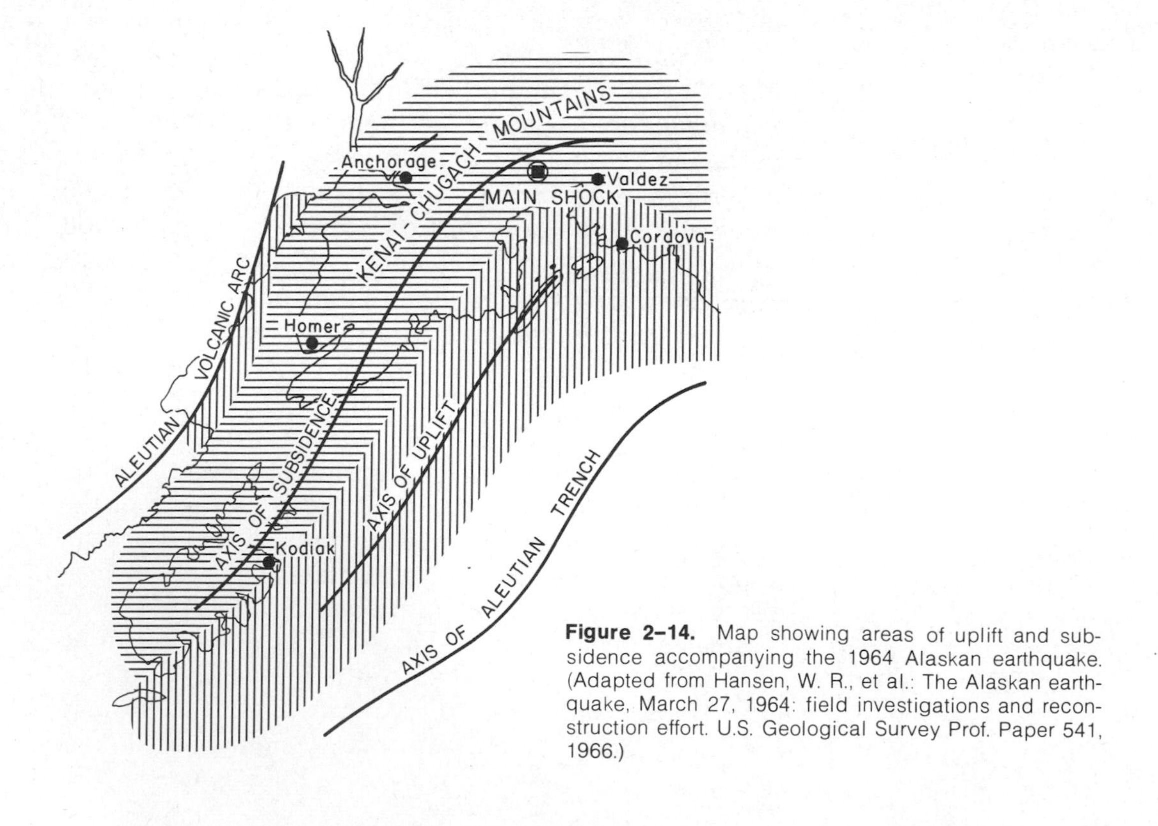

Figure 2–14. Map showing areas of uplift and subsidence accompanying the 1964 Alaskan earthquake. (Adapted from Hansen, W. R., et al.: The Alaskan earthquake, March 27, 1964: field investigations and reconstruction effort. U.S. Geological Survey Prof. Paper 541, 1966.)

Figure 2–15. Kelp (seaweed) wilts in the sun following uplift of the sea floor during the 1964 Alaskan earthquake. Uplift exposed and killed many marine communities, but most reestablished themselves within a few years. (U.S. Geological Survey photo.)

1964. Many marine organisms were killed, but most marine communities re-established themselves within a few years. Salmon spawning areas were damaged by uplift, and subsidence allowed salt water to invade coastal fresh-water marshes and lakes. The commercial clam habitat was badly damaged. On the other hand, uplift reduced flood damage hazards to nesting areas of ducks, geese and trumpeter swans.

Vibration Damage

Vibration damage is a complex phenomenon. It depends on both the vibration resistance of a structure and on seismic wave characteristics. The degree and extent of vibration damage varies widely. For example, the Peruvian earthquake of 1970 (M = 7.7) devastated much adobe construction (Fig. 2–16) in Huaraz, Peru, about 100 miles from the epicenter. The much stronger 1964 Alaskan earthquake (M = 8.3) heavily damaged buildings in Anchorage, 80 miles from the epicenter, but the city was not devastated as was Huaraz. Anchorage suffered relatively lighter damage because its buildings, for the most part, were of stronger construction than buildings in Huaraz.

Main Factors in Vibration Damage. Two factors, inertia and horizontal strength, stand out among those

Figure 2–16. Destruction of adobe buildings in Huaraz, Peru, after the 1970 earthquake. (U.S. Geological Survey photo.)

contributing to the vibration resistance of a house, building, dam or other man-made structure.

There is an old party trick in which dishes, glasses of water and other paraphernalia are placed on a table covered with a cloth. If the tablecloth is pulled away sharply enough, the objects will not be disturbed. *Inertia* keeps them stationary when the tablecloth is removed. The law of inertia states that objects at rest tend to remain at rest, and objects in motion tend to continue to move in the same direction unless acted on by an external force. In the 1886 Charleston earthquake, the ground moved, the brick foundations of buildings tried to move with the ground, and the buildings themselves tended to stay put—like objects on the tablecloth. The result was that the foundations fractured, and the houses fell to the ground.

During the 1971 San Fernando earthquake, many freeway overpasses collapsed. The ground moved, the piers of the overpasses tried to move, and the bridge sections of the overpass tended to remain stationary. The vibration broke the piers loose from the upper bridge portion, resulting in either an unsafe overpass or collapse (Fig. 2–17).

Overcoming differences in inertia is the fundamental problem of earthquake-resistant construction design. The structure, whether building, dam or bridge, must be strong, and it must be braced, anchored and bonded so that all elements move together. Horizontal strength is another important factor. Examine the classroom walls around you, or the walls of a typical modern building. Unless designed to resist earthquakes, the walls are strong vertically but weak horizontally. Support of walls, roof and upper storys requires vertical strength. Without vertical strength a building would collapse under its own weight. Horizontal strength is recognized to be necessary in certain regions, where strong winds apply horizontal force to buildings; but in most areas, it is not considered an important factor—until an earthquake strikes.

High amplitudes and rapid vibrations of S-waves applied horizontally, where buildings are weakest, form a potent combination. Horizontal shaking due to

Figure 2–17. Collapsed freeway overpasses near San Fernando, California, caused by the 1971 earthquake. (U.S. Geological Survey photo.)

S-waves threw adobe blocks from the top of the wall on the left in Figure 2–16.

In addition to inertia and horizontal strength, seismic wave height (amplitude), rapidity of vibration (frequency), duration of shaking and the type of rock or soil beneath the structure (geologic foundation) also affect the extent of damage.

With the possible exception of the role of the geologic foundation, the importance of each of the above factors is probably self-evident: The amplitude of a seismic wave determines the amount of motion; we are all subjected to low vibration levels daily without being aware of it; and the ground shakes continuously, primarily as a result of air currents, rainfall, tidal motion and animal (including human) activity. Damage occurs only when the height of the vibratory waves is great, as during an earthquake.

Earthquake waves diminish as they spread out from their source. The decrease in amplitude results from the spreading of earthquake energy through a much greater volume of rock. This spreading, combined with an absorption of energy by rocks, limits the extent of severe damage from earthquakes. Earthquake waves cause grains in rocks to rub together, generating frictional energy, which remains in the rocks as heat. Some energy is absorbed by a frictionlike mechanism within the crystalline structure of individual mineral grains, a principle exploited in microwave ovens, where high frequency sound (P-waves) heats a sandwich or cooks a roast by excitation of molecules inside the food. Conversion of sound energy to heat warms the food in an amazingly short time. Although absorption of microwaves and earthquake waves is similar in some respects, earthquakes do not cause rocks to be perceptibly hotter. The amount of heat transferred to a given rock is small; it is the volume of rock absorbing small amounts of earthquake energy that is tremendous. If rocks did not absorb the energy, the earthquake waves would travel around and around the earth.

Absorption and spreading of earthquake energy limit the extent of severe damage from an earthquake. The 1897 Indian earthquake caused serious damage over an area with a mean diameter of 600 miles. Areas of serious damage resulting from twentieth-century earthquakes have not been as great. The average radius of perception (that is, how far away the earthquake was

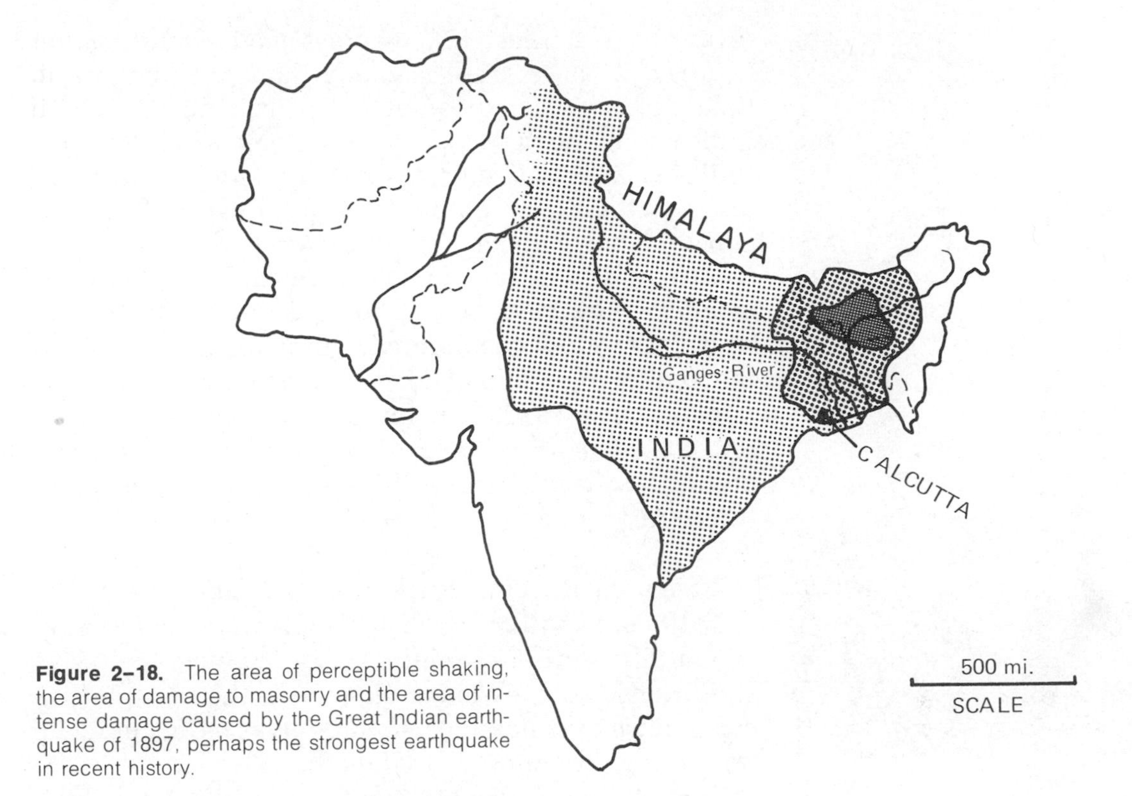

Figure 2–18. The area of perceptible shaking, the area of damage to masonry and the area of intense damage caused by the Great Indian earthquake of 1897, perhaps the strongest earthquake in recent history.

felt) for the 1897 Indian earthquake was 900 miles (Fig. 2–18).

If a building is close to an active fault during a major earthquake it generally does not make much difference whether the building is one mile or ten miles way. Vibration will be intense everywhere in the immediate vicinity of the fault. Buildings in the fault zone, however (Fig. 2–19), run the double risk of damage from shaking and damage from surface movement of the fault.

Figure 2–19. Housing developments obliterate the surface expression of the San Andreas fault near San Francisco. The solid line traces the axis of the fault. (U.S. Geological Survey photo.)

The role of duration of shaking is simple: the longer a building is shaken, the greater the chance that it will suffer major damage or collapse. A relatively long duration of shaking intensified damage of modern buildings in Anchorage, Alaska, during the 1964 earthquake (Fig. 2–20).

Certain geologic foundation materials, such as water-saturated uncemented granular soil or sediment, magnify earthquake wave amplitude. Solid rock is a far safer foundation material in earthquake-prone areas. Vibration damage in San Francisco during the 1906 earthquake clearly shows the influence of geologic foundation (Fig. 2–21). Buildings in old stream beds and on land reclaimed from the bay or swamps by

Figure 2–20. Earthquake vibration severely damaged this apartment building in Anchorage. Relatively long duration of strong shaking contributed to the damage. (U.S. Geological Survey photo.)

filling were hardest hit, whereas many buildings on bedrock hills were not severely damaged, in spite of their proximity to the fault (less than 10 miles). As indicated in the figure, damage was great near the Ferry Building, which is on the northeast side of San Francisco, farthest from the fault. A building partly on unconsolidated soil and sediment and partly on solid rock may be severely damaged because of differences in amplitudes in rock and unconsolidated sediment. Buildings on unconsolidated sediments may be safe if the sediment is adequately compacted prior to construction. However, in many older areas, construction on fill and sediment preceded legislation of adequate building codes. These buildings risk great damage in future earthquakes.

Effects on Specific Structures. Seismic waves affect different structures in different ways. Houses respond differently from tall buildings, dams, highways and bridges. A knowledge of some of the peculiarities of response by different structures gives us an appreciation of the problems of earthquake defense.

Wood frame construction resists earthquake shaking far better than masonry construction. Its wall sheathings and wall panels brace the frame against horizontal vibration. Like a reed bending and surviving the great storm which blows down the sturdy—and rigid—oak, the wood frame house yields to earthquake vibration without breaking. Brick, concrete block and other masonry collapse more readily because they do not yield.

However, not all wood frame houses resist earthquake shaking equally well. In the Charleston earthquake of 1886, brick chimneys and foundations of wood frame houses failed, causing extensive damage. Plaster frequently cracks; windows break, especially those with large glass panes. Modern post and beam construction lacks much of the bracing found in conventional construction. Post and beam houses and buildings require careful design to resist earthquakes. Conventional house design, although more resistant than most to earthquake damage, should provide for bracing, anchoring and bonding of all elements to insure that the house vibrates as a unit. It does little good to have the strongest house and the strongest chimney in town if they are free to bang against one another during earthquakes.

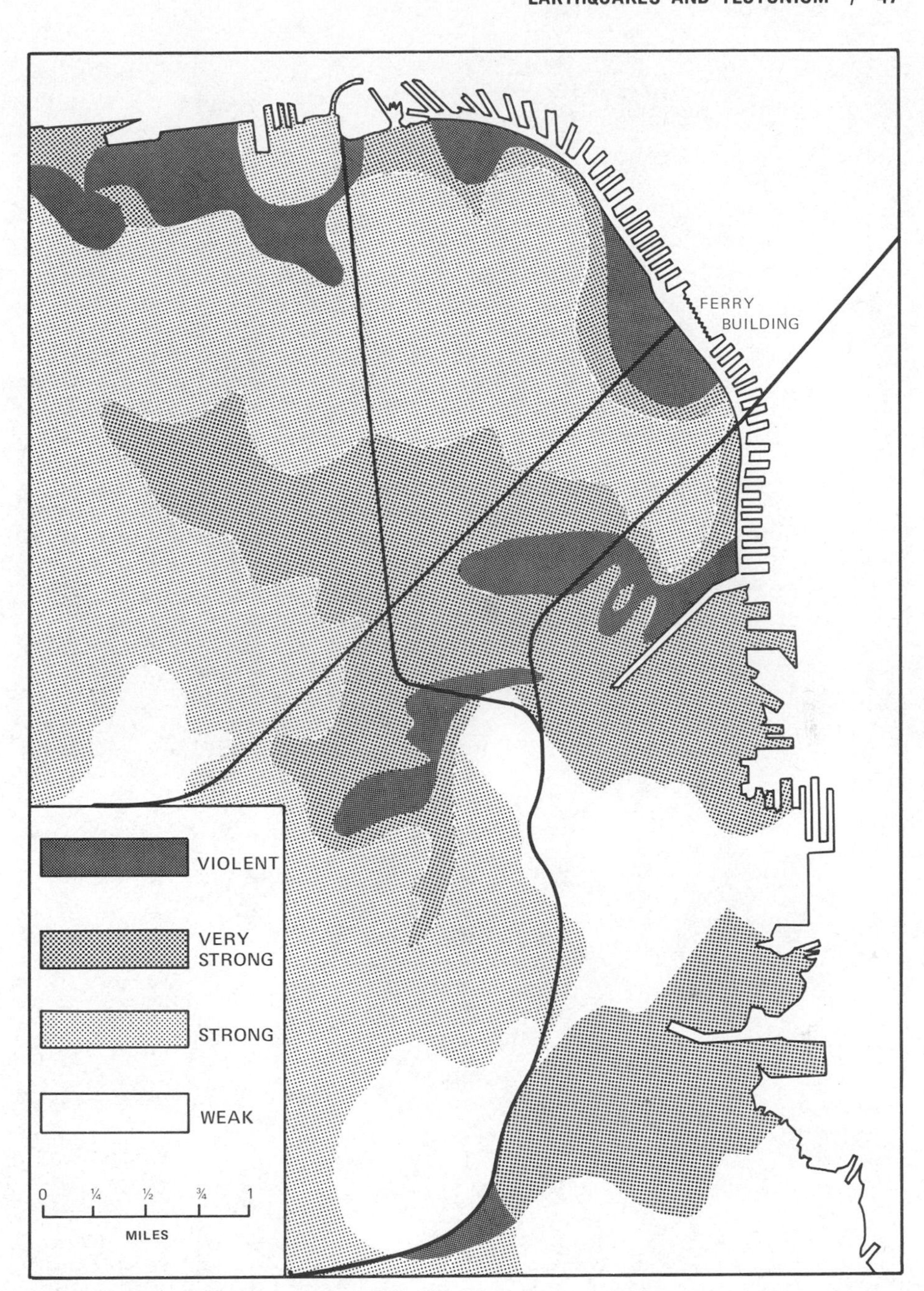

Figure 2–21. Distribution of the damage due to the 1906 San Francisco earthquake. Land recovered from the bay, reclaimed and filled swampland, river bottoms and other areas of thick unconsolidated soil experienced the most damage. Buildings on bedrock experienced relatively little damage.

Adobe (a mixture of mud and straw) is probably the material least resistant to earthquake damage, and primitive stone and mortar construction is little better. The abundance of these two types of construction in primitive countries accounts for the great loss of life during earthquakes. For example, 12,000 died in Aga-

Figure 2-22. Damage to the Lower Van Norman Dam caused by the San Fernando earthquake of 1971. The dam barely held, averting disastrous losses of life and property damage. (U.S. Geological Survey photo.)

dir, Morocco, in 1960; 12,000 died in northwestern Iran in 1963 – but only 100 died in the Taiwan, Formosa, earthquake of 1963. Although Formosa is densely populated, the predominant wood construction prevented extensive loss of life.

Tall buildings are usually built strongly and solidly. Nevertheless, they sometimes fail owing to vibration or other causes. In addition, they vibrate *resonantly*, which complicates their design. Go to the top of a tall building when the wind is blowing. The building will be swaying back and forth in the wind. The *period* of the sway (the time it takes for the building to sway from one side to another and back again) depends on several factors, one of the most important being the height of the building. If the period of the back and forth motion of surface waves happens to coincide with one of several natural frequencies of vibration of a particular building, serious damage can result.

Dams occasionally fail during earthquakes. An earthfill dam gave way in the 1925 Santa Barbara (California) earthquake. No dams failed in the 1906 San Francisco earthquake, although faulting ruptured and displaced one dam on the fault. The 1971 San Fernando earthquake severely damaged the Lower Van Norman Dam (Fig. 2–22), which overlooks the heavily populated San Fernando valley. Failure of the dam would have

Figure 2–23. The Twentymile River Bridge fell into the river, driving some of the wood pilings through concrete spans. The adjacent steel railroad bridge received little damage from the Alaskan earthquake. (U.S. Geological Survey photo.)

Figure 2–24. Rails skewed away from their bed as underlying soil shifted toward the river during the 1964. Alaskan earthquake. (U.S. Geological Survey photo.)

caused much damage and many deaths in the valley below.

Earthquakes damage highways and railroads in many ways. The Twentymile River Bridge near Anchorage collapsed into the river as wooden pilings (vertical supports beneath the bridge) were driven through the reinforced concrete deck (Fig. 2–23). Rails

Figure 2–25. Crumpled, overturned and smoke-darkened oil tanks in tank farm in Seward, Alaska, still smoking shortly after the 1964 earthquake. (U.S. Geological Survey photo.)

Figure 2–26. Oil storage tanks in the San Francisco Bay area. These tanks are constructed on low-lying fill near Hayward fault. What are their chances of survival during the next large earthquake?

are frequently twisted (Fig. 2–24) and roadbeds rupture owing to vibration, fault movement or slumping of underlying rock and soil. Damage to highways and railroads during a severe shock is usually extensive.

Oil storage tanks located in dock areas (Figs. 2–25 and 2–26) may rupture as they did at Seward, Valdez and Whittier, Alaska, in 1964. In each of these ports, the oil caught fire and added to the damage.

Secondary Effects

In many earthquakes, damage due to secondary effects such as fire, landslides or seismic sea waves (*tsunami*) far exceeds the damage due to direct effects. In 1906 and 1923 respectively, fire devastated San Francisco and Tokyo after great earthquakes broke water mains and tipped over heaters. A seismically triggered landslide killed thousands in the 1970 Peru earthquake, and damage from *tsunamis* is almost always cataclysmic.

Slumps, Slides and Fissures. In the zone of intense shaking that surrounds an earthquake epicenter, strong vibrations may cause loose sandy soil to liquefy or become "quick" (as in quicksand). This happens because cohesiveness of loose soils depends on frictional contact between adjacent grains of soil. If soil is shaken

Figure 2–27. Shaking by the 1964 earthquake liquefied a clay layer beneath the Turnagain area of Anchorage, causing overlying soil, houses and streets to slide and slump downhill toward the nearby bay. (U.S. Geological Survey photo.)

violently, grains lose contact with one another. If the soil is also saturated with water, grains "float" for a fraction of a second, bump into other grains and float again. During the time of floating, the soil is "quick." It lacks strength and acts as a liquid. In some instances, liquefaction causes heavy objects to sink into the soil. This was the case during violent shaking in the 1934 Bihar-Nepal (India) earthquake, in which an observer reported that his car sank to the axles. In other instances, the liquefied soil flows and causes slumping and landsliding. A spectacular landslide occurred during the 1964 Alaskan earthquake, when a clay layer beneath the residential area of Turnagain Heights in Anchorage liquefied. Overlying soil and rock fractured and began to slide downhill toward Cook Inlet. Damage to property was extensive (Fig. 2–27).

Compaction and Settling. Fill a glass with a granular material such as fine sand, flour or sugar. Tap the side of the glass lightly. Vibration causes the grains to pack together more closely, and the surface settles slightly. Soil is a granular material and will settle when shaken, just as the sand, flour or sugar settles in the preceding experiment. Now add water to the glass. Notice that the sand settles even more quickly. When shaken, the grains rotate and move slightly, which causes them to pack together more tightly. During an earthquake, ground surface may settle several feet, as it did during the 1970 Peruvian earthquake, when a residential area built on low-lying swampy ground in Chimbote subsided and had to be abandoned.

Tsunamis and Seiches. A tidal wave 50 feet high sweeps across the sea, destroying everything in its path—at least that's the way the movie version usually goes. Actually, *tsunamis* or seismic sea waves have nothing to do with tides. Tsunamis are usually caused by earthquakes but may result from volcanic eruptions or occasionally from hurricanes. Hollywood is right in one respect, however—tsunamis can be extremely destructive.

Tsunamis are not true seismic waves (which travel through water at speeds of almost a mile per second) but are gravity-controlled waves that travel through the open ocean at speeds up to 0.2 miles per second. In the open sea, tsunami wave heights are no more than a few feet, and wave crests are far apart, so true tsunami waves are imperceptible to ships at sea. As tsunami waves approach the shore, the shallower water causes the waves to slow down and increase in height. If the wave enters an inlet that narrows rapidly, the energy of the wave is compressed into a smaller volume, which increases the height and destructive power of the wave. Such narrow inlets are common in Japan, where the term *tsunami* originated.

Most tsunamis are caused by earthquakes in subduction zones. As a result, tsunamis are common along the coasts of Japan, Peru and Alaska. Hawaii suffers so much from tsunamis originating in these and other earthquake zones surrounding the Pacific that an alarm system was devised to provide a warning at least four hours in advance of arrival of a tsunami originating along the Pacific margin.

The Alaskan and nearby coasts have been hit often and hard by tsunamis. A tsunami that followed the 1964 earthquake caused extensive damage along the Alaskan coast and 16 deaths by drowning in Oregon and California. In 1946, a tsunami wave over 100 feet high demolished a large lighthouse on Unimak Island in the Aleutians. The base of the lighthouse was 45 feet above sea level.

Tsunamis acting in concert with other earthquake phenomena are particularly destructive in ports and harbors, which are vulnerable to virtually every earthquake hazard. Tsunamis batter docks and port facilities, sometimes depositing vessels hundreds of yards inland (Fig. 2–28). The ground on which port facilities are built usually consists of unconsolidated sediments, which magnifies earthquake wave heights. The low elevation and nearby water practically insures that the soil is water saturated. Water saturation increases the likelihood of the soil's liquefying or becoming quick and rupturing and sliding. Submarine landslides destroy cables and pipes laid on the seafloor.

Although most tsunamis (and earthquakes) originate in the Pacific, an occasional tsunami sweeps across the Atlantic, such as the one generated by the Lisbon earthquake of 1755, which sent waves 50 feet high crashing into the Portuguese coast. Waves 15 feet high were reported in the West Indies. After the Charleston earthquake of 1886, a small tsunami flooded rice fields along the Cooper River near Charleston.

Figure 2-28. Tsunamis caused by the 1964 earthquake washed vessels into the center of Kodiak, Alaska. (U.S. Navy photo.)

A *seiche* is another kind of wave, set up in an enclosed body of water such as a lake or pond. A seiche can be easily generated experimentally in a pan or glass of water. Partially fill a container with water, then tilt the container slowly back and forth. A wave will form on the surface and move from side to side. Tilting the container back and forth "in time" with the wave will quickly intensify the wave, and may cause it to flow over the edge of the container. This wave is a miniature seiche.

During an earthquake, long slow surface waves tilt a pond or bay gently from side to side. If the rate of tilting is the same as a *seiche period* of the body of water, a seiche will occur. Seiches occurred along the Texas coast following the 1964 Alaskan earthquake. Waves as much as six feet high damaged small craft. Water was agitated in swimming pools in Texas and Louisiana. Seiches following the 1971 San Fernando earthquake sloshed water from pools throughout Southern California.

Damage to Utilities. Few of us realize the extent of our society's dependence on water, electricity and liquid fuels—at least until a catastrophe denies us the use of these utilities. The disruption of utilities exposes us to fire, cold and illness, post-earthquake factors that often have caused far more misery than the earthquake itself.

Rupture of water and gas mains and collapse of power lines occur during most severe earthquakes. Fire is a serious hazard. In addition to the danger of fire itself, roads and bridges are damaged, making it difficult for firemen to mobilize. Rupture of water mains impairs their ability to fight the fire. Fires following the 1906 San Francisco earthquake and the 1923 Tokyo earthquake caused far more damage than all other effects combined.

In 1923, charcoal burners were the principal source of heat in most Japanese homes (this is still the case in rural Japan). In the great Tokyo earthquake of that year, knocked-over burners started fires throughout Tokyo. Firemen, hampered by debris and broken water mains, could not control the fires raging through the wooden houses and apartments that comprise most of Tokyo. Tokyo literally burned to the ground. An estimated 143,000 people died. In the United States, fires following the 1906 earthquake in San Francisco caused more property loss than the earthquake proper.

Cold or rainy weather following an earthquake may cause deaths by exposure. Exposure is a more serious problem in underdeveloped countries than in industrialized nations. Other health hazards develop after longer interruption of utilities.

3 FOUR EARTHQUAKES

The San Fernando earthquake of 1971, the Charleston, South Carolina, earthquake of 1886, the Denver earthquakes of the 1960's and the Managua earthquake of 1972 were not monster earthquakes. The magnitude of the San Fernando earthquake was 6.6, that of the Charleston earthquake was about 7, the Managua earthquake was 6.2 and the largest magnitude observed in the Denver sequence was between 5¼ and 5½. A few dozen people were killed by the San Fernando and Charleston earthquakes; no one was killed during any of the Denver earthquakes; and an estimated 4000 to 6000 people were killed by the Managua earthquake.

Each of the following earthquake case histories illustrates one or more important points. The San Fernando earthquake shook our complacency and gave a hint of what is to come when a really strong earthquake hits a California metropolitan area. The Denver earthquakes were triggered by man's activity. They revealed much about the mechanism of earthquakes. The Managua earthquake reminds us that people in many parts of the world still live in circumstances such that even modest-sized earthquakes can cause great damage and loss of life. Finally, a discussion of the Charleston earthquake is included as an example of a major earthquake in the United States elsewhere than California or Nevada. Strong earthquakes have occurred within the recent past in Missouri, South Carolina, the Atlantic Ocean off New England and many of the western mountain states, and smaller earthquakes are even more widespread.

San Fernando Earthquake of 1971

Early in the morning of February 9, 1971, the sun was rising over the San Gabriel Mountains north of Los Angeles, sending the first light into the San Fernando Valley. On that morning, many residents would not awaken to the ringing and buzzing of an alarm. At 6:01 AM, an earthquake jolted them out of bed and shook them violently for about 60 seconds.

The old Veteran's Hospital, built before enactment of stringent building codes, collapsed, killing 45 patients and staff members. Altogether the death toll would rise to 64. The initial estimate of damage was $500 million, and the final figure may eventually reach $1 billion.

Large cracks appeared in the Lower Van Norman Dam. The dam, built in 1918, appeared in imminent danger of collapse. Eighty thousand people living in the valley below the dam were evacuated. Quick removal of water pressing on the weakened dam probably averted the largest single disaster in the earthquake history of the United States.

What monster earthquake created such havoc and caused so much damage? Actually, it wasn't a monster earthquake at all but an earthquake with a magnitude of 6.6. Southern California averages an earthquake of equal or greater magnitude every 4 years. However, most of these occur in sparsely populated regions; this earthquake occurred immediately north of the San Fernando Valley, a densely populated section of the Los Angeles metropolitan area. Imagine what would have happened had this been a magnitude 8.3 or 8.4 earthquake with ground amplitudes 60 times greater.

San Fernando Valley and Vicinity

A few miles north of Los Angeles, the San Andreas fault intersects the Garlock fault (Fig. 2–29) in the region of the "great bend." Here the fault motion differs from that of the rest of the San Andreas. The southwestern block slides beneath the northeastern block, forming a local minisubduction zone. Crumpled and uplifted rocks of the overriding northeastern plate form the Transverse Ranges, so called because they trend east-west. (Most other ranges in California trend northwest-southeast.) In the great bend, depression of underthrusting rocks of the southwestern plate has formed valleys filled with sediments eroded from surrounding highlands. The San Fernando Valley is one of these valleys. The San Gabriel Mountains, which border the valley on the north, are part of the overthrusting northeastern plate.

Numerous faults have been identified in the border zone of the San Gabriel Mountains and San Fernando Valley, but no activity had been observed before 1971. People in the San Fernando Valley thought they were as safe from earthquakes as anyone who lives on the California coast.

The Earthquake

Eight miles below the San Gabriel Mountains and nine miles north of San Fernando, the earth shuddered. The underthrusting block slipped a few feet down and to the left. The rupture spread, moving outward and upward. The rupture broke the surface in a zone nine miles long and 300 to 1000 feet wide (Fig. 2–30). Uplift of the northeastern block formed a *scarp* (small cliff) up to 3 feet high in some areas (Fig. 2–31).

The earthquake triggered more than 1000 landslides. The largest slide occurred in an area where soil liquefied and slumped down a gentle 1½ to 3 degree slope. The slump zone was ¾ mile long, and total movement

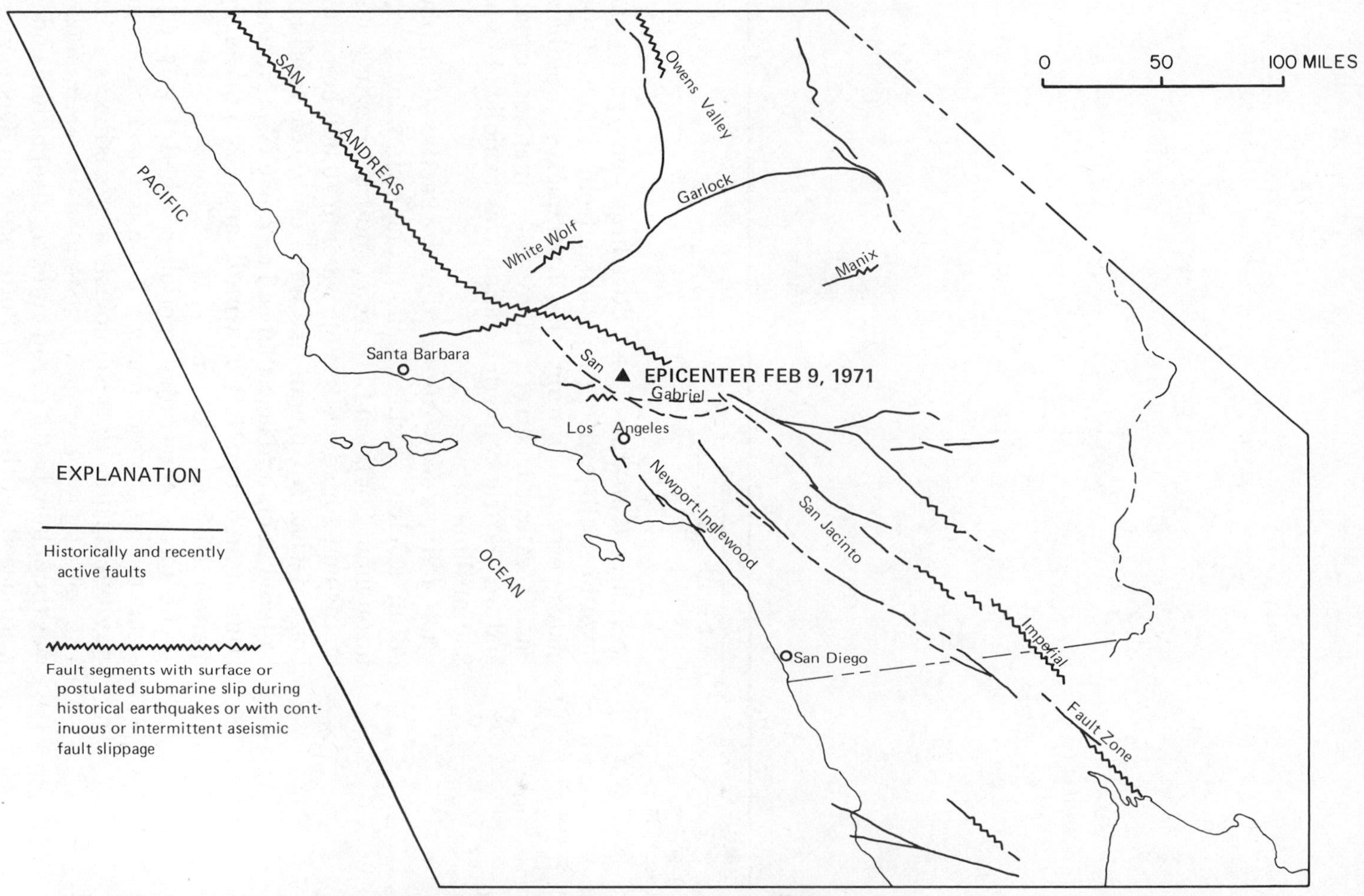

Figure 2–29. Major faults historically active in Southern California. A zone of east-west treading faults offsets the otherwise straight San Andreas in the "Big Bend" north of Los Angeles.

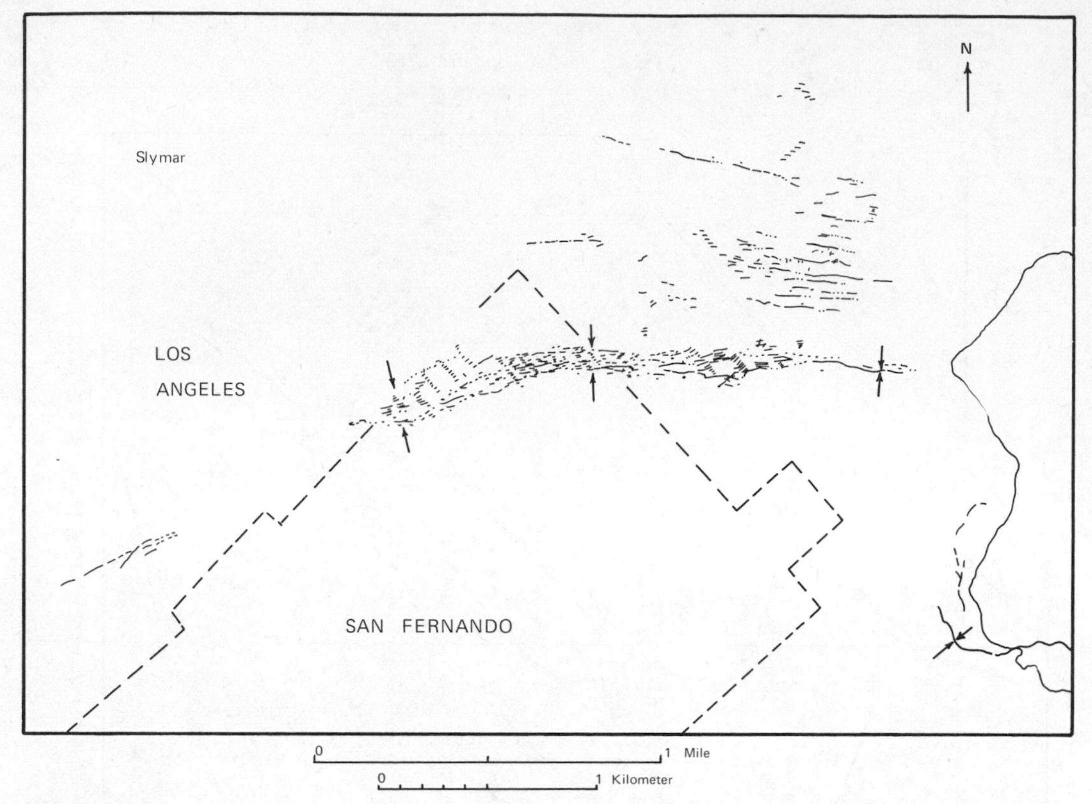

Figure 2–30. Surface faulting in Los Angeles and San Fernando during the 1971 earthquake. Surface displacement was distributed across a zone rather than in a single break. (After U.S. Geological Survey Staff: The San Fernando, California earthquake of February 9, 1971. U.S. Geological Survey Prof. Paper 733, 1971, p. 74.)

was less than three feet. Resulting damage to the Juvenile Hall (a detention center from which 80 inmates escaped during the earthquake), Southern Pacific railroad, streets, highways, pipelines, canals and an electric converter station was calculated to be $30 million.

Oil and gas seeps appeared in the Pacific off Malibu Point; seiches sloshed water from swimming pools throughout Southern California; and fractures in water mains and aqueducts cut off water to 10,000 homes in Los Angeles. A tsunami-like wave up to two feet high battered dams and lake shores. The Lower Van Norman Dam shifted two feet downhill and, as previously mentioned, nearly failed.

The force of recorded motion reached 1 g (1 g is equal to the force of gravity; 0.1 g is 1/10 as strong as gravity), surprising seismologists and engineers alike. The largest force usually generated by earthquakes was thought to be 0.1 g, and building designs and codes were based on this figure. Portions of the Olive View Hospital collapsed (Fig. 2–32), and damage rendered the hospital unfit for use.

Figure 2–31. Fault scarp near San Fernando. The fault lies downward and away from the observer. (U.S. Geological Survey photo.)

The Fire Department was able to extinguish several fires before serious damage resulted. It will be some time before the full extent of damage to sewers and other underground facilities is fully known.

Figure 2–32. Second floor of Olive View Hospital Psychiatric Unit rests on collapsed first floor following failure of support columns during the 1971 San Fernando earthquake. This was a new building. (Photo by D. F. Moran.)

Lessons of the Earthquake

It would be difficult to convince people in the San Fernando Valley whose homes were severely damaged that the earthquake was a good thing, but long-term benefits to Southern California could easily exceed earthquake damages if the lessons of the earthquake are learned and applied.

The San Fernando earthquake occurred near the center of the permanent seismograph net operated by the California Institute of Technology. Federal, state and university researchers moved quickly. Mobile instruments were on the scene recording aftershocks within a few hours of the main shock. About 250 *strong motion* seismographs had been installed previously, making Southern California the best-covered area in the United States. Engineers recorded the actual forces applied to structures by earthquake waves, and found them to be as great or greater than previously observed.

Data collected during and after the San Fernando earthquake clearly indicate that three things must be done to minimize the impact of future earthquakes in Southern California. First, we must learn to recognize potentially dangerous faults. Prior to the earthquake, faults in the San Fernando Valley and southern San Gabriel Mountains were thought to be relatively inactive. Close examination of the area after the earthquake revealed that past activity was greater than suspected.

Second, safety criteria for older structures must be re-examined. More than half the deaths in the San Fernando area occurred when the old pre–building code Veterans Hospital partially collapsed. Damage to the old Lower Van Norman Dam menaced thousands of people. Fortunately the water level of the lake behind the dam had recently been lowered, perhaps making the difference between a minor inconvenience and a major disaster. Vibratory forces recorded during the earthquake were far larger than imagined. Safety considerations may dictate replacement of many older structures, especially hospitals, schools and other public buildings where collapse would endanger many people simultaneously.

Third, re-examination of design criteria for new buildings is needed. The old design limit of 0.1 g is clearly inadequate. Forces of 0.5 to 0.75 g were common in the valley, and up to 0.4 g was recorded in buildings in Los Angeles more than 20 miles away. It was also discovered that vibration was stronger on upper floors than lower floors of many tall buildings, a fact which must now be considered by building code experts. Most new buildings survived the earthquake with minimal damage, but some, such as the Olive View Hospital, were severely damaged. Reasons for the damage must be established and building design criteria modified accordingly. Some building code changes have already gone into effect. Los Angeles County now requires additional wall bracing in new construction, including anchoring of chimneys and water heaters. A new Veterans Hospital in another area of Los Angeles County will replace the old, severely damaged hospital.

Managua, Nicaragua, Earthquake of December 23, 1972

The Managua earthquake of 1972 contrasts sharply with the San Fernando earthquake of the previous year. Its magnitude (6.2) was slightly less than the 6.6 magnitude of the San Fernando earthquake. But Nicaraguan buildings are not constructed to resist earthquakes. As a result, the earthquake devastated Managua and killed an estimated 4000 to 6000 people. With limited resources, these people and others in underdeveloped earthquake-prone regions of the world have little choice but to rebuild their adobe homes and wait passively for the next earthquake.

Geologic Setting

Managua lies near the western edge of the Nicaragua Trough, a depression which runs through much of Central America. The depression probably formed as a result of differential motion of the Caribbean and Pacific plates. The Caribbean plate appears to be moving eastward, dragging eastern Nicaragua with it, while the Pacific plate is moving northward, sliding obliquely beneath western Nicaragua. In the process, eastern and western Nicaragua are separating; the Nicaragua Trough is the surface expression of the zone of separation.

Active volcanoes in and around the Nicaragua Trough have filled the trough with lava and ash. The upper 1.4 kilometers of the trough consist mainly of unconsolidated ash and lake bed deposits, a factor that tends to intensify earthquake damage.

The Earthquake

On the morning of December 22, 1972, two small earthquakes rattled Managua. But small earthquakes are frequent, and no particular attention was paid to these foreshocks. At 12:31 AM the following morning, the main shock hit, trapping many residents in bed beneath the rubble of their collapsing homes made of *taquezal*, a combination of wood and adobe. Aftershocks strong enough to be felt continued for several weeks.

Lessons of the Earthquake

Approximately 400,000 people—20 per cent of the total population of Nicaragua—lived in Managua. Managua was the seat of government and the major industrial center of Nicaragua. Between 4000 and 6000 people were killed and another 20,000 injured. Roughly 57,000 buildings were destroyed or severely damaged, and between 200,000 and 250,000 people were left homeless. Among the survivors, poor health will probably shorten the life span of a significant number. Recovery of Nicaraguan industry will require several years.

Following the earthquake, there was some discussion of relocating Managua in a safer area, or at least rebuilding on bedrock rather than the soft alluvial soil of the trough. As this is being written, it seems more likely that Managua will be rebuilt much as it was. As with most developing countries, Nicaragua lacks the resources to build many steel-reinforced buildings capable of withstanding vibration. The average citizen similarly lacks resources necessary to deviate from taquezal construction.

The fault responsible for the earthquake goes through the city of Managua. It is probably part of the same fault system responsible for the devastation of Managua in 1931. For most people of Managua, the only available earthquake protection is prayer that another 40 years will elapse before Managua is again destroyed.

Charleston Earthquake of 1886

Earthquakes in parts of the United States outside "earthquake belts" in California and Nevada are not as rare as one might think. A series of earthquakes struck the central Mississippi River Valley in 1811–12. Judging from reports of effects, magnitudes of several of the earthquakes exceeded 8, and one earthquake magnitude may have exceeded the 8.3 level estimated for the San Francisco earthquake of 1906.

Numerous earthquakes have occurred along the Appalachian Mountains, the St. Lawrence River, around Charleston, South Carolina, and off the coast of New England. In 1929, an earthquake in the Atlantic Ocean south of Newfoundland (the Grand Banks earthquake) sent a tsunami against the Newfoundland coast, causing much damage and several deaths. The earthquake triggered submarine landslides that broke 12 transatlantic cables. The slides carried away large segments of the cables, necessitating expensive repairs.

The frequency of earthquakes of varying magnitudes in certain parts of the world indicates a relationship between the number of small earthquakes and the number of large earthquakes in a given area. The data suggest that in earthquake-prone areas, a small but predictable percentage of the earthquakes will be "monster" earthquakes. Stated another way, the question is not *whether* a magnitude 8+ earthquake will rock the Charleston (or Grand Banks) area but *when* the earthquake will occur. Monster earthquakes can be expected in all earthquake-prone areas. They merely occur more often in some areas (such as California and Nevada) than others (such as Charleston or New England).

Geologic Setting

Following the breakup and drifting apart of the American, Eurasian and African plates 150 to 200 million years ago, the coastal zone of the southeastern United States sagged gently so that "hard" rock of the underlying original continental crust is several thousand feet beneath present-day Charleston. Surface and near-surface "rocks" consist of sand, silt, mud and clay eroded from distant Appalachian highlands and carried by rivers to the sea. Some of the sediment is deposited in the sea; other sediment is being deposited in sounds and bays along the coast. Unconsolidated rock and soil make the Charleston area especially vulnerable to earthquake damage.

The Earthquake

Between 9:51 PM, August 31, 1866, and 8 AM the following day, ten shocks shattered the calm of the old southern city of Charleston. The first and strongest shock consisted of two periods of vibration separated by one or two seconds. This P-S interval indicates that Charleston was approximately 10 to 15 miles away from the focus of the earthquake. (An interesting bit of trivia is apparent in times reported for the shocks. Virtually all clocks in Charleston were pendulum clocks.

Earthquake vibration stopped the clocks during the first shock. For this reason, the time of the initial shock of earthquakes occurring during the eighteenth and nineteenth centuries is almost always known precisely. Times of subsequent shocks are almost always estimated.)

Approximately 90 per cent of the brick buildings in the city were damaged, almost all chimneys collapsed and damage to plaster and foundations was widespread. Altogether 102 buildings were declared unsafe and pulled down, and a number of others were pulled down because of the excessive cost of repair. Officially 27 people were killed outright. Injuries and exposure raised the number of deaths to 83. A small tsunami in the Cooper River inundated neighboring rice fields. Cracks and fissures opened in the ground, spewing mud and water, and an odor of sulfur gas was noticed.

The epicenter of the earthquake was 14 miles northwest of Charleston, near the railroad station of Tenmile Hill. In Summerville, the settlement nearest the epicenter, two people were killed. Wood frame buildings in Summerville were badly damaged, in contrast with

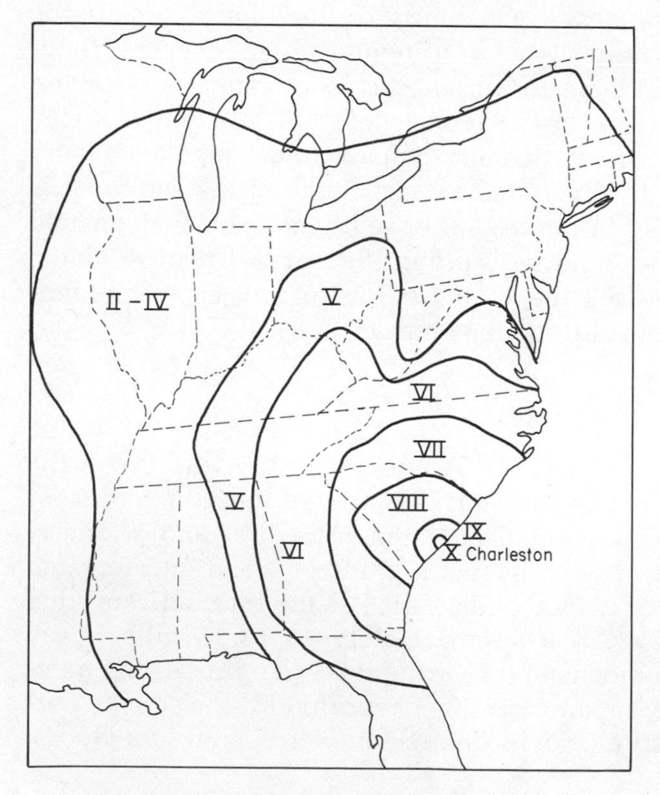

Figure 2–33. Intensities observed during the 1886 Charleston earthquake, the largest earthquake striking the east coast of the United States during recorded history.

the slight damage to wood frame buildings in Charleston.

Water rose in wells just before the shock, and one foreshock had caused concern on the preceding Friday, August 27. In Langley, South Carolina, some distance away, a millpond dam broke. The resulting flood washed out railroad tracks below the dam, wrecked two trains and drowned two firemen. Lighthouses rocked in New York, chimneys fell in Kentucky and Ohio and people felt the earthquake in Arkansas, Wisconsin, and Ontario. Intensities are shown in Figure 2–33.

Prognosis

Small earthquakes continue to occur in a zone extending inland from Charleston. Little is known about the nature of the fault or faults causing the earthquakes. The technical ability needed to establish the regime and periodicity of earthquakes in this area is available. Hopefully, this knowledge will be utilized before the next severe earthquake occurs.

Denver Earthquakes of the 1960's

The Denver earthquakes caused very little damage. Some plaster cracked, tremors disturbed people and a few chimneys were damaged. A number of earthquakes were felt, but except for three in 1967 that had magnitudes between 5 and 5½, all had magnitudes less than 5. Why, then, are the Denver earthquakes important?

Man triggered the Denver earthquakes, but this is not the primary reason for their importance. Man had triggered earthquakes before. Over 600 earthquakes accompanied the filling of Lake Mead in Arizona. Filling of the Koyna Reservoir in India triggered an earthquake with a magnitude of 6½, resulting in 200 deaths and widespread destruction. Underground nuclear explosions frequently trigger earthquakes in southern Nevada. Rocks in many areas are deformed and near the fracture point, and a small disturbance in such an area can cause an earthquake.

Fluid injection (pumping fluids into rocks) triggered the Denver earthquakes. Oil companies use this process routinely to fracture oil-bearing rocks around an oil well. When properly done, the technique increases the flow of oil into a well, but fluid injection alone does not trigger magnitude-5 earthquakes.

In terms of the important information gathered, Denver was a fortuitous location for the earthquakes. Had they taken place in an earthquake-prone area, they probably would not have been recognized as being man-made. Had they occurred near most other cities outside earthquake-prone areas in the United States, the instrumentation and scientific expertise necessary to solve the riddle of the earthquakes would not have been so readily available. In Denver, however, seismologists of the U.S. Geological Survey had been recording earthquake waves generated by underground nuclear and dynamite explosions. These explosions took place in many widely scattered areas, so the seismologists, led by L. C. Pakiser and J. H. Healy, had mounted their seismographs on trucks. They could move into an area and set out detectors within a few hours. In nearby Bergen Park, Colorado, Colorado School of Mines had recently established a seismograph station. Without the U.S.G.S. and School of Mines facilities, the following story might never have been told, and we might not have realized for many years that man may be able to control earthquakes by fluid injection.

Geologic Setting

Denver is located on the Great Plains at the foot of the Rocky Mountains (Fig. 2–34). The area was not always as high as it is now. For millions of years during the Mesozoic era (Age of Dinosaurs), the crust in the area subsided slowly. At times a vast sea covered the area, at other times it was above sea level and much of the time the area was swampy. As the crust subsided, sediments carried in by rivers filled the basin. During much of the time, sediments were predominantly fine clay. The clay compacted into a thick, widespread rock unit now called the Pierre Shale. Altogether over 12,000 feet of sediments were deposited in the Denver basin. Formation of the Rocky Mountains signaled the end of sedimentation in the basin. The entire area began to rise, and rocks formed from basin sediments began to erode. The original land surface of the region

Figure 2–34. The Denver skyline partially buried in early morning smog and haze.

is still below sea level. We know little about rocks of the prebasin crust except where wells such as the Rocky Mountain Arsenal well penetrate the old rocks. These rocks are thought to be similar to ancient crustal rocks exposed elsewhere.

Rocky Mountain Arsenal Well

The Army was in a bind. Since 1942, they had manufactured chemical weapons at the Rocky Mountain Arsenal north of Denver (Fig. 2–35). Until 1961, waste water evaporated in nearby open pits; then waste from the pits began seeping into the ground, contaminating the ground water. Something had to be done.

The Army decided to drill a well into the ancient crustal rock beneath the sediments. Waste liquids could be pumped into the rock. The depth of the well insured that liquids could not endanger the near-surface water supply. It seems like a good plan. The well, 12,045 feet deep and 100 feet into the ancient crustal rock, was completed in September, 1961. Pumping of waste into the disposal well began in March of 1962.

The Earthquakes

About seven weeks after pumping began, people in the area around the arsenal began to feel earthquakes.

Figure 2–35. The Rocky Mountain Arsenal main gate. The Rocky Mountains rise in the distance. Denver lies to the left rear of the observer.

More than three years elapsed, however, before it was realized that the well might be related to the earthquakes.

In November, 1965, a consulting geologist, David Evans, presented data that demonstrated a correlation between volume of injected waste and earthquake frequency (Fig. 2–36). Evans pointed out that recent work by a graduate student at Colorado School of Mines

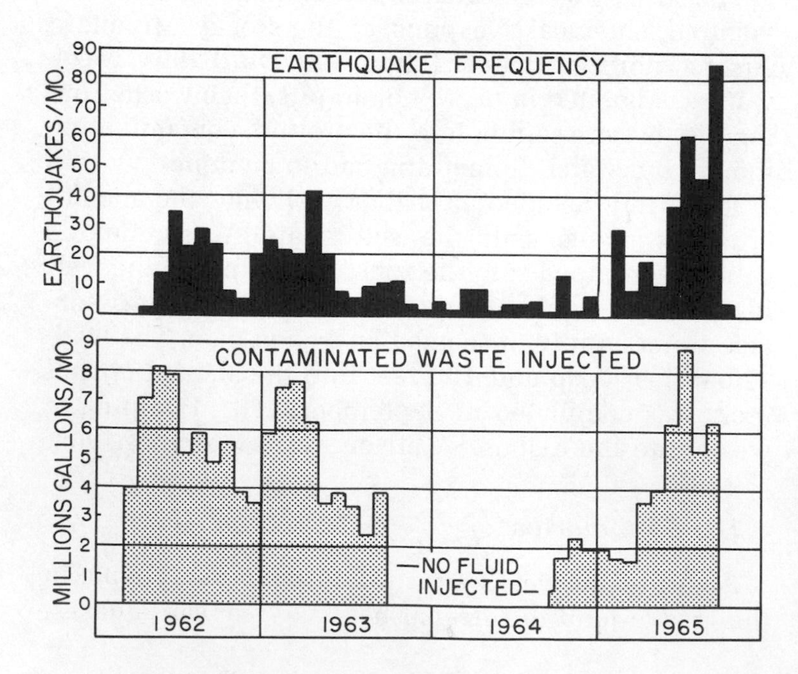

Figure 2–36. David Evans's correlation of fluid injected and earthquake frequency. Evans's data precipitated a crash study of the Denver earthquakes, which resulted in cessation of pumping of water into the ground at the Rocky Mountain Arsenal.

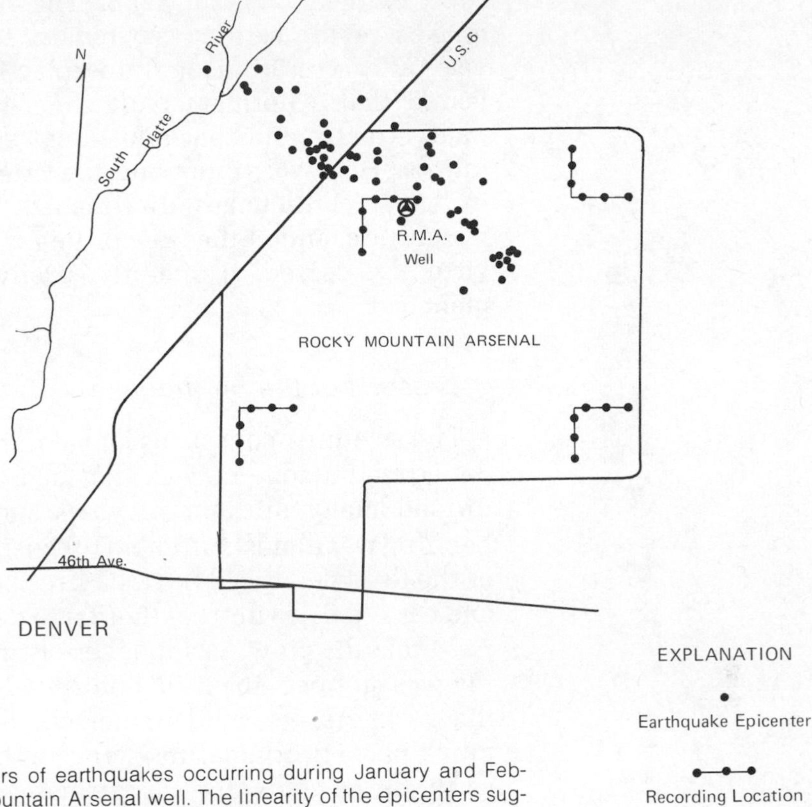

Figure 2–37. Epicenters of earthquakes occurring during January and February near the Rocky Mountain Arsenal well. The linearity of the epicenters suggests motion of a northwest-southwest trending fault. (From U.S. Geological Survey data.)

showed that the earthquake epicenters formed a rough oval, with the well near its center.

At the request of the U.S. Army Corps of Engineers, U.S.G.S. seismologists quickly established seismographs in the vicinity of the well. During January and February of 1966, U.S.G.S. seismographs recorded 62 earthquakes (Fig. 2–37). Earthquake foci clustered in a northwest-southeast trending zone 6 miles long, 2 miles wide and between 3 and 3½ miles deep. The center of the earthquake zone was just below the bottom of the well.

Newspaper accounts and historic records were searched for evidence of previous earthquakes in the area, as were records from a seismograph station operated in Boulder, Colorado, between 1954 and 1959. The search produced no evidence of previous earthquakes in the area of activity.

Realizing the probable relationship of well and earthquakes, the Army permanently stopped injection of fluid on February 20, 1966. The three largest earthquakes of the series occurred in 1967, more than a year after cessation of fluid injection. Seismologists feared that a dormant fault zone had been activated inadvertently and that additional earthquakes would follow. However, following the brief burst of activity in 1967, earthquakes diminished in frequency and magnitude, and at the time of this writing, activity has virtually ceased. Hopefully, Denver is no longer menaced.

Lessons of the Earthquake

The scientific consensus is that the well at the arsenal penetrated a zone of rock that had been bent and deformed many millions of years ago. The amount of bending was almost sufficient to cause rupture. Injection of the fluid decreased the rock's resistance to fracturing and caused new ruptures (faults) to develop in the rock. As the faults grew, earthquakes occurred. After pumping was stopped, the fluid continued to spread through the rock. As the fluid dispersed, the rocks regained much of their original resistance to fracturing, and the earthquakes gradually diminished.

The most important result of the Denver earthquakes is the intriguing suggestion that *man might be able to eliminate disastrous earthquakes in some areas by deliberately causing many small earthquakes*. Large earthquakes occur when opposing crustal blocks stick instead of slipping. If the rocks are severely deformed before rupture and slip occur, the earthquake will be severe. If, on the other hand, the rocks are weakened and rupture more easily, then earthquakes in the area will be less severe. If rocks are artificially weakened in areas where severe earthquakes are certain to occur, many small earthquakes could theoretically replace the occasional severe earthquake and cause a great deal less damage.

Postscript

The role of fluid pressure in earthquakes is an area of active research. As this is being written, J. H. Healy and C. B. Raleigh are continuing their research on this problem. They are working in the Rangeley oil field in a remote area of northwestern Colorado. They have been able to "turn on" earthquakes by pumping water into a well and then "turn them off" by pumping the water out. It is a very exciting study.

4 WHAT ABOUT THE FUTURE?

Control

As we have seen, investigation of the Denver earthquakes and continuing research at the Rangeley oil field suggest that man eventually may be able to control earthquakes to some extent. Although this is an intriguing possibility, we must remember that research is still in a very early stage.

Meanwhile, many questions remain unanswered. The San Andreas fault is hundreds of miles long. How many wells must be drilled in order to eliminate earthquakes with magnitudes greater than 7? How much will the wells cost? A deep oil well may cost millions of dollars. State and federal governments have been unwilling to provide adequate support for earthquake research; is it likely that they will underwrite expensive earthquake control wells? Can underdeveloped countries afford the costs? Costs of earthquake control might constitute a disproportionately large amount of the gross national product of a poor country. When all these factors are considered, it becomes clear that many years may pass before large-scale earthquake control is a reality.

Prediction

Earthquake control seems far in the future, but earthquake prediction is feasible within the next 5 to 10 years. Japanese scientists already issue bulletins, similar to storm warnings in the United States, notifying residents of the possibility of severe earthquakes in selected areas.

Earthquake predictive techniques fall into two rather broad categories: long-range predictions (years) and short-range predictions (hours, days or months). The history and periodicity of activity along the San Andreas fault in California, for example, indicates that the fault will probably rupture within the next few decades and that the rupture will probably occur somewhere between Los Angeles and San Fran-

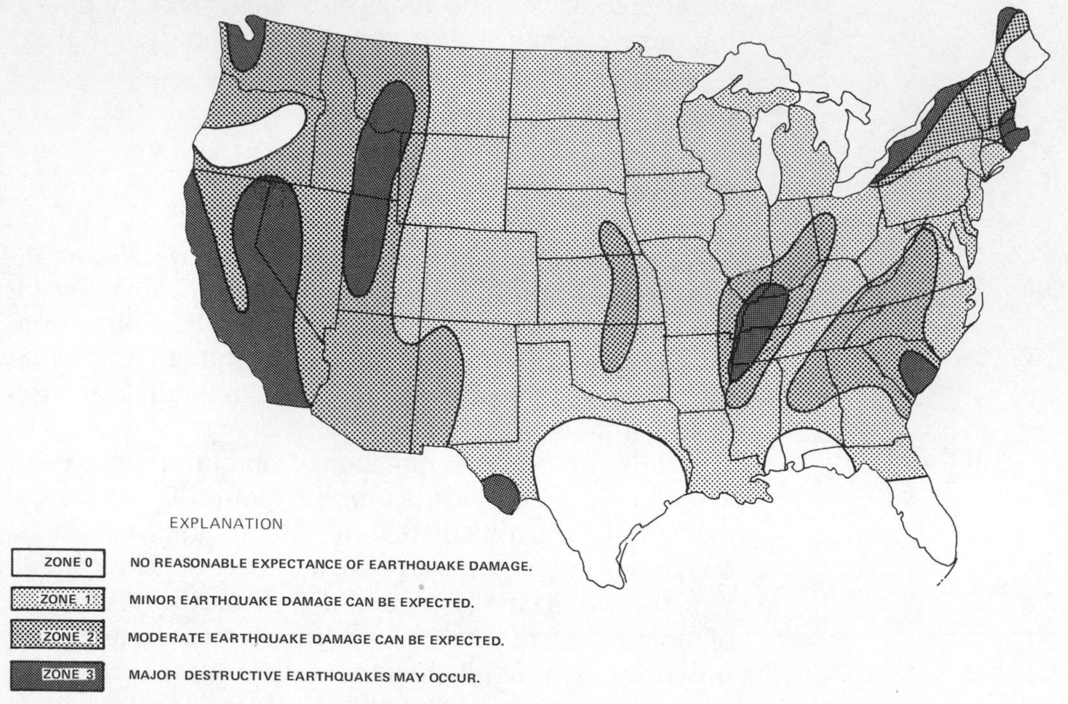

Figure 2–38. Earthquake risks in the U.S. (National Oceanographic and Atmospheric Administration.)

cisco. In this area the fault has "locked" since an earthquake in 1857. Crustal plates on either side of the fault are moving at a rate of about 1.6 inches per year. Accumulated strain is therefore about 15 feet, or enough to cause a large earthquake (a shift of 15 to 20 feet accompanied the 1906 earthquake). Damage will run into billions of dollars, and thousands of people will be killed or injured. This prediction is not fiction. It is a view widely held by seismologists and has been publicized on nationwide television.

Seismologists have divided the United States into zones based on likelihood of earthquake damage (Fig. 2–38). Certain earthquake-prone regions have been further subdivided into seismic risk zones and areas of major faults. Earthquake risk maps constitute a crude form of long-range prediction.

Vagueness of long-range predictions limits their usefulness. Most people will not worry about nebulous future dangers when present-day problems press. Short-range prediction would be more effective. Short-range predictions are based on rock behavior in the laboratory and subsequent confirmation in field studies. In the laboratory, it was learned that heavily stressed rock yields slowly. For a relatively short time before rupture, the rate of yielding speeds up. Changes in electrical, magnetic and other physical properties of the rock seem to occur during the period of faster yielding. Seismologists are trying to use these changes to predict earthquakes a few hours or days in advance.

Japanese seismologists measure tilt of the ground surface, changes in distance between markers embedded in the ground, number and frequency of very small earthquakes and small changes in the earth's magnetic field near fault zones. Significant changes in some or all of these phenomena occur before major shocks. Swarms of very small earthquakes have been particularly good indicators of forthcoming major shocks. However, the Japanese scientists report that not everyone appreciates their efforts. In particular, those engaged in tourist-related businesses were opposed to the publicized warnings. The problem, therefore, is not an exclusively scientific one.

At the time this book is being written, monitoring of the P-wave—S-wave velocity ratio offers the best method of predicting earthquakes. This method consists of repeated measuring of P-wave and S-wave velocities in earthquake-prone regions, dividing the P-wave velocity by the S-wave velocity and plotting the ratio. The ratio decreases a few weeks or months before a major earthquake, and the magnitude of the earthquake is roughly proportional to the duration of the anomalous ratio. That is, the longer the low ratio is observed, the larger will be the earthquake. The graph in Figure 2–39 shows the behavior of the velocity ratio before and after the San Fernando earthquake.

The mechanism causing the change in seismic wave velocities is probably similar to the mechanism that caused the Denver earthquakes. Simply described, this mechanism operates as follows: The Denver earthquakes resulted from the increased pressure of fluids in the pores of the rock. Increased pressure caused by pumping weakened the rocks around the well to a point at which "fossil" stresses frozen in the rock for millions of years caused fracturing, probably along a pre-existing fault. Pumping fluid out of wells strengthens the rocks, and earthquakes cease.

A similar natural phenomenon apparently causes the anomalous behavior of the P- and S-wave velocity ratio immediately prior to some earthquakes. In these cases, the rocks begin to bend somewhat more rapidly a few

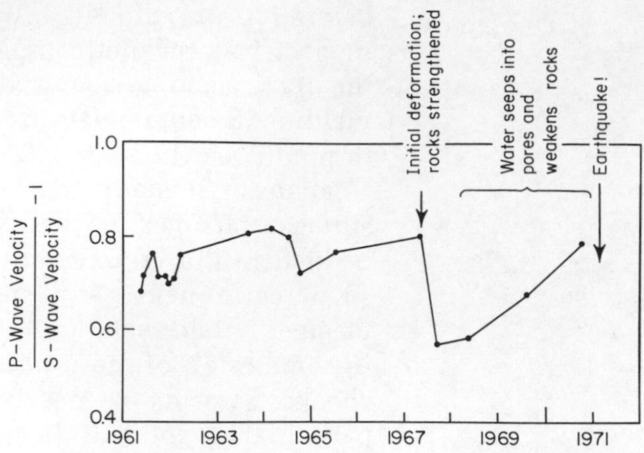

Figure 2–39. Compressional wave–shear wave velocity ratio anomaly preceding the 1971 San Fernando earthquake. (Adapted from Whitcomb, J. H., et al.: Earthquake prediction: variation of seismic velocities before the San Fernando earthquake. Science *180*:632–635, 1973.)

weeks or months before the earthquake. The bending causes expansion in some parts of the rock mass. Most of the expansion occurs in the form of increased pore volume. Water in the pores is now under less pressure because of the expansion. Often *there is not enough water to completely fill the pores.* This lack of water has the same effect as pumping water out of the wells. The rock is *strengthened* and becomes *more resistant* to fracturing. At the same time the rock becomes *more compressible* and the P-wave (which is a compressional wave) velocity drops. This is because any body, rock or otherwise, is easier to squeeze or compress if it has air pockets in it than if its holes are filled with water, which is hard to compress.

The greater strength of the rock is temporary, however. Fluids, mainly pore water, seep into the undersaturated region. The seepage, like the pumping of water into wells, gradually weakens the rock until it fractures and slips, causing an earthquake.

The time duration of the anomalous velocity ratio is a function of the volume of deformed rock. A larger volume of deformed rock means that (1) a longer time is required for the water to seep in and fill the partially empty pores; and (2) a proportionally larger earthquake will occur.

Results of United States and Japanese studies indicate that an effective warning system capable of predicting earthquakes a few hours or days in advance of a large shock is technically feasible. Establishment of the warning system will require substantially more money than is presently allocated. Not only must potentially dangerous fault zones be instrumented and continuously monitored but, as indicated in the San Fernando case history, many potentially dangerous faults remain undiscovered. They too must be located and monitored. In California, the San Andreas system is well monitored in spots, but much remains to be done. With the exception of the San Andreas system, most potentially dangerous faults in the United States are monitored only by conventional seismograph stations usually too far away to provide useful prediction data.

Damage Prevention

Seismologists often look at the gloomy side of earthquakes and earthquake damage, seeing only those things needing to be done and forgetting what has already been accomplished. For example, comparison of the death toll of the San Fernando earthquake with the death toll of the Managua earthquake, in which thousands died, shows the effectiveness of earthquake-resistant building design in the United States. Earthquakes cause relatively few deaths in the United States, and with proper preventive measures, most of these could be prevented.

Building codes have gradually improved. New information, such as the very strong forces observed during the San Fernando earthquake, will result in even better codes. However, improved building codes usually apply only to new construction. Older buildings and structures such as the Veterans Hospital and Van Norman Dam in the San Fernando area are usually exempt from re-evaluation of damage potential. Hopefully, either something will be done about unsafe older structures or an early warning system will allow evacuation of questionable structures before the next severe earthquake.

3

VOLCANOES AND IGNEOUS PROCESSES

The San Francisco Peaks near Flagstaff, Arizona, in the spring. Northern Arizona Indians thought that demigods lived on the mountain. The peaks are on the rim of an extinct volcano.

1 INTRODUCTION

Eruption of Mt. Pelée, May 8, 1902

On the afternoon of May 7, 1902, the French colonial governor, Louis Mottet, his wife and the administration principal secretary, Edouard L'Heurre, boarded a coach in Fort-de-France, Martinique, and left for St. Pierre (Fig. 3–1), where they met the mayor and several prominent citizens. The discussion that afternoon centered around two things—the intermittent eruptions of nearby Mt. Pelée and the election scheduled for May 10, three days later. The eruptions had begun on April 23 and had been increasing in size and frequency. St. Pierre was at a standstill. Ash blanketed everything, reaching a depth of one foot in some places. There were reports that lava flows and mudflows had killed people living on the mountainside. How much did Mt. Pelée actually endanger St. Pierre? Should the city be evacuated?

The people attending the meeting in St. Pierre that afternoon were leaders of the incumbent Progressive

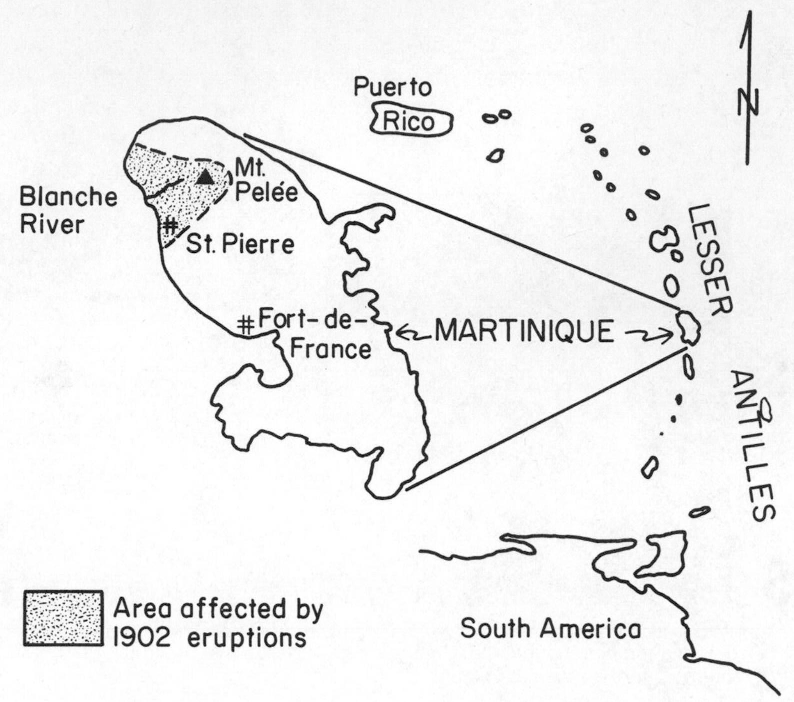

Figure 3–1. Lesser Antilles and Island of Martinique.

Party. The opposition Radical Party was vigorous and had intelligent leadership. The election next Sunday would be close. The city would probably go to the Progressives, the countryside to the Radicals. Evacuation of the city would cost the Progressives votes and possibly the election. The meeting continued through the afternoon, breaking up late. The governor and his wife decided against the bumpy ride back to Fort-de-France that evening. L'Heurre returned alone to await an expected cable from the French government in Paris. This action would save his life.

Louis Auguste Ciparis didn't care who won the election or whether or not the city was evacuated. He was locked in a cell that was especially small and uncomfortable on the morning of May 8. For several days now, smoke and ash from the rumbling volcano had been filtering into the cell, filling the air with fine grit. Ash piled up outside the door, cutting off what little light and air might otherwise seep in, but at least the heat kept away the jailers.

As Ciparis cursed the heat, the *Roraima* arrived and tied up at the St. Pierre docks. Crewmen of the *Roraima*, who could plainly see the smoking volcano in the clear morning air, wanted to unload and get away as quickly as possible. They wouldn't make it.

At 7:50 AM, May 8, 1902, four explosions ripped the mountainside. Two clouds billowed out, one rising vertically, the other roaring laterally out of the mountain and devastating a large area southwest and west of the summit (Fig. 3–1). St. Pierre, a gay waterfront city, the "Paris of the West Indies," was no more.

Governor Mottet and his wife died as hot gases swept through open screenless windows of the tropical city. Leon Compare-Leandre, a shoemaker, survived the blast, then ran to safety through four miles of hot ash. A rescue party probing the smoking ruins two days later discovered Ciparis trapped in his prison cell. Ash piled around openings had blocked the entrance of hot gases that might have killed Ciparis, who was later pardoned. The *Roraima*, on fire, missing boats, bridge, mast and funnel and with 28 of her 47-man crew dead or dying, remained afloat. Of 19 vessels in the St. Pierre harbor on May 8, only two survived. The eruption killed an estimated 30,000 people. Ciparis and Compare-Leandre were the only survivors in St. Pierre.

A *glowing avalanche* had overwhelmed St. Pierre. This type of eruption, characterized by red-hot ash and rock enclosed in a rapidly flowing cloud of gas, is the most violent of the several classes of eruptions recognized by volcanologists. The heat of the avalanche softened glass, carbonized fresh fruit and set wood, fabric and paper on fire. Hot gases easily penetrated the buildings in St. Pierre, which, like those in most tropical cities, lacked glass in their windows. Death probably came quickly from inhalation of hot (over 600° Centigrade) gases.

Large amounts of gas dissolved in the magma were responsible for the glowing avalanche. All lavas contain dissolved gas, but in most cases the gas merely forms harmless bubbles in the lava. Inside Mt. Pelée, however, the weight of the overlying rocks held the gas in solution. As the rising magma neared the top of the volcano, the weight of overlying rocks decreased until gas pressure exceeded rock pressure. At this point, the gas-charged magma exploded violently, and the mixture of gas and hot fragments of lava flowed down the mountainside. Several glowing-avalanche eruptions followed the May 8 eruption of Mt. Pelée, and one on August 30, 1902, killed an additional 2000 people.

Eruption of Kilauea, Hawaii, 1959–60

The placidity of the 1959 eruption of Kilauea on the island of Hawaii contrasts strongly with the violence of the 1902 eruption of Mt. Pelée. The Kilauea eruption had no glowing avalanche and resulted in no great casualty list and only light damage. The relatively tame Kilauea eruption is particularly interesting, however, because volcanologists' instruments followed moving magma from deep within the earth, first to the crater where an eruption occurred, then back underground toward the site of the final eruption from the flank of the volcano.

The Hawaiian Islands lie at the southeastern end of a great mountain chain extending more than 1500 miles from Hawaii to Midway Island and beyond. The mountain chain (except the island of Hawaii) consists of extinct volcanoes formed as magma seeped through cracks in the earth's crust. Many geologists believe that Hawaii and its active volcanoes overlie a *hot-spot* located in the upper mantle and that movement of the sea floor over the hot-spot formed the new volcanoes. According to this hypothesis, the northern end of the Emperor Seamount chain was originally above the hot-spot (Fig. 3–2).

During the period 43 to 70 m.y. (million years) ago, the Pacific plate drifted slowly northward as the Emperor Seamounts formed. Approximately 43 m.y. ago, the direction of plate motion changed from northward to westward, and the speed of drift increased. The Hawaiian chain formed during the following period, 43 m.y. ago to the present. The lithosphere cools and contracts as it drifts away from the hot-spot, causing subsidence and submergence of older volcanic islands. Hence, the Emperor Seamounts and many of the western mountains of the Hawaiian chain are now beneath the surface of the sea.

Two volcanoes in the chain, Kilauea and Mauna Loa, are active. They are located on Hawaii, the southeast-

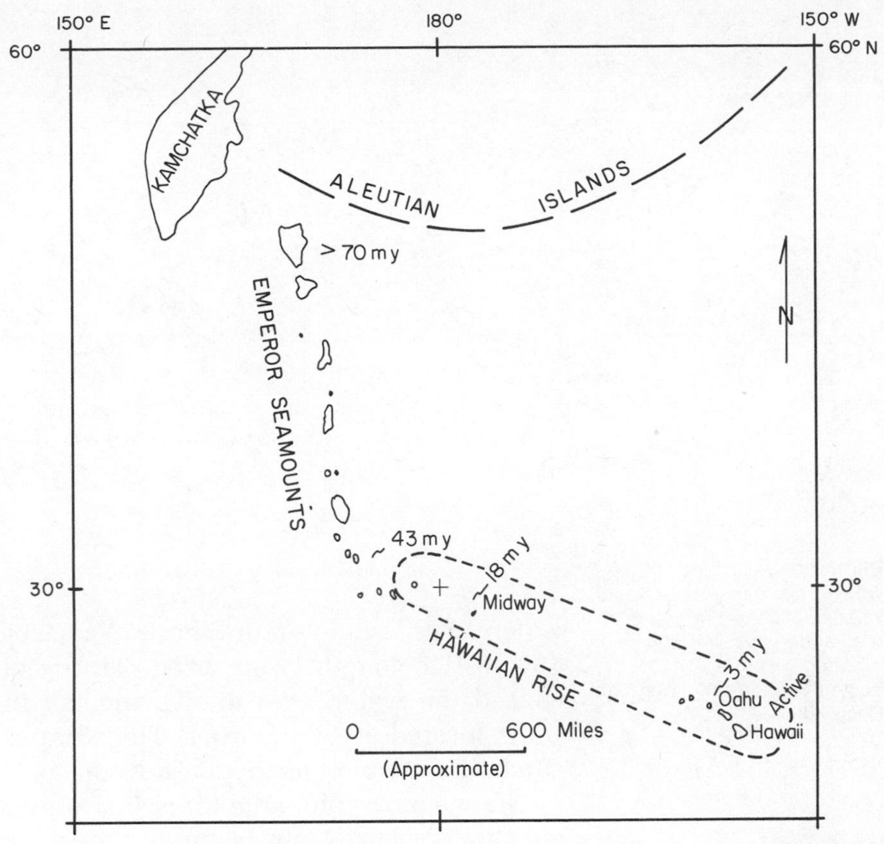

Figure 3–2. The northern Pacific Ocean, the Emperor Seamounts, the Hawaiian rise and selected ages of volcanic rocks collected from the region.

ernmost island of the mountain chain. The summit of Mauna Loa rises more than 13,000 feet above sea level and more than 30,000 feet above the sea floor of the Pacific. Measured from base to summit, Mauna Loa is the highest mountain on earth. Kilauea, which rises over 4000 feet above sea level on the eastern tip of the island (Fig. 3–3) has been the more active volcano during recorded history.

The 1959–1960 eruption consisted of two major outpourings of lava. The first, a summit eruption, occurred between November 14 and December 21, 1959; the second, a flank eruption, occurred between January 13 and February 20, 1960.

Volcanologists observing the eruption relied on two instruments to tell them what was happening underground prior to and during the eruption. Tiltmeters (a device for measuring very small changes in the inclination or tilt of the earth's surface) consisting of

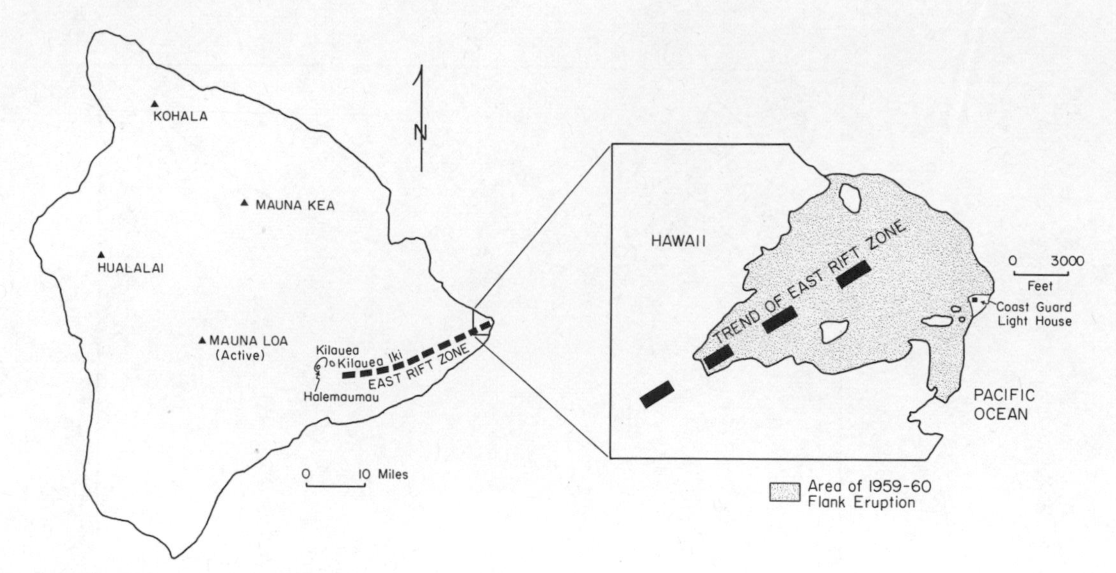

Figure 3–3. Island of Hawaii. Summit eruption of Kilauea occurred in Halemaumau and Kilauea-Iki. The flank eruption occurred near the eastern end of the East Rift Zone.

water tubes laid out in the shape of a triangle, each side 150 feet in length, were used. Tilting of the ground caused the water level to rise and fall in observation ports located at each corner. This simple device indicates tilts as small as .00001 degree.

The seismograph is the other instrument used by the volcanologists to study underground migration of magma. Two aspects of magma movement generate earthquakes. In order for the magma to move through the ground, old cracks must be reopened and new cracks opened. Each time the magma forces open another crack, the rocks shift slightly and a small earthquake occurs. As the magma moves through underground cracks, its movement causes a "gurgling" type of vibration, which the seismographs record. Seismologists distinguish two gurgling vibrations, a "spasmodic tremor" when magma is moving underground and a more periodic "harmonic tremor" when the lava erupts and flows from its vent. Vibration caused by opening of cracks and those caused by movement of magma and lava have distinct and identifiable characteristics on seismograms.

The first clue that an eruption might be forthcoming was noticed in February, 1959, when tiltmeters indicated that Kilauea's crater was swelling and tilting outward. In May, the swelling subsided somewhat, but it began again in early August.

In the middle of August, seismographs on the island recorded a great swarm of earthquakes whose foci were

30 to 35 miles deep and a few miles northeast of the Kilauea summit. Cracks deep beneath the earth opened during these earthquakes. The magma had begun its slow ascent to the surface, accompanied by spasmodic tremors recorded on the seismographs.

By September, upward-moving magma was close enough to the surface to cause a series of shallow, light earthquakes beneath the northern rim of Halemaumau (Figs. 3–3 and 3–4). On November 1, a series of shallow, very light earthquakes began, with as many as 1000 per day being recorded. During the second week of November, the rate of swelling increased to three times the rate observed in previous months. On November 14, the number and size of shallow earthquakes intensified even more as the magma moved the remaining distance to the surface. For five hours, the entire summit of Kilauea shuddered, and at 8:08 PM, lava erupted in Kilauea-Iki crater (Fig. 3–3) immediately east of the main *caldera* (a large crater usually formed by inward collapse after eruption). The shuddering ceased abruptly, and a strong harmonic tremor appeared on the seismographs.

Between November 14 and December 21, 17 distinct phases of activity were recorded. In the first and longest phase, 30 million cubic meters of lava poured into the crater of Kilauea-Iki. This phase lasted for a week.

Figure 3–4. Steam and hot gases rising from Halemaumau, the main summit crater of Kilauea. (See Figure 3–3 for exact location.)

During the eruption, tilting subsided slightly but later increased still further. Successive eruptions decreased in duration and amount of lava. Eruption from the summit ceased altogether on December 21, the same day that tilting reached its peak.

During the final week of December, an earthquake swarm occurred in the previously inactive East Rift Zone (Fig. 3–3). Cracks opening along the rift allowed magma to flow laterally underground along the rift zone. In early January, frequency of earthquakes in the rift zone increased. The center of earthquake activity moved progressively as the magma inched its way eastward.

On January 13, 1960, a harmonic tremor announced the eruption of lava along a fissure in the East Rift Zone about 25 miles east of Kilauea's summit. The fissure was a little more than 1/2 mile in length. During the succeeding five weeks, the fissure disgorged 100 million cubic meters of lava, which flowed down the gentle slope of Kilauea into the sea (Figs. 3–3 and 3–5).

The summit of Kilauea subsided sharply in January, 1960, and continued to sink until the summer of that year. The withdrawal of magma from beneath the summit left summit rocks poorly supported, and near the end of January, many small earthquakes occurred as the summit rocks cracked and sank. On February 7, the floor of Halemaumau (Fig. 3–6) sank 150 feet. A

Figure 3–5. Lava erupting from the East Rift Zone flowed down a gentle slope toward the sea. A tongue of lava stopped a few feet from a lighthouse near the beach. Location of the lighthouse is shown in Figure 3–3.

Figure 3–6. Floor and wall of Halemaumau. Condensation of sulfur in gases rising from beneath the floor causes white crust. Note layering in walls. Each layer represents a single lava flow.

smaller area near the center sank 200 to 350 feet but was partially refilled by lava.

Igneous Activity

Eruption of volcanoes such as Kilauea and Mt. Pelée are surface manifestations of *igneous activity*, a term used by geologists to denote all aspects of magma formation and movement. Volcanic activity is a small but obvious part of igneous activity. Less visible but far more important are other aspects of igneous activity. Some important forms of igneous activity take place far beneath the ground and will not affect man until millions of years of erosion have removed overlying rocks. Other important geologic processes are currently taking place but occur so slowly that their relationship to igneous activity is not immediately apparent.

A large part of the world's population lives in areas subject to volcanism. The densely inhabited islands of Japan, the Philippines and Southeast Asia are good examples. In this area, much of the adjacent heavily populated mainland provides evidence of Tertiary (0 to 60 m.y.) volcanic activity. The islands of Greece are remnants of an island arc formed by northward subduction of Mediterranean sea floor; the volcanoes of

Italy are well known; Mexico, Central America and western South America contain numerous active volcanoes; and rocks in much of the western United States are capped by lava flows or other volcanic ejecta. Clearly, an understanding of volcanism is important on the basis of the number of people living in or near past or present volcanic provinces. Few of these people realize the extent to which volcanism has affected their lives. For example, volcanic soils are among the world's richest. Minerals and elements brought to the surface with the lava help crops to grow as *weathering* (see Chaps. 5 and 10) releases these nutrients from the igneous rocks.

Lava flows alter the shape of the land. The courses of rivers change. Some areas become fertile and habitable, while others become uninhabitable. Volcanic mountain ranges deflect moisture-bearing winds (see Chap. 4) and control the distribution of rainfall. These and other effects of volcanism persist for millions of years after the last cinder spews from the crater and the last drop of lava solidifies.

The engineering properties of lavas and igneous rocks exposed by erosion differ significantly from the properties of many other rocks, and these differences affect construction costs. Volcanic terrains are seldom flat and rolling; more often they consist of rugged mountains and steep-sided valleys. The shape imposed on the land by volcanism is thus a factor in highway, railway and tunnel design.

Igneous processes provide heat that can be converted into electricity. *Geothermal energy*, or energy from hot rocks, is discussed in Chapter 8.

Many minerals are obtained directly or indirectly from *igneous rocks* (rocks which have been molten or fluid at some point in their history). Copper is mined from ancient lavas in the Great Lakes region and elsewhere; gold in the fabulously rich Witwatersrand of South Africa is thought to have been weathered out of ancient lavas, then concentrated in sands and gravels deposited in a nearby sea. The United States obtains much of its aluminum ore from Jamaica, where it is produced by weathering of igneous rocks. These are a few examples of the many ways that igneous activity shapes our environment.

Finally, we cannot ignore the esthetic contributions of igneous processes as we stand and gaze at the distant cone of a large volcano such as Mt. Fuji in Japan or a massive mountain range such as the Sierras of California and Nevada. Our lives are richer as a result of the topographic expression of igneous processes.

In this chapter, we will discuss volcanoes and their direct effects, the origin and distribution of heat within the earth, the three great families of volcanic rocks and volcanic phenomena including flows, vents, and volcanic mountains, concluding with a discussion of volcanoes and man. Indirectly related topics such as soils, geothermal energy and minerals will be discussed in later chapters.

2 WHAT CAUSES VOLCANOES?

The eruption of a volcano like Kilauea is the final chapter in a long book. The book begins about five billion years ago when the earth was first taking form, attaining its present size and shape and trapping large quantities of heat in its interior. After its initial formation, the earth's interior continued to gain internal heat from decay of radioactive elements, tidal dissipation and possible chemical reactions.

Heat has been escaping as well as accumulating. Every day heat "seeps" through rocks and into the sea or the atmosphere on its way into space. This process is known as conduction. *Convection* is another means of transferring heat from the earth's interior to its surface. The process of convection causes molten rock to rise to the surface because it is lighter than surrounding solid rocks, while cooler rocks tend to sink. Some geologists believe that convection causes sea floor spreading and subduction. They perceive the spreading-subduction process as a gigantic, continuously rotating cell in which hot rock rises along mid-ocean ridges, releases some of its heat, then moves laterally away from the ridge, cooling as it goes. The cooler, denser rock then sinks into the subduction zone, mixes with mantle rock and moves back toward the ridge along mantle pathways. This model, sometimes called the "conveyor belt model," is controversial, but most geologists agree that some sort of convection takes place in the earth's interior.

The distribution of heat within the earth is another

factor in the evolution of a volcano. Heat is not uniformly distributed within the earth. Heat flow investigations indicate that the earth's crust close to mid-ocean ridges and in marginal seas between island arcs and continents, as well as near inferred hot-spots, tends to be abnormally hot.

Pressure plays an important role in volcano formation. Increasing pressure tends to raise the melting temperature of rocks. The interaction of the *geothermal gradient*—the increase of temperature with depth inside the earth—and the pressure gradient is probably responsible for the existence of the low velocity zone (see Chap. 2).

Molten magma must escape to the surface of the ground in order for a volcano to erupt, and since most volcanoes are associated with faults or fracture zones, it would seem that the lava rises to the surface via pre-existing cracks formed by the constant shifting of the crust and lithosphere. The most active parts of the earth's crust, and therefore the areas with the greatest concentration of faults and fractures, are the plate boundaries. Thus it is no surprise to discover that active volcanoes are concentrated along plate boundaries. The volcanoes of Hawaii are some of the few that are active in plate interiors.

Sources of the Earth's Heat

Gravity

Although the temperature of the matter that first formed the earth is not known, we do know that these particles—possibly of atomic size, possibly actual pieces of rock—were floating in space under virtually no pressure. When the particles came together and formed the earth, their mutual gravitational attraction pulled them tightly together. As a result, immense pressures developed inside the earth. Rocks near the center of the earth were squeezed into a much smaller volume, a process that released tremendous amounts of primeval heat, much of which is still trapped inside the earth. It is impossible to duplicate in the laboratory the high pressures known to exist deep within the earth. Without being able to duplicate these pressures, we cannot measure the heat given off by the squeezed rock, but it may have been enough to melt the primeval earth, especially if matter forming the earth was already hot.

Radioactivity

Many radioactive elements occur in rocks, but only a few are sufficiently abundant to contribute significant amounts of heat to the earth's interior. The three most important radioactive elements are uranium, thorium and potassium-40. In the early history of the earth, these elements were probably distributed uniformly throughout the earth. Today, they are concentrated in the crust and upper mantle rocks, where they contribute some of the heat needed to melt the rocks that ultimately flow onto the earth's surface as lava.

Earth Tides

Each day two small bulges move around opposite sides of the earth as a result of the gravitational attraction of the moon. Each day the earth slows down by an almost immeasurably small degree as a result of friction created by the moving bulges, and each day tidal friction heats rocks inside the earth. The amount of heating due to tidal friction is relatively insignificant today, but it may have been important during the distant geologic past, when the moon was much closer to the earth and tides were larger. A significant fraction of the heat now trapped within the earth may be fossil heat originally generated by tidal friction.

Chemical Reactions

Originally the earth's interior was probably homogeneous, but now rocks are separated into three concentric zones. Lighter rocks are at the surface, and denser rocks are in the middle of the earth. Many chemical reactions must have taken place as rocks melted, solidified and shifted about. A net release of heat probably accompanied these reactions and added to the total heat in the earth's interior.

Distribution of Molten Rock within the Earth

Low Velocity Zone

Temperature measurements in mines and wells indicate that rocks become hotter with increasing depth.

The rate of temperature increase varies from place to place, but averages almost 2° F. per 100 feet outside volcanic areas. If the temperature of rocks increased uniformly at this rate, the melting temperatures of common crustal rocks would be reached at 20 miles or less, depending on water content. However, all rocks are not molten at depths greater than 20 miles because 1) pressure of overlying rocks increases their melting points; and 2) the rate of temperature increase observed in surface rocks appears not to persist at depth. The average depth at which rocks begin to melt is about 40 miles beneath oceans and is somewhat greater beneath the continents.

The way in which rocks melt differs from the way in which ice, for example, melts. Ice melts at precisely 32 degrees F., whereas rocks melt over a range of temperatures. A rock may begin to melt at 2000 degrees F. and not become completely molten until the temperature reaches 2200 degrees F. At temperatures between 2000 degrees F. and 2200 degrees F., solid and liquid components occur together. In addition, if increasing temperature were the only factor, rocks would begin to melt at a certain depth and be completely molten at some greater depth. However, a number of other factors affect the melting of rocks inside the earth. The most important other factors are the increase of pressure with depth (which causes the melting point of rocks to rise with depth), and water content (which lowers the melting point). Beneath the sea floor, the pressure effect dominates, and mantle rocks at depths greater than about 180 miles are entirely solid, no longer containing a molten component. Rocks at intermediate depths contain a molten component, but the percentage of molten material probably does not exceed a few per cent except locally.

Most of our information about the zone of partial melting derives from studies of seismic waves traversing this region. Molten rock in this zone slows seismic shear waves and, to a lesser extent, slows compressional waves traversing the zone. Shear wave velocities are especially sensitive to melting. Molten rock amounting to a few per cent of the whole is sufficient to account for a decrease in seismic wave velocities. This relatively small fraction (but large amount) of molten rock probably constitutes the reservoir in which much lava originates.

Hot-Spots

Not all magma originates in the low velocity zone. Some magma apparently originates in local hot-spots such as the one beneath Hawaii, which results either from a local accumulation of radioactive elements or from other poorly understood factors. Some hot-spots are above the low velocity zone; others seem to be within it.

During *orogenesis* (mountain-building), the lower crust is heated to the point at which large quantities of low-density rock form by melting and resolidification. Effects of this type of heating will be discussed in a later section.

3 THREE GREAT FAMILIES OF VOLCANIC ROCK

Geologists divide rocks into three groups; igneous, metamorphic and sedimentary. Igneous rocks form through melting of pre-existing rocks or through solidification of lava and magma; metamorphic rocks form through alteration by heat and/or pressure of pre-existing rocks; and sedimentary rocks form through deposition of rock-forming materials such as sands, shales, or silts or by chemical precipitation of rock-forming materials such as salt or calcium carbonate in bodies of water.

Igneous rocks are further divided into two classes, intrusive and extrusive, on the basis of whether the rock solidifies inside the earth or extrudes onto the surface of the earth and solidifies there. Lava is a type of extrusive rock; magmas that do not reach the surface (and there are many instances of this) solidify and form intrusive rocks.

Both intrusive and extrusive rocks are further divided on the basis of their chemical composition. Three great chemical families of rock encompass the vast majority of extrusive rock; the basalts, the andesites and the rhyolites. These divisions are based primarily on the percentage of SiO_2 (silicon dioxide, or quartz in its pure form) in the rock. Typical basalts contain about 50 per cent SiO_2, typical rhyolites about 75 per cent and andesites contain between 50 and 75 per cent SiO_2.

Differences in the nature of lava flows, the nature of

material ejected from volcanic vents, the shapes of volcanoes and the history of volcanic eruptions are directly related to differences in rock type. A brief discussion of the major families of igneous rocks will facilitate later discussions of volcanic phenomena and of volcanoes and man.

It should be pointed out, however, that the chemistry of magmas and their physical properties are complex topics. Many combinations of different elements occur. In some cases, seemingly small changes in composition cause relatively large changes in physical properties. The following discussion is therefore necessarily general. The student should consult an advanced text for further information about this subject.

Basalts

The most abundant lavas are the basalts. Virtually all lavas occurring on the oceanic islands are basalts, including the lavas of Hawaii, Iceland and other islands of mid-oceanic ridges. Many continental lavas are basalts as well, including the vast lava plateaus such as the Columbia of the northwestern United States and Deccan of India.

Most basalts originate through melting of mantle

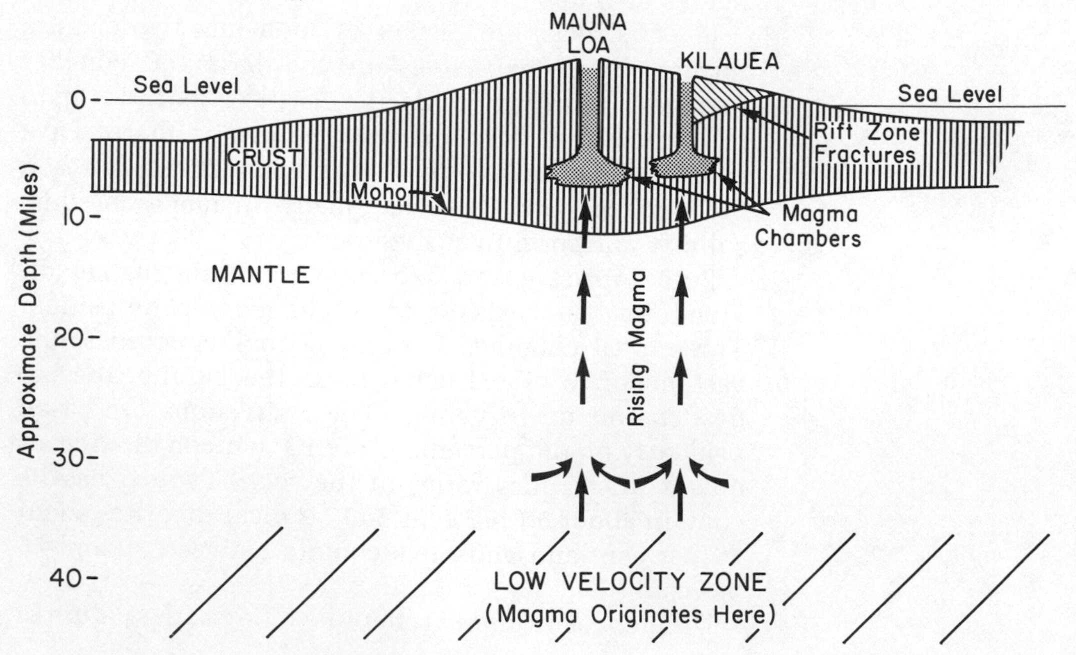

Figure 3–7. Inferred crustal structure of Kilauea and Mauna Loa volcanoes, Hawaii. Lava probably originates near the top of the low velocity zone, rises via fractures, then accumulates inside the volcano, causing the volcano to swell. When the lava erupts through the surface, swelling subsides. In the 1959–60 eruption, the lava flowed underground through rift zone fractures and erupted again on the flank of Kilauea. (From Macdonald, G.A., and Hubbard, D.H.: Volcanoes of the national parks in Hawaii, 5th ed. Hawaii National Park, Hawaii National Historical Association, 1970.)

and if a large percentage of the whole is melted, magma composition will be basaltic. (A similar phenomenon occurs when basalt forms from partial melting of upper mantle rocks, which contain less SiO_2. This phenomenon, differentiation of igneous rocks, is a fascinating subject but is largely outside the scope of this work.)

Rhyolites

Rhyolitic magma can be formed from melting of basalts, upper crustal igneous and metamorphic rocks or from melting lavas and sedimentary assemblages, which are commonly found in great, elongate, slow-sinking troughs called *geosynclines.* Geosynclines form early in the mountain-building sequence, filling slowly with sediments carried from surrounding high-

Figure 3–8. Distribution of the earth's active volcanoes, world rise system and deep ocean trenches.

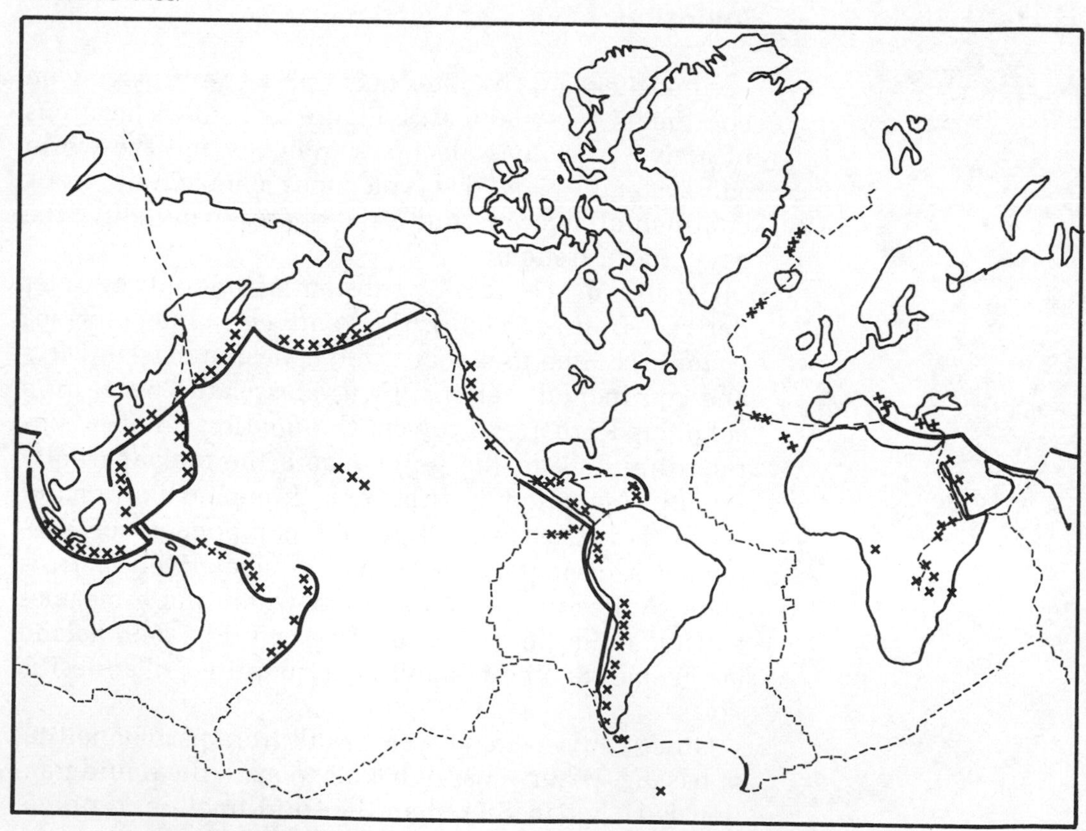

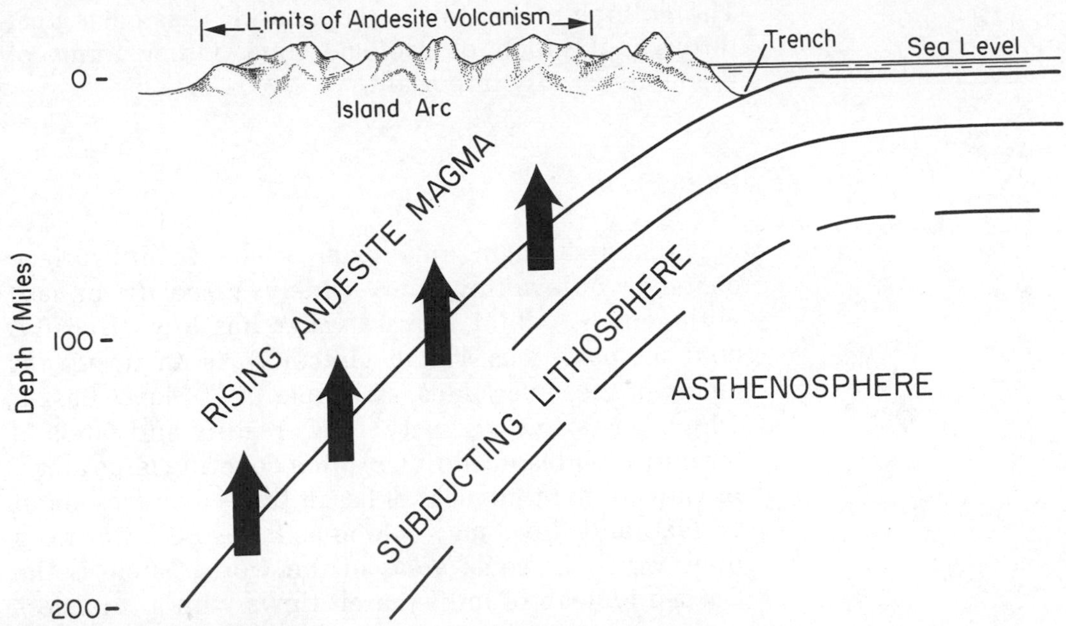

Figure 3–9. Relation of depth of subduction and formation of andesite magma. (Adapted from Hatherton, T., and Dickinson, W. R.: The relationship between andesitic volcanism and seismicity in Indonesia, the Lesser Antilles and other island arcs. J. Geophys. Res. *74*:5301–5310, 1969.)

land. Up to 60,000 feet of sedimentary and volcanic rock may accumulate before the geosynclines stop sinking. The Gulf Coast of the United States is a good example of a young geosyncline. Rocks in the lower parts of geosynclines often melt, forming huge bodies of rhyolite magma, some of which escapes to the surface, where it erupts, often with tremendous violence. Sheets of ash and debris are spread over many square miles. The ash sheets, called *ignimbrites,* are far more voluminous and widespread than the glowing avalanche of Mt. Pelée, for example, although the explosive nature of the two types of eruption is similar.

Each type of extrusive volcanic rock has its intrusive equivalent. The equivalents of basalt and andesite are gabbro and diorite, respectively. The intrusive equivalent of rhyolite is granite, a name widely and often improperly applied to many igneous and metamorphic rocks.

4 VOLCANIC EJECTA

In this section we will discuss lava flows and other forms of volcanic ejecta. Lava flows account for more than half of the volume of material ejected from volcanoes, but other forms of ejecta are also important.

The volume of ash flows, for example, has been great throughout much of geologic time. Other forms of ejecta are locally important.

Lava Flows

Thickness, length, areal extent, velocity and surface character of lava flows vary widely. *Viscosity*, or ease with which a fluid moves (water has low viscosity; cold molasses has a high viscosity), is an important factor in these variables, as is quantity of lava. Basalt, which has a low viscosity, flows readily and tends to form thin widespread flows, often covering large areas. In Hawaii, the thickness of basalt flows averages about 15 feet, and flows may extend as far as 35 miles from their vents. In some areas of the world, such as the Deccan Plateau of India, basalt flows with a thickness of 200 feet or more occur, but these tend to be exceptional. More siliceous (rich in SiO_2) lavas are more viscous and form thicker, shorter lava flows. One rhyolitic flow in the Mono Craters area of eastern California is a mile long and more than 700 feet thick (Fig. 3–10).

Lava flows travel at widely varying speeds. Glowing avalanches and ash flows may travel at speeds of 100 miles per hour or more. Very fluid basalt flows, such

Figure 3–10. Southern Coulee of the Mono Craters, California. The coulee (a thick rhyolitic flow) extends toward the viewer and is approximately 1 mile long and 700 feet thick, making it one of the thicker known flows.

Figure 3–11. Pahoehoe lava in the main caldera of Kilauea, Island of Hawaii. Fluid lava flowed down the small gulley extending toward the viewer.

as the one erupted by Mauna Loa in 1950, travel at rates of several miles per hour, whereas more viscous lavas travel at speeds measured in feet per day. Except for glowing avalanches and ash flows, human beings are in very little danger of being overtaken by lava flows.

Pahoehoe, Ää and Block Lava

Surfaces of lava flows vary from smooth to rough, depending largely on the viscosity of the lava. Extremely fluid (low viscosity) lavas flow smoothly, forming relatively level surfaces (Fig. 3–11) that sometimes have a "ropy" texture. This ropy textured lava, called *pahoehoe* (a name originated in Hawaii, where pahoehoe lava occurs in abundance), is limited to basaltic lavas, because the other lavas are too viscous.

Large areas of lava surfaces in Hawaii and other basaltic terrains consist of broken, angular fragments resembling "clinkers," the residue left after coal is burned. The surface is rough and knobby (Fig. 3–12), and the lava contains abundant gas bubbles. Hawaiians named this lava *aa*. It is often mistakenly called "blocky" lava, but examination of the surface of an ää flow shows that the "blocks" are actually plates and not equidimensional. Individual plates are stacked

Figure 3–12. Ää lava. The object in the center of the picture is a light meter, which is 3 to 4 inches long. Note the rough, knobby, "clinkerlike" texture of the surface.

Figure 3–13. Surface of an old ää flow, Garibaldi volcano, Ethiopia. Note the rough surface of the lava.

chaotically on and against one another, forming a very rough surface (Fig. 3–13).

Viscosity of ää lavas is slightly greater than that of pahoehoe lavas. In Hawaii, the part of a flow closest to the vent may be pahoehoe and the distant part, ää lava. Gases coming out of solution as the lava flows away from the vent increase the viscosity of the lava, causing the change in texture. Like pahoehoe, ää texture is

Figure 3–14. Surface of S.P. lava flow in the San Francisco volcanic field, northern Arizona. The surface of much of the lava flow consists of blocks similar to those in the foreground. (U.S. Geological Survey photo.)

Figure 3–15. Tree molds in Hawaii. A lava flow as thick or thicker than the height of the mold overran a forest in this area. Lava adjacent to the trees cooled and solidified, remaining behind when the main mass of the flow drained away.

limited to basaltic lava flows. Basaltic block lava is not unknown, but true blocky-textured lava generally occurs in more viscous lavas such as the andesitic flows in areas of the western United States (Fig. 3–14). Individual blocks are roughly equidimensional, with diameters ranging from several inches to a foot or more.

Initially, lava flows are inhospitable badlands. The length of time required for rough, rocky flows to weather into rich soil depends on climate. In the warm humid air of Hawaii, a few decades to a few centuries is sufficient. In the cold dry heights of New Mexico and northern Arizona, several thousand years may be required.

Tree Molds

Formation of tree molds requires extremely fluid lava (Fig. 3–15), which hardens upon contact with the relatively cool wet wood of trees. The main mass of the lava flows on, leaving behind a smoking tree trunk surrounded by several inches of hardening lava. In time, the unburned wood decays, leaving a hollow pillar in the lava. In many cases the interior mold perfectly preserves the shape of the bark and small limbs.

Pillow Lavas

Some lavas explode violently upon contact with water, others, which contain little gas or which are extruded into very deep water, do not. Those that do not explode often solidify into globular masses called *pillows* (Fig. 3–16). Pillow lavas are abundant in geosynclinal rocks, where they seem to have extruded onto the floor of the geosyncline. They are subsequently uplifted during orogenesis and exposed by faulting and erosion.

Vesicles, Scoria and Pumice

All magma contains dissolved gas. When the lava reaches the earth's surface, reduced pressure allows the gas to come out of solution. In the case of Mt. Pelée, the gas exploded violently, but in most instances, the gas merely forms *vesicles* (gas bubbles) in the lava (Fig. 3–17). Moderately vesicular lava is called *scoria;* extremely vesicular, gassy lava (Fig. 3–18) is

Figure 3–16. Pillow lavas exposed on a hillside north of San Francisco are thought to have formed in deep water. They were lifted to their present position during mountain building along the California coast.

called *pumice*. Pumice is almost always rhyolitic; scoria is generally less siliceous. All pumice and some scoria float, because of the large volume of gas bubbles. In consequence of its light weight, pumice is used as a decoration stone—large boulders can be lifted and moved about easily. Pumice with very small vesicles is widely used as an abrasive. Scoria is used for construction and also for decoration because of its distinctive texture.

Figure 3–17. A core 2¼ inches in diameter taken from the Kana-a flow in northern Arizona. Note abundant vesicles. Vesicles are concentrated near tops and bottoms of flows. (U.S. Geological Survey photo.)

Figure 3–18. Close-up of pumice, showing network of vesicles.

Fragmental Flows

Not all volcanic flows consist of lava. Some are composed of fragments and particles creeping along slowly. Other fragmental flows move rapidly, as particles are rafted along on a cushion of compressed air. Glowing avalanches and ash falls, the two main types of fragmental flows, have caused much damage in the past, devastating St. Pierre (as we have seen), Pompeii in 79 AD and many other areas.

Pyroclastics

Volcanoes in eruption eject a variety of solid and semisolid particles. Red-hot fragments often shoot several thousand feet into the air, sometimes causing lightning. When the fragments land, they cover the ground with a deposit of variable thickness. Collectively, solid particles ejected by a volcano are called *pyroclastics* (pyro=Greek *fire*; klastos=Greek *broken*).

Fine pyroclastic material ejected from volcanoes is classified according to size. Small fragments with diameters between 2 and 64 millimeters (approximately 0.1 to 2.5 inches) are called *lapilli* or, more commonly, *cinders*, and frequently accumulate around vents (Fig.

Figure 3–19. Cinder cones in northern Arizona. The cones are formed by cinders spewing from vents and are easily eroded because the cinders are loose and uncemented. Commonly, as is the case here, a row of cinder cones marks the line of a fault or fracture that served as a conduit for the molten magma.

3–19). Particles with diameters of less than 2 millimeters are called *volcanic ash.*

Glowing Avalanches

Macdonald recognizes three varieties of glowing avalanche based on their modes of origin. In *Peléean* eruptions, hot material is ejected laterally. The force of ejection combined with the force of gravity sends particles flying down the mountainside with great force and at high velocity.

At the time of the eruption of Mt. Pelée, Soufriere Volcano on nearby St. Vincent Island erupted also (on May 7, 1902), forming a glowing avalanche similar to the one formed at Mt. Pelée. However, at Soufriere, the hot gas and particles erupted skyward, fell back to earth and were pulled down the mountainside solely by the force of gravity. The heat of the avalanche and its velocity and force diminished somewhat during the initial upward flight of the particles.

Merapi Volcano in Java pushed out a dome of hot, semiconsolidated lava. As the dome grew, large pieces fell off the edge of it and broke, sending avalanches of glowing rock fragments down the mountainside. This type of glowing avalanche is more docile than the Soufriere type.

Hot ash may also settle on steep sides of a volcano,

only to be dislodged by an earthquake or eruption. The resultant avalanche may have many of the characteristics of a glowing avalanche because of heat retained in the slowly cooling ash.

In most avalanches, solid particles ride on a cushion of air trapped beneath the avalanche. Cold air trapped beneath a glowing avalanche expands, increasing the buoyancy of solid particles in the avalanche. Lava particles may continue to emit gas during the avalanche, further increasing the fluidity of the fast-moving mass. The air-and-gas cushion permits the avalanche to attain great speeds as it races down the mountainside.

Ash Flows

Large areas of California, New Mexico, Oregon and Nevada are covered with layers of ash, which is usually rhyolitic but occasionally andesitic. Thicknesses of individual layers range from near zero to several dozen feet. Some ash flows originate from a common vent, while others seem to come from multiple vents in a restricted area. Multiple-vent eruptions may cover areas of as much as 12,000 square miles and have volumes of 500 cubic miles. One ash flow extended at least 70 miles from its source.

Very large volume ash flow eruptions are associated with multiple calderas or large-scale depressions. Ash flows from multiple calderas have had volumes of as much as 2000 cubic miles. Smaller scale flows originate from fissures, craters and domes. Glowing avalanches are a form of small-scale ash flow originating from a crater or dome.

The mechanism responsible for eruption and flow of ash flows is similar to the mechanism of the glowing avalanche. However, the quantity of matter ejected, the temperature of the ash cloud and the amount of gas is greater in an ash flow. The erupted gas forms a cushion and is responsible for the fluidity of the flows. The force of gravity and the quantity of material cause the ash to flow downhill, covering minor terrain irregularities.

Many ash flow deposits consist of granular unconsolidated particles. Others were originally hot enough and thick enough to "weld" the particles together. Rock formed in this manner is called *welded tuff.*

Figure 3–20. Bluff comprised of ash flow deposits north of Bishop, California. Hard welded tuff on top prevents erosion of unwelded underlying ash flow deposits.

Welded tuff covers the flat-topped low plateau in Figure 3–20. The hard welded tuff cover resists erosion and prevents water from washing away the less resistant underlying ash flow deposits.

Mudflows

Mudflows are one of the most destructive phenomena associated with volcanic eruptions. A boiling mudflow buried a sugar refinery prior to the eruption of Mt. Pelee; another overwhelmed Herculaneum during the eruption of Mt. Vesuvius in 79 AD. The mud at Herculaneum subsequently hardened to rock, making archaelogical excavation more difficult than in ash-covered Pompeii.

Initially, mudflows usually consist of fine particles of volcanic origin. As the mudflow travels downhill, it incorporates soil, rocks, boulders, trees and other movable objects. Mudflows usually follow stream valleys on the mountainsides and spread out at the base of the mountain.

Two of the largest known mudflows originated at Mt. Rainier, Washington (see also p. 124). The Electron mudflow occurred about 500 years ago, and the larger Osceola mudflow occurred 5000 years ago. A modern repetition of either of these mudflows would result in extensive damage and many deaths in cities and towns now occupying sites of the old mudflows. Such a mudflow would greatly exceed the Buffalo Creek disaster (see Chap. 14).

Fumaroles, Geysers and Hot Springs

All magmas contain small amounts of dissolved gas, usually carbon monoxide, carbon dioxide, sulfur dioxide and steam (ground water coming into contact with the magma may contribute additional water vapor).

Vents emitting volcanic gases are called *fumaroles* and are associated with many active and *dormant* (inactive for a long period of time but not extinct) volcanoes. Fumaroles may persist for hundreds or thousands of years after the final eruption as hot magma inside the earth slowly cools. Deposition of mineral components around a vent often results in colorful encrustations. Sulfur, a common deposit around fumarole vents, may collect in mineable quantities.

Geysers and hot springs occur when ground waters encounter hot rock within the earth. Hot water, like hot air, is lighter than its cold equivalent and therefore rises after being heated by rocks. Sinking cold water replaces the hot water, becomes heated and rises convectively. The waters of hot springs often include significant amounts of dissolved mineral matter, including radioactive substances. The alleged benefits of drinking and bathing in mineral waters are widely advertised. Rock alteration and deposits of unusual minerals are common around hot springs.

Geysers resemble hot springs in that water, mainly ground water, coming into contact with hot rock or volcanic steam at depth, is heated and rises convectively to the surface. Geyser water, however, is hot enough to form steam periodically. Sudden release of steam pressure causes the geyser to erupt. Volcanic gases may also contribute to the eruptive mechanism in geysers, but steam derived from ground water is the major force.

The mechanism causing geyser eruptions resembles the mechanism that caused the eruption of Mt. Pelée. In each case, weight of overlying material (rock at Mt. Pelée, water in a geyser tube) prevents formation of a gas phase (volcanic gases in Mt. Pelée, generally steam in a geyser) until the liquid (lava and water, respectively) approaches the top of the conduit. When pressure diminishes sufficiently, a gas phase forms suddenly and explosively. The explosion clears the vent temporarily, but water refills the vent and the process begins anew.

Hot springs, geysers and fumaroles have potential as sources of heat and power (see Chap. 8).

5 THE EXITS

Lava emerges from the interior of the earth through many openings. Some exits or *vents* are located on tops of imposing mountains; others lie cradled within encircling rims of vast calderas; and still others consist of small fissures, each contributing its quota to a lava flow or ash flow.

Summitry

Summits of volcanoes come in a wide variety of forms. Some are so broad and flat that the vent can be found only with difficulty. Pronounced indentations and multiple concentric rims distinctly identify other volcanic summits.

Craters

The simplest and perhaps most common volcanic vent is a *crater*. Craters can be formed by explosion or by collapse following removal of magma. A single volcano may have one crater or numerous craters, all of which have erupted at one time or another. The older literature abounds with tales of volcanoes "blowing their tops." Like much volcano mythology, however, the dimensions of the story greatly exceed the dimensions of the truth. Diameters of true explosion craters seldom exceed one mile, and most larger "craters" are not craters at all but calderas formed by collapse. Craters formed by violent explosions are limited to cases in which water (ground water, sea water or river or lake water) came into contact with molten magma. In these cases, the combined effects of explosion and collapse can produce a large crater or caldera.

Halemaumau, the main summit crater of Kilauea in

Figure 3–21. Crater of Garibaldi volcano in Ethiopia. Dark area in lower right is a recent lava flow. Vegetation covers the remainder of the crater floor.

Hawaii, is 3000 feet across and was formed largely by collapse (Figs. 3–4 and 3–6). Part of the crater of Garibaldi Volcano in Ethiopia is shown in Figure 3–21.

Calderas

Large, more or less circular depressions formed by collapse are called *calderas*. Kilauea caldera, in which Halemaumau lies, is about 2½ by 2 miles across. Dimensions of Ngoro-Ngoro caldera in Tanzania are more than twice those of Kilauea, and La Garita caldera in southern Utah has a diameter of between 25 and 30 miles.

As mentioned previously, early volcanologists examining volcanoes that were missing much of their summits (Fig. 3–22) assumed that summit rocks had been blown outward in a violent eruption. It was a natural assumption to make, but later more careful examination of the ejecta showed that the ejecta surrounding large volcanic depressions or calderas consist almost entirely of new lava. Pieces of summit rocks, which should have been abundant, were almost nonexistent. The small amount of summit and wall rock in the ejecta led scientists to conclude that calderas form mainly from inward collapse. Careful measurements of the quantity of ejected lava confirm this conclusion.

Ash flow eruptions create the largest calderas. La

Figure 3–22. In much of the old literature, volcanoes such as Fantale in Ethiopia were said to have "blown their tops." Actually, large central depressions of Fantale and other volcanoes were formed by collapse. Only Fantale's rim remains, rising approximately 3000 feet above the surrounding plain.

Garita, the Timber Mountain caldera of southern Nevada (8 by 12 miles) and the Valles caldera of northern New Mexico (13 miles across) all resulted from ash flow eruptions.

Formation of calderas occurs late in the life of a volcano, usually after an extended period of quiescence. It is thought that during the quiescent period, basaltic or andesitic magma in the conduit melts part of the siliceous crust and combines with it, forming a rhyolitic magma, causing eruption of an extensive ash flow that forms the characteristic collar of ash and tuff surrounding the caldera.

Necks, Pipes and Kimberlites

With the exception of welded tuff, lava and pyroclastic debris forming the sides of a volcano usually are not strongly cemented. Rainwater rushing down the sides of volcanoes easily erodes their flanks. The material in the neck of the volcano often consists of a single block of hardened magma, which resists erosion. The result of the differential resistance to erosion of flank and conduit rock is often a *volcanic neck* (Fig. 3–23). Volcanic necks also form when the conduit traverses soft erosion-susceptible rock. At the Castle Buttes in Arizona, cracks radiating from the neck subsequently

Figure 3–23. Volcanic necks in the Simien Mountains of central Ethiopia crowd against one another. Some of the necks are 13,000 feet in elevation.

filled with magma. Hardened magma in the cracks is now exposed (Fig. 3–24).

Pipes are round intrusions. Some, but not all, originally fed magma to volcanoes. Others lacked surface openings and were entirely intrusive. Pipes tend to change shape with depth, generally grading into elongate fissures, although in some areas they grade

Figure 3–24. Hardened magma exposed by erosion at the Castle Buttes, in Arizona. Similar features occur in many older volcanic areas. Lava that passed through conduits like these now caps the broad flat-topped mesas in the background.

into *diatremes* (a type of vent produced by the explosive energy of gases in magmas). A volcanic neck is a portion of a pipe stripped of its enclosing rock walls by erosion.

Pipes composed of a mineral assemblage called *kimberlite* sometimes contain diamonds. Kimberlite originates deep within the mantle and rises toward the surface as a mixture of fluid gases and hot particles.

6 MOUNTAINS

Most people think of volcanoes as beautiful cone-shaped mountains (Fig. 3–25), but this is only one of many shapes assumed by oozing lava and fiery ash accumulating near a vent. Some volcanoes form broad plateaus, others form stubby domes and some consist merely of fissures in the ground, which cannot be seen because of the cover provided by erupted lava and ash.

Shapes of volcanoes depend on the amount and composition of their lavas. Fluid basaltic lava spreads readily, building broad gently domed mountains, andesitic lavas form beautiful cones and rhyolitic lavas usually build short, thick flows. Very fluid basalts may build no mountains at all but may form broad plateaus instead (rhyolitic lavas also build broad

Figure 3–25. Mount Fuji, Japan. Fuji is a classic example of a composite volcano built of alternating layers of ash and lava. In this view, lights twinkle on the far side of a lake dammed by lava flows. The cloud on top results from condensation of moist air flowing up the sides of the mountain.

plateaus, but this is because the viscous lava erupts as ash flows). Eruption of small amounts of cinders and ash build *cinder cones* (Fig. 3-19).

Fissures

Although lava conduits of most volcanoes are roughly circular in outline, many lava flows do not erupt from circular conduits but from networks of *fissures* (cracks) in the earth's surface (Fig. 3-24). Fissure eruptions come in all sizes, some extruding only small amounts of lava while others form part of a vast outpouring of lava, which eventually may cover a large area with an extremely thick layer of lava.

Fissure eruptions may occur on the flanks of volcanoes. Earlier in this chapter, we discussed the eruption of Kilauea, where lava bubbled up temporarily into Kilauea-Iki crater, escaped into cracks in the East Rift Zone and finally erupted through fissures 25 miles east of the summit. Lava extruded through the fissures eventually covered an area of several square miles.

Flank eruptions similar to the 1959-60 Kilauea eruption are common. Rifts or other zones of weakness cross many volcanoes. Lava seeking an outlet at a level lower than the summit forces open the cracks and flows through them onto the surface somewhere on the flank of the volcano.

A large eruption occurred when a fissure near Mt. Katmai, Alaska, spewed out a rhyolitic ash flow in 1912. The smoking ash covered an entire valley 14

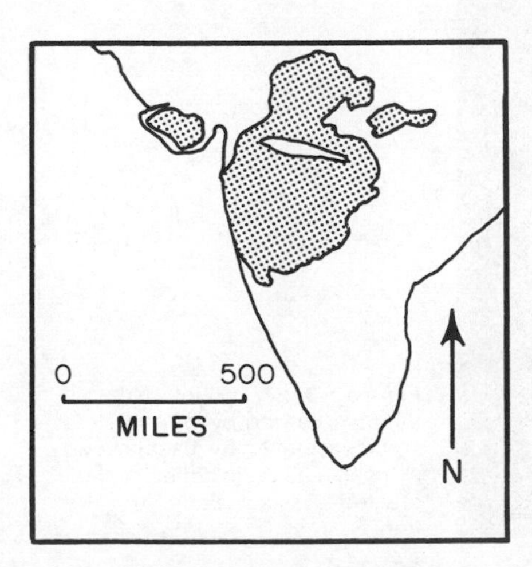

Figure 3-26. Extent of Deccan plateau basalts in India. (Adapted from Macdonald, G. A.: Volcanoes. Englewood Cliffs, N.J., Prentice-Hall, 1972.)

rocks in the low velocity zone. Molten magma inches its way toward the surface through cracks in the mantle and crust. When cracks open during passage of the magma, earthquakes occur, and magma movement can be traced with some accuracy. However, because rocks of the low velocity zone are plastic and yield by flowing rather than by brittle fracture (which is essential for earthquakes), it is impossible to tell at precisely what depth the magma originates. Figure 3–7 shows a cross-section of the main island of Hawaii and the possible location of the sources of lava. Magma appears to be stored temporarily in a magma chamber about five miles underground until it can break through the overlying rocks and extrude upon the surface. Although oceanic volcanoes are not identical, basic structures are probably grossly similar to that shown in Figure 3–7.

Andesites

Andesites and rhyolites occur almost exclusively on continents and island arcs. Figure 3–8 shows locations of active volcanoes, deep-sea trenches and the world rift systems. Note that volcanoes concentrate along continental margins and island arcs paralleling the deep ocean trenches.

Ocean crust containing substantial amounts of water derived from crystallization and entrapment of sea water in sediments is subducted beneath the island arcs and continental margins. Heat releases the water from subducted crust and lowers the melting point of surrounding rock to the point where the rocks partially liquefy. Liquefied mantle rock, liquefied portions of subducted crust and liquefied components of subducted sedimentary rocks rise in a relatively narrow zone. Andesitic volcanoes form where these magmas extrude onto the surface of the earth (Fig. 3–9). Island arcs usually contain significant quantities of andesitic lavas.

Other andesites appear to result from partial melting of basalts. When a basalt begins to melt, the liquid fraction is richer in SiO_2 than the solid fraction. If only a very small amount of the whole is melted, magma composition will be rhyolitic; if a greater percentage of the whole is melted, magma composition will be andesitic;

Figure 3–28. Drakensberg Mountains of South Africa. The Drakensberg range is composed of a sequence of thick basalt flows. Some flows are visible as layers in the mountainside.

Most vents supplying basalt to lava plateaus and plains are buried by the lava oozing from the vent. Occasionally, a row of spatter cones marks the location of a fissure. Otherwise the overlying lava securely hides vent locations until time and erosion expose them.

In most cases, lava of plateau basalts consists of a very fluid pahoehoe type, although ää lavas also occur. The lavas are so fluid that they flow down slopes of less than 1 degree. Individual flows are typically be-

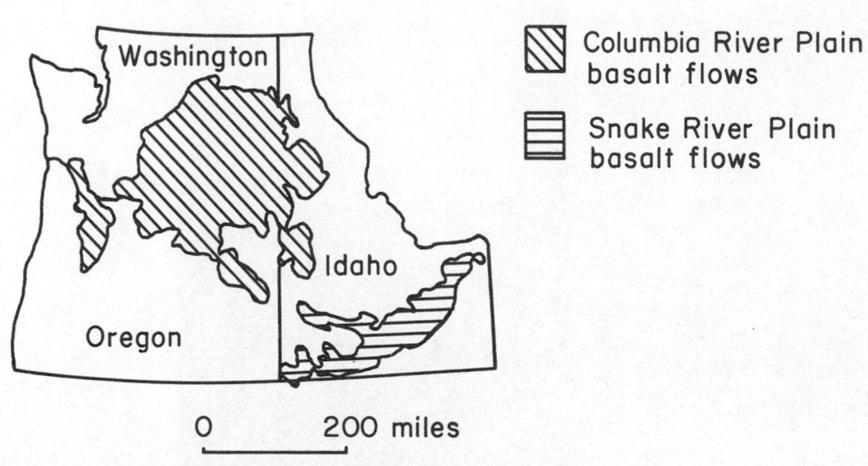

Figure 3–29. Location of the Columbia River Plateau and Snake River Plain. Plain basalt flows in Washington, Oregon and Idaho. (Adapted from Macdonald, G. A.: Volcanoes. Englewood Cliffs, N.J., Prentice-Hall, 1972.)

tween 15 and 30 feet thick but may reach thicknesses of 400 feet.

Basalt plateaus seem to form in areas where the lithosphere is cracked and spreading. This is true in Iceland, which straddles the crest of the Mid-Atlantic Ridge, in East Africa, where fluid basalts mantle either side of the rift valleys and in the eastern United States, where tension cracks formed during the initial phase of separation of North America and the Eurasian-African lithospheric plates. Lithospheric tension opens numerous cracks, which extend into the low velocity zone and serve as conduits for the rising magma.

Shield Volcanoes

Fluid basalts erupted from individual or closely grouped vents form *shield volcanoes*. The name derives from the profile of the volcano (Fig. 3–30), which resembles a round shield lying face up on the ground. Shield volcanoes have gentle slopes that are typically less than 10 degrees. Shield volcanoes consist almost entirely of hardened lava; pyroclastics comprise an insignificant fraction of the whole. Repeated thin flows gradually build these volcanoes.

Rift zones radiate from the summits of some shield volcanoes, and in certain areas elongation can be

Figure 3–30. Mauna Kea, Hawaii. Mauna Kea is one of five coalescing volcanoes that form the Island of Hawaii. Note the broad, gently convex profile reminiscent of a shield placed face up on the ground.

Figure 3–31. San Francisco Peaks, north of Flagstaff, Arizona. The Peaks, rising to a height of over 12,000 feet, surround a caldera in which an Ice Age glacier nested. The mountain at the foot of the peaks consists of thick flows of silicic lava.

traced to the quantity of lava erupted along the rift zone. The rift zones of the Hawaiian volcanoes, for example, converge toward the central caldera of each volcano.

Bottom profiling and dredging indicate that shield volcanoes are abundant on the sea floor.

Composite Volcanoes

Volcanoes comprised of alternating strata of pyroclastics and lava are known as *composite volcanoes* or stratovolcanoes (Figs. 3–25 and 3–31). They are magnificent soaring mountains, which may have base diameters of 20 miles or more. Mount Fuji in Japan rises more than 12,000 feet above sea level.

Composite cones may erupt lavas of every type from basalt to rhyolite, with textures ranging from pahoehoe to the fine ash of a glowing avalanche. Typically, however, the average composition of the lavas is intermediate. Composite volcanoes usually erupt from a central vent, but the last eruption of Fuji, for example, blasted a hole in its southeastern side.

Devastating eruptions are historically associated with composite volcanoes such as Vesuvius in Italy, Kilimanjaro in Tanzania, Mayon in the Philippines and Pelée in Martinique.

Domes and Coulees

Rhyolitic and other siliceous lavas too viscous to form extensive individual flows and too explosive to form a thick accumulation of flows generally result in ash flows and ash flow plains. In a few places, short stubby rhyolite domes flanked by thick short lava flows called *coulees* occur (Fig. 3–10).

7 VOLCANISM AND MAN

Volcanic Histories of Three Areas

The following brief histories of three areas illustrate the diverse nature of the impact of volcanoes on man. The eruption of Krakatoa in 1883 ranks with the greatest eruptions in history. The sound of the explosion was heard 3000 miles away and a tsunami (see Chap. 2) generated by the eruption swept over low-lying areas of nearby islands, killing more than 36,000 people.

The eruption of Thera in 1500 BC obliterated an island in the Mediterranean that was populated by a culturally advanced people. Side effects of the eruption may have contributed to the collapse of the Minoan civilization on nearby Crete, thus changing the course of history.

Volcanoes of the Cascade Range of Washington, Oregon and northern California have erupted sporadically during the past few million years. Lassen Peak in California was active most recently, erupting during the period 1914 to 1921. Long periods of quiescence interrupted by violent eruptions characterize these volcanoes. They have now been quiet for many years, but like strain slowly accumulating near a fault, these volcanoes will erupt someday, possibly with disastrous results for people living nearby.

Krakatoa, Indonesia

Krakatoa is one of the many volcanic islands of Indonesia. Prior to 1883 its eruptions had been sporadic, the last previous eruption consisting of andesitic pitchstone.

The volcano began to erupt on May 20, 1883, and continued intermittent activity. Without any warning, on August 20, the volcano erupted with great violence. People 100 miles away heard the sound. At about 2 PM an even greater explosion shot ash and gas to a height of 17 miles. On August 21, falling ash from the frequent eruptions turned day into night. On the nearby island of Sumatra, the darkness lasted 2½ days in the area closest to the volcano.

On August 27, an eruptive climax ejected a cloud of ash and gas 50 miles into the atmosphere. Two great explosions followed within hours of the first. Ash falling from the sky blanketed an area of 300,000 square miles. Australians more than 2000 miles from Krakatoa and inhabitants of Rodriguez Island nearly 3000 miles away heard the noise of the main explosion.

The caldera of Krakatoa was partly underwater before the eruptions began, which probably contributed to the formation of the enormous tsunamis that followed each eruption. The tsunamis, which reached heights of 130 feet, swept inland as far as ten miles in low-lying areas of nearby islands. Deaths due to the tsunamis exceeded 36,000.

Rafts of floating pumice ten feet thick accumulated in the straits between Java and Sumatra; falling ash in southeastern Sumatra burned people's hands and faces; fine ash fell on ships 1600 miles away; during the following year ash in the atmosphere reduced sunlight reaching the earth by 13 per cent; and equatorial skies glowed at night for several years.

The eruptions shot out at least 3.8 cubic miles of hot magma and rock fragments. During or following the eruption, the summit of Krakatoa collapsed, forming a large caldera.

Thera, Greece

Thera (or Santorin as it is now known) belongs to the Cyclades, a group of islands in the Aegean Sea between Greece and Crete (Fig. 3–32). Rich volcanic soil and wealth from trading contributed to the rise of the ancient Minoan culture centered on nearby Crete, which flourished from about 3000 to 1100 BC. Thera, whose cities formed part of the Minoan world, prospered also, until about 1500 BC, when the huge multistoried palaces of Knossos, the Minoan capital in Crete, were suddenly destroyed. The island of Crete was devastated, and a primitive but powerful Greek culture filled the vacuum left by the collapse of the Minoan civilization. The influence of Greek civilization reaches through the ages, affecting our lives even today. The English language and western philosophy, for example, might be much different had not the Minoan civilization collapsed.

Excavation at Knossos and Thera indicates that two natural disasters, an earthquake and a tsunami, either singly or together contributed to the sudden collapse of the Minoan civilization at its height. In this chapter we will consider only the tsunami.

Thera is composed of five islands. Three of these consist of pieces of a single caldera rim projecting out of the water, and pieces of a central cone within the caldera form the other two islands (Fig. 3–33). The caldera rim islands formed as the result of an eruption or eruptions similar to the one on Krakatoa, and the islands of the central cone formed 1000 years later. Ash from the original eruption buried the Minoan towns and villages to depths of as much as 200 feet, and sediment sampling by marine geologists shows that prevailing winds carried ash 600 miles to the southeast.

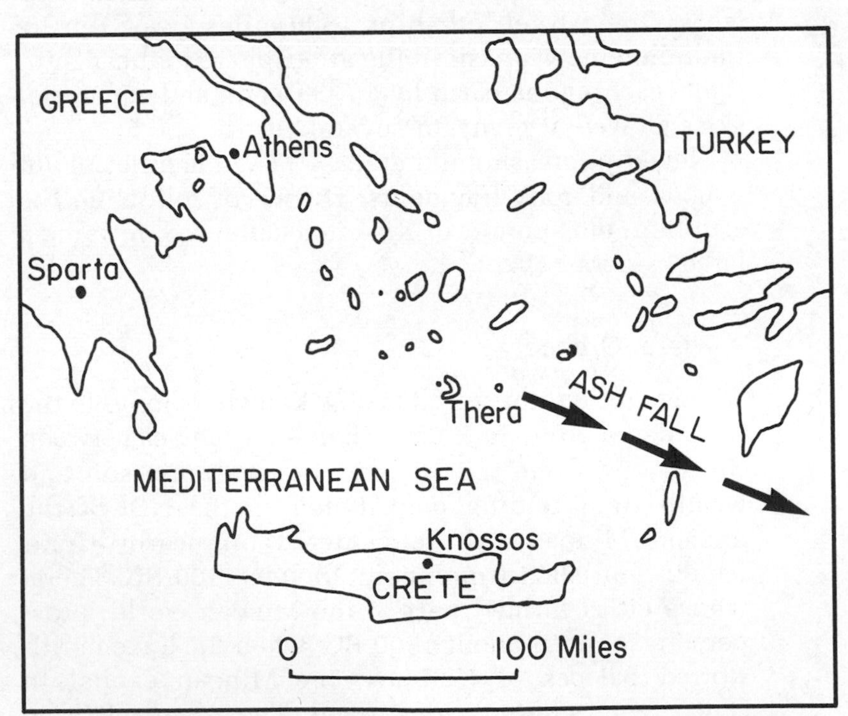

Figure 3–32. Thera (Santorin) and surrounding lands of the eastern Mediterranean. The ash fall following the 1500 BC eruption extended 600 miles to the southeast.

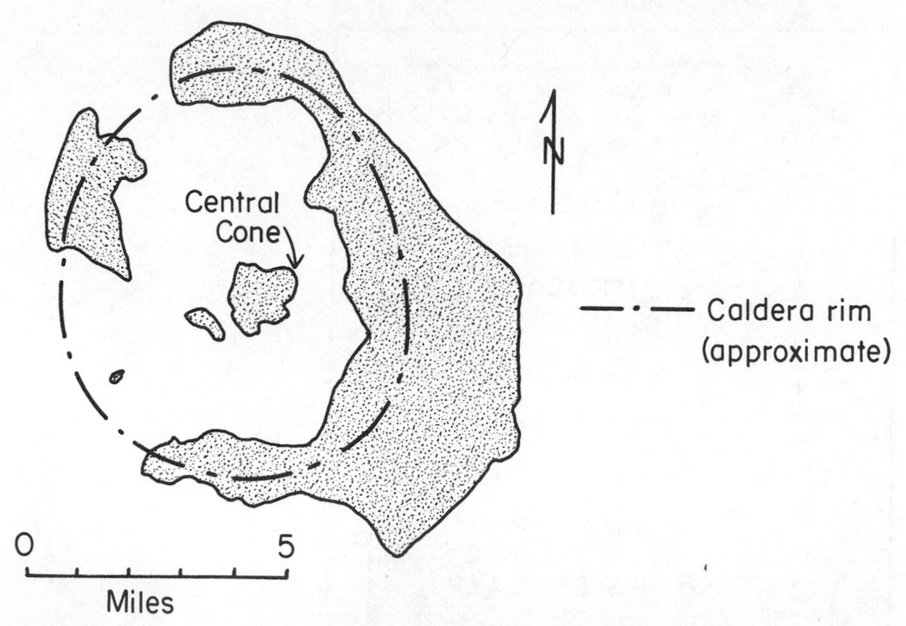

Figure 3–33. Islands comprising Thera (Santorin) today consist of pieces of the caldera rim and central cone sticking up through the water. The caldera floor is as much as 1300 feet below sea level.

Some scientists suggest that a tsunami like those accompanying the Krakatoa eruptions ravaged Knossos and the northern coast of Crete. It is certainly true that the devastation of Crete occurred at about the same time that Thera erupted, but it remains to be proven that a tsunami was the principal factor in the sudden eclipse of Minoan civilization. If a tsunami did form following the eruption, it would have severely damaged low-lying cities of northern Crete.

Thera continues to be slightly active, and an earthquake rocked the islands as recently as 1956.

Cascade Range, Western United States

Volcanoes of the Cascade Range extend from Mt. Baker near the Canadian border in northern Washington to Lassen Peak in northern California (Fig. 3–34). Seven volcanoes—Mt. Baker, Mt. Rainier, Mt. St. Helens, Mt. Hood, Mt. Shasta, Cinder Cone and Lassen Peak—have erupted since 1800, but only one eruptive episode, the Lassen activity between 1914 and 1921, was carefully observed and reported. In fact, the Lassen Peak episode is the only eruption within the conterminous United States (i.e., excluding Alaska and Hawaii) that was observed by non-Indian inhabitants.

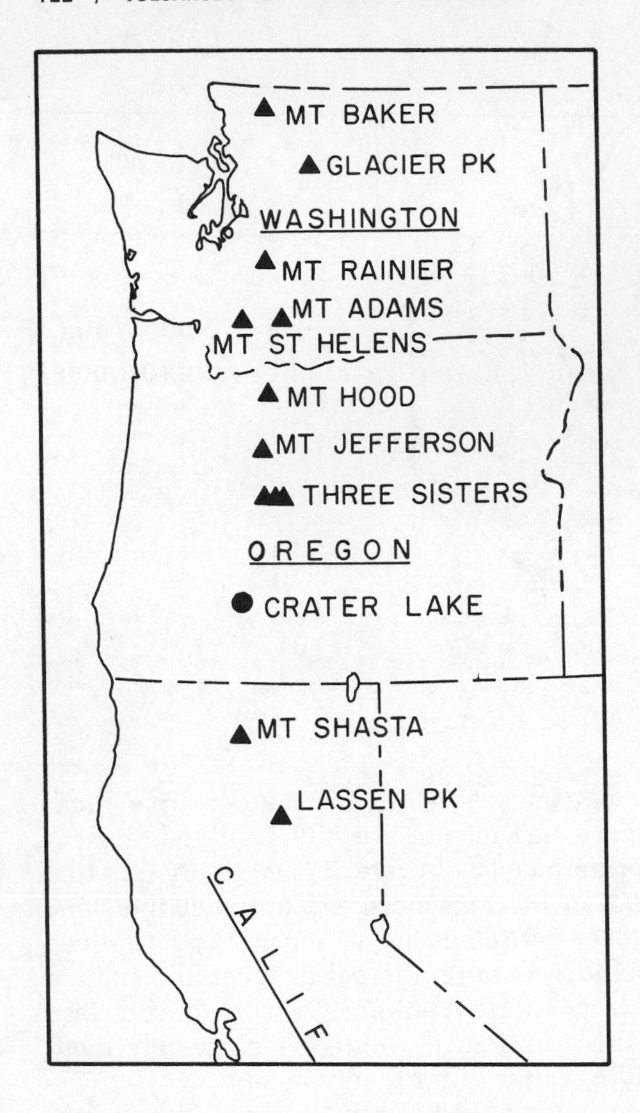

Figure 3–34. Major volcanoes of the Cascade Range.

The Lassen Peak episode began on May 30, 1914, when fissures and a small crater appeared ¼ mile from the summit. Sand, boulders and mud erupted over an area 200 feet in diameter. During the early summer of 1914, successive explosions extended the crater and spewed ash over larger and larger areas. A major eruption, which sent an ash cloud up 11,000 feet on July 18, climaxed the early summer phase of the episode.

The period of quiet that followed lasted seven weeks, after which eruptions continued throughout the summer and fall. Forest Service observers reported 69 individual eruptions during the year. Snow, cold, clouds and generally bad weather hindered observation during the winter. The first exploration party reached the summit in late March, 1915, and found that the crater had expanded to a width of about 1000 feet.

On May 19, 1915, the forces inside the volcano pushed a sticky agglomeration of glowing andesitic lava into the crater. As the viscous mass rose above the crater, rim, incandescent blocks of andesite broke off and rolled down the mountainside. The entire mass rose and fell periodically, then finally receded into the crater.

On the opposite side of the peak, unknown to observers of the andesite blob (who were watching through field glasses at a safe distance of 21 miles), a horizontally directed blast ripped a gash in the side of the mountain. The explosion scoured the surface, sweeping away trees and boulders. Hot gases melted large snowbanks on the mountainside, and a mixture of pyroclastics, trees, boulders and water roared down into the valleys, washing away four ranches. The mudflow eventually dissipated and stopped more than 20 miles from its source. Another eruption and mudflow followed on May 22.

During the remainder of 1915, the peak subsided and activity diminished, and 1916 passed with only minor activity.

Activity increased in early 1917. An eruption on May 18 lasted 18 hours and threw out a cloud 12,000 feet high. In May and June, 1917, the volcano erupted 34 times, abruptly ceasing activity on June 29. No activity was reported in 1918, and only minor activity was reported in 1919–21. Thereafter the volcano became dormant.

It is interesting to note that activity surged in May of 1914, 1915 and 1917, when snows were melting in the crater. Investigators of the eruption hypothesized that meltwater reaching hot magma via cracks and fissures triggered the larger eruptions and that no major eruptions occurred in May, 1916, because the summit had been too hot during the previous winter to permit accumulation of snow.

There are many similarities between the Lassen Peak eruption and the 1902 Pelée eruption. Each was a composite volcano of predominantly andesite composition. Eruptions consisted mainly of clouds of small incandescent particles and included horizontally directed glowing avalanches, and both volcanoes sent mudflows rolling down their flanks. Differences in

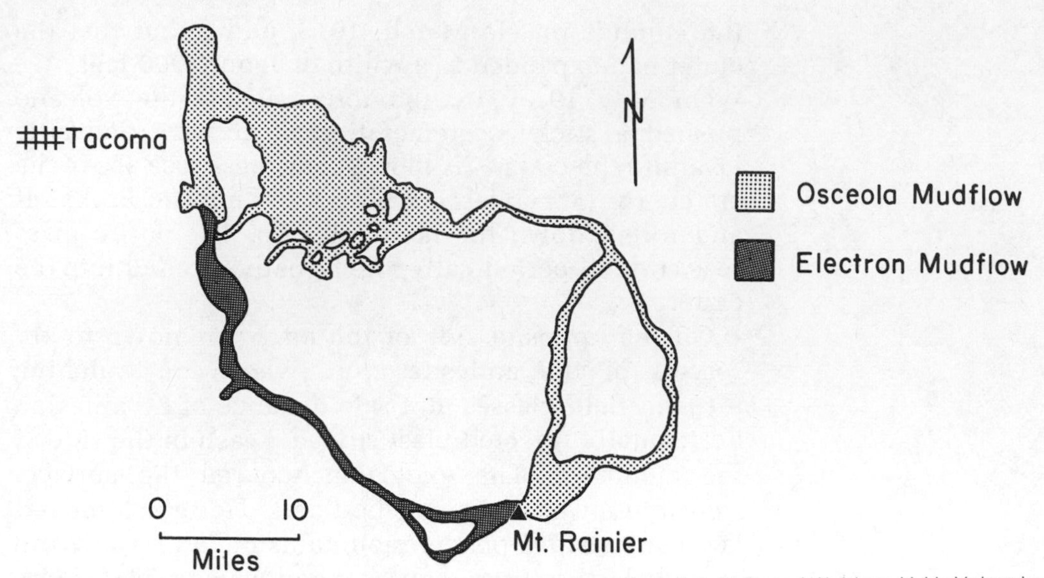

Figure 3–35. Extent of Osceola and Electron mudflows. (From Crandell, D. R., and Waldron, H. H.: Volcanic hazards in the Cascade Range. *In* Olsen, R., and Wallace, M., eds., Geologic Hazards and Public Problems: Proc. Conf., 1969.)

population density were mainly responsible for differences in loss of life and damage. People living near the Cascades, like people along the San Andreas fault, live in the shadow of disaster. When Crater Lake volcano erupted 7000 years ago, ash fell over a wide area, with six inches or more falling as far away as 100 miles. A comparable contemporary eruption of one of the volcanoes in southern Oregon might affect cities as far north as Seattle, Tacoma or Portland.

Mudflows accompanying eruptions constitute a serious danger. Two mudflows, the Electron (ca. 1500 AD) and Osceola (ca. 3000 BC), originating near the summit of Mt. Rainier, streamed down the Puyallup and White River Valleys, respectively (Fig. 3–35), smothering vegetation and wildlife in the Puget Sound lowland. Today, dams block two valleys heading on Mt. Rainier. The reservoirs fill only during times of heavy rainfall and are empty during much of the remainder of the year. Either dam could contain a mudflow the size of the Electron, but neither would stop one the size of the Osceola. Mudflows occurred at Pelée and Lassen following periods of heavy rainfall and heavy snow accumulation. Water in a reservoir would intensify a similarly timed mudflow at Mt. Rainier.

Hazards

Volcanoes present five main hazards:

1) Lava flows
2) Ash falls (including glowing avalanches)
3) Mudflows
4) Noxious gases
5) Tsunamis

Because all of these have been discussed in previous sections of this chapter, a brief summary only is provided.

Lava flows principally endanger property. The fastest flowing lava moves so slowly that a person in good health can *usually* escape it by walking. Property damage from lava flows can be extensive, however.

Ash falls and glowing avalanches create a more serious problem. Once the eruption begins, the cloud of hot gases and ash moves so rapidly that people cannot easily evade the onslaught. Inhalation of these gases causes death by lung damage or suffocation. Immediate evacuation is required (though this is not always possible, as we have seen in the case of Mt. Pelée).

Mudflows are most dangerous when they descend narrow valleys. On broader expanses, they can spread out and dissipate. Water trapped by a lake in the crater of Pelee and water formed by melting snow on the flank of Lassen Peak contributed greatly to the 1902 and 1915 mudflows. Ground water probably contributes to mudflows also, but it is not clear whether it alone can fuel a major mudflow. As in the case of imminent ash flows, immediate evacuation seems to be the best policy.

Poisonous components of volcanic gas include carbon monoxide and sulfur dioxide. In addition to deaths resulting from these noxious fumes, oxygen is undoubtedly dangerously depleted in glowing avalanches. Gases issuing from fissures can cause deaths of animals and birds as well as extensive damage to vegetation. As in the case of industrial pollution, volcanic gases may contribute to a higher incidence of respiratory disease.

Tsunamis caused by eruption are rare, but they can devastate low-lying coastal areas and cause thousands of deaths, as we have seen. Volcanic regions, with their rich soil and promises of abundant harvests, attract settlers then turn on them, overwhelming them with ash, lava, water and mud. But memories are short and the need for arable land is great, so the cycle begins anew.

Benefits

Volcanoes benefit man in many ways. Nutrients in volcanic soil enhance plant growth. The porous and vesicular nature of lava and ash speeds the weathering process. In hot humid regions, where nutrients leach quickly from soils, addition of fresh ash and lapilli revitalizes the soil. Rich volcanic soils have contributed to the rise of many civilizations and cultures, including ancient Thera, ancient Rome and modern-day Indonesia and Japan.

Economically valuable minerals accompany magma as it rises toward the surface. Copper deposits of northern Michigan occur in ancient lava flows, and in other regions diamonds occur in pipes rooted deep within the mantle. Pumice is employed as an abrasive and as decorative stone. Cinders are combined with ash and asphalt to surface roads, as well as being a component of concrete blocks. The Deccan Plateau would be uninhabitable were it not for the presence of spring and well water found in vesicular flows.

All islands in the deep ocean basins originated as volcanoes. The Aleutian islands traversed by forebears of the American Indians en route to North America are volcanic. Viking explorers launched their voyages to North America from the volcanic island of Iceland. Volcanic islands, island arcs and plateau basalts add to the amount of land above sea level and compensate for land lost by erosion.

Geothermal steam presently powers electric generators in many areas of the world (see Chap. 8), and this nonpolluting and underutilized energy source can be used to supply a significant fraction of our imminent power needs. In some areas, it may be able to replace conventional power sources to a great extent.

The esthetic beauty of symmetrical volcanic cones has fascinated men for thousands of years. Historic and prehistoric Indians of the southwest believed that their gods lived on the top of the San Francisco Peaks. The Japanese call Mt. Fuji "Fujisan" or "Mr. Fuji," and their folklore is rich in myths of activities in, on and around Fuji. Thousands of tourists visit the volcanoes of the Cascades and the geysers and hot springs of Yellowstone National Park. The quality of our lives is enhanced by the soaring beauty of volcanic mountains and the wide reaches of volcanic plains.

Prediction

Although it is now possible to predict eruptions of a few well known frequently erupting basaltic volcanoes, eruption prediction research is still in a very early stage. Techniques developed for use at these volcanoes may be adequate to predict the less frequent eruptions of andesitic volcanoes, but much work remains to be done in the area of detailed predictive studies of non-basaltic volcanoes.

Techniques which show promise include the following:

1) Earthquake monitoring to trace underground movements of magma and shifting of large segments of rock beneath volcanoes;

2) Monitoring of swelling and deflation (see section on the Kilauea eruption, earlier in this chapter);

3) Fumarolic activity monitoring of amount and temperature of water and composition of gases:

4) Monitoring of changes in magnetic and electric fields near the volcano caused by temperature changes in rock inside and beneath the volcano; and

5) Observation of variations in surface temperatures through infrared photography.

Every major earthquake in the United States stimulates interest in earthquake prediction and flow of money into earthquake prediction research. Human nature being what it is, however, a volcanically related disaster will probably be required to prod us into action. If this is true, let us hope that it is a long time before the disaster arrives.

4 OUTSIDE THE SOLID EARTH

The Blue Planet.

1 INTRODUCTION

He was a violent man in a violent age. His kinsmen banished him from his homeland for manslaughter, forcing him to move to another country. Now, after a second and third outbreak of fighting and killing, he was again being exiled, this time for three years. Forced to leave Iceland in 982 AD after having been banished from his native Norway, Erik the Red, Viking, turned his eyes westward.

The Icelanders knew that land lay to the west. Fifty years earlier a Norwegian sailor named Gunnbjorn had arrived in Iceland and told of a land he had seen after being blown to the west and south of Iceland. Erik now sailed for this land. During his three years of exile, he explored the southern margin of the western land, marking future sites of farms, homes and fishing and hunting grounds. Setting a precedent followed by subsequent colonizers in the New World, he named the land Greenland in contrast to the starkly named Iceland.

Vikings occupied the last of Iceland's arable land by 930 AD. A few years later (ca. 975), Iceland suffered a severe famine. With land scarce and memories of starvation fresh, 25 ships loaded with Icelanders returned with Erik to Greenland in 986. Their settlement near the southern tip of Greenland eventually grew to include 190 farms, 12 churches, a cathedral and a monastery.

Bjarni Herjolfson arrived in Iceland from Norway to discover that his father had sold his farm and gone to Greenland with Erik the Red. Bjarni obtained as much information concerning the location of Erik's new settlement as he could, then hoisted his sails for Greenland. Fog closed in on him three days out of Iceland. Days later, when the fog finally lifted, Herjolfson sighted a low, well-forested land. He sailed north along this strange coast for two days, then continued southwest for three days before sighting the rugged, glaciered coastline of Greenland. Four more days of sailing brought him to Erik's settlement. Bjarni Herjolfson had become the first European to see the North American mainland.

Fifteen years later, Leif Erikson, son of Erik the Red, bought Bjarni's ship, enlisted some of Bjarni's crew and sailed Bjarni's course in reverse. Leif landed (Bjarni had not) first on Baffin Island, then on Labrador and finally on Vinland (or Wine Land). Like his father, he assigned enticing names to his discoveries. After wintering in Vinland, Leif returned to Greenland, loudly praising Vinland, its grapes, grass, timber, fish and mild winters.

Erik's death thrust the responsibilities of government on Leif and thus prevented his return to Vinland. At this time, Thorfinn Karlsefni followed Bjarni's and Leif's trails, leading a colonizing expedition of men, women and livestock. Jealousy over women, hostile Indians, insufficient manpower and other factors caused a great deal of unrest in the Vinland settlement. Because of dissatisfaction among the colonists, Karlsefni led his band back to Greenland after three winters in Vinland. Greenlanders tried once more to colonize Vinland, but this second effort failed for much the same reasons as the first.

With the attempted colonization of North America,

Norse exploration of the New World reached its apex. Thereafter, the Norse settlements in Greenland gradually decayed. Greenlanders continued to sail to North America until at least 1347, when winds blew to Iceland a ship attempting to return to Greenland from the North American mainland. The Greenland Norse survived only a few decades longer.

The Norse settlements in Greenland failed for a number of reasons, including social unrest in Norway, outbreaks of plague in northern Europe and increased resistance of northern Europeans to Viking predations.

In addition, the exportable resources of North America and Greenland consisted of fur, timber and perhaps a little fish. The value of these resources on the European markets was too small to justify the risk of the sea voyages to America. Conversely, the desire for South and Central American gold impelled the Spanish into numerous expeditions of conquest and plunder two hundred years later. European merchants eagerly accepted all the gold the conquistadors brought from the Americas. During this same period, the sixteenth century, the English made no significant progress in colonizing North America, contenting themselves with raiding, plundering and piracy. Not until the late seventeenth century did colonization of North America reach significant proportions.

The most important reason for the failure of the Norse settlements, however, was a general cooling of the earth accompanied by an advance of glaciers in Greenland and sea ice in the north Atlantic (Fig. 4–1). The Norse colonization of Iceland and Greenland during the ninth and tenth centuries coincided with the end of a relatively warm epoch, in which mean

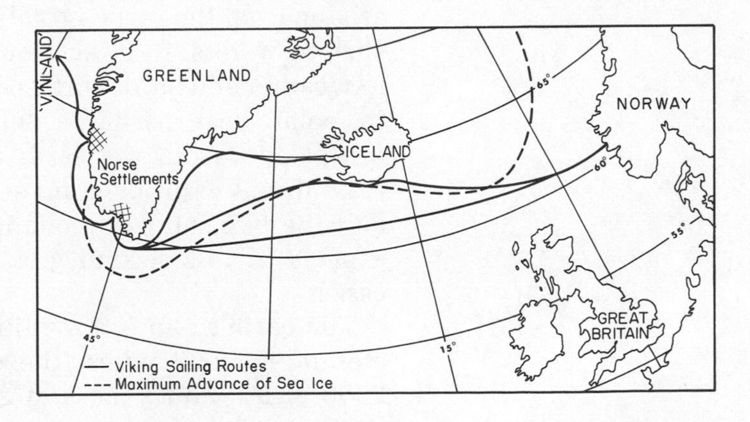

Figure 4–1. Norse settlements in Greenland, Norse sailing routes and maximum advance of sea ice during the eighteenth and nineteenth centuries. (From Denton, G. H., and Porter, S. C.: Neoglaciation. Sci. Am. *222*(6):106, 1970.)

temperatures (see Fig. 4–19, p. 159) had already peaked earlier in the first millennium.

As summers became shorter and colder, Greenlanders found it increasingly difficult to raise enough food for their small colony. It is also possible that inbreeding necessitated by lack of immigrants led to physical and mental deterioration of the settlers. With sea ice closing their harbors and glaciers advancing into their valleys, the last Norse settler died sometime around 1500 AD, a time ironically coincident with Columbus's voyages of discovery far to the south.

2 THE SUN

There are two great families of geologic processes; those deriving their heat from within the earth and those deriving their heat from without the earth, that is, from the sun. The Viking settlers of North America were victims, at least in part, of the latter process.

In Chapters 2 and 3 we discussed sea floor spreading, earthquakes and volcanoes, all of which belong to the family of earth-shaping geologic processes originating from forces acting mainly within the solid earth. Heat released by radioactive decay of certain isotopes in rocks, gravitational compression, tidal friction (this source is outside the earth, but the heating occurs within the earth) and chemical reactions drive a vast machine that melts rock, carries molten rock to the earth's surface, compresses other rock into high mountains and causes earthquakes.

Solar heat, the driving force of the second great family of geologic processes, enters the atmosphere, heating it and driving the winds (with help from forces arising from the earth's rotation). Solar heat evaporates surface waters, forming clouds that provide rainwater to weather and erode surface rock. Long-term variations in solar heat partially determine whether glaciers sculpt and scour the earth or dry desert winds push sand dunes grain by grain across desert wastes. In this and the two following chapters we will discuss some aspects of this second great family of geologic processes.

The earth's sun is a medium-sized star with a diameter of 865,000 miles (the earth's diameter is about 8000 miles) and a mass of 2×10^{33} grams (the earth's mass is about 6×10^{27} grams or about 1/300,000 that of the sun). The sun rotates about its axis in the same direction that the planets move. The visible surface layer of the sun, the *photosphere,* consists of gases at temperatures ranging from 7700°F. near the outer boundary to about 11,000°F. at the base of the photosphere. Pressures and temperatures in the solar interior or *nucleus* become very great.

The sun is a huge nuclear fireball that has been burning for 5 billion years or more. It radiates huge amounts of energy into space, which it produces by converting hydrogen into helium in its interior. Scientists divide solar radiation (and other radiation as well) into groups based on the rapidity of vibration (frequency) of the waves. Very rapidly vibrating waves are called *gamma* rays, slightly less rapidly vibrating waves *x-rays;* then descending the frequency scale we encounter ultraviolet light, visible light, infrared light, microwaves and radio waves.

The earth intercepts less than 0.0025 per cent of the sun's total radiation. The earth's acceptance of solar radiation varies from place to place, creating temperature gradients that constitute important elements in global climate, rainfall, weathering and erosion equations. We will return to these aspects of solar radiation later in this and succeeding chapters.

Turbulence in the photosphere occasionally sends charged gas particles streaming into space. The charged gas particles or *ions* may strike the earth, some becoming trapped by the earth's magnetic field. If the stream of gas particles is great enough, it creates fluctuations in the earth's magnetic field strong enough to disrupt radio communications. Periods of high solar ion bombardment are called *magnetic storms.* Severe storms, which average less than one per year, may cause the earth's magnetic field to vary as much as 5 per cent. Ions trapped in the earth's magnetic field produce the *aurora borealis* or *northern lights.*

3 THE ATMOSPHERE

The gray, dry, lifeless lunar landscape terminating abruptly against a black sky contrasts strongly with the blues, greens and browns of the earth's landscape gently silhouetted against the pale blue of the earth's atmosphere. A unique combination of size and distance from the sun provides the earth with its insulating blanket of air, which not only retains the heat necessary to warm this planet's life forms but also transports and disperses oxygen and water essential to survival of its flora and fauna. If the earth were much larger, a dense atmosphere rich in methane and ammonia would inhibit evolution of life as we know it; if much smaller, its gravitational field would not pull strongly enough and would allow essential gases to escape, leading to development of a dry and arid landscape such as those of Mars and the moon; if it were much closer to the sun, the earth would have a thick hot cloud of carbon dioxide like that of Venus; and if it were much farther from the sun, lessened solar radiation would soon convert it into a glacier-covered wasteland.

Composition

The blanket we call our atmosphere lies for the most part (97 per cent) within 18 miles of the earth's surface. The absolute "top" of the atmosphere is rather difficult to define, because the gases of the atmosphere become gradually thinner with increasing distance, but six thousand miles is a sufficiently accurate approximation. Nitrogen, which comprises 78 per cent of dry air (Fig. 4–2), acts primarily as an inert component, although certain bacteria extract it and form compounds essential to plant growth. Oxygen makes up 21 per cent

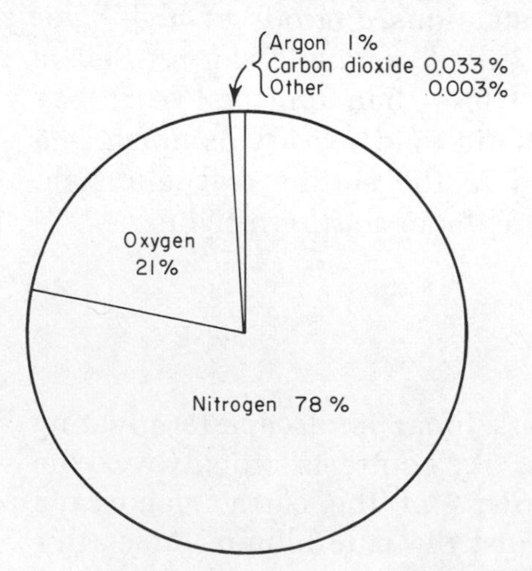

Figure 4–2. The composition of the earth's atmosphere. (Adapted from Glueckauf, E.: Compendium of Meteorology. Boston, American Meteorological Society, 1951.)

of the atmosphere and is very active, combining readily with rock-forming materials, with organic materials in the decay process, with iron to form rust and with carbon in the animal metabolic cycle. Plants separate carbon and oxygen given off as animals breathe and use the carbon (with water, oxygen and other elements and compounds) to build plant tissues and cells. Oxygen given off by plants returns to the atmosphere to be used again by animals. Argon makes up most of the remaining 1 per cent of the air. A number of gases are present in amounts of less than one per cent, but carbon dioxide is the only one of these that is present in amounts greater than 0.01 per cent in dry air. All of the gases just described occur in uniform proportions throughout the atmosphere.

Water vapor can comprise as much as 4 per cent by weight of the air we breathe (note that preceding percentages for nitrogen, oxygen, argon and carbon dioxide were for *dry air*) but usually comprises a smaller percentage, sometimes as little as 0.1 per cent in cold dry polar air. Amounts of water vapor in the atmosphere vary greatly with time and place.

Man introduces significant amounts of sulfur dioxide, nitrogen dioxide, nitrous oxide and ammonia into lower levels of the atmosphere. Forest fires, winds and volcanic eruptions contribute dust. Evaporation of sea waters contributes minute quantities of salt. These components, which concentrate in lower levels of the atmosphere and may be totally absent at higher levels, are the *air pollutants.*

Energy Budget

The earth's atmosphere maintains a delicate balance between incoming solar energy and energy radiated outward from the earth into space, insuring a comfortable temperature range for animal and plant growth. The complex branching process whereby heat from the sun enters the atmosphere, heats the land and sea, evaporates water and returns it to earth as rain and then returns to space is called the earth's *energy budget.*

Both heat from the sun and heat escaping from within the earth warm the earth's surface. Of these two heat sources, the sun is by far the most important. At the

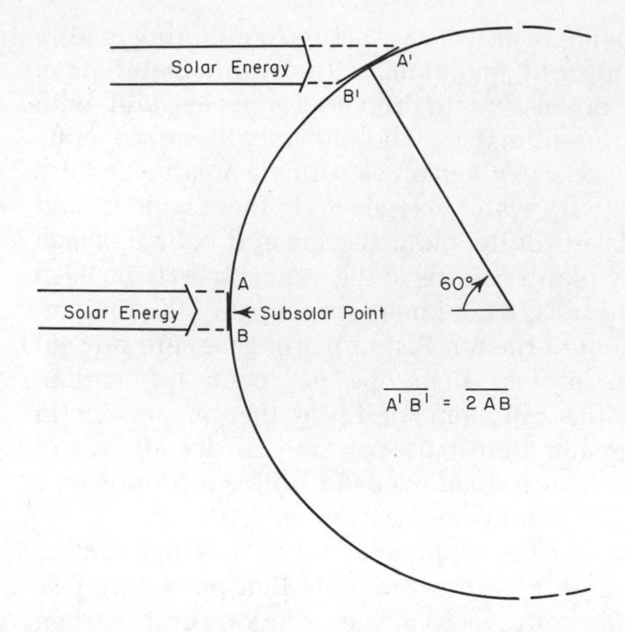

Figure 4–3. Effect of latitude on intensity of radiation per unit area. At 60°N latitude, a given area on the earth's surface receives only half as much energy per unit of area as received at the equator because of the angle of incoming sun's rays.

subsolar point (the point on the earth's surface located on a line between the center of the earth and the center of the sun) the sun heats the earth at a rate of about two calories per square centimeter per minute (cal/cm²/min). Heat from within the earth amounts to less than 1/10,000 this amount.

Although the sun provides 2 cal/cm²/min, the amount reaching the surface of the earth may be much less. Latitude, season of the year, cloud cover, reflectivity of the surface material (called the *albedo*) and scattering, absorption and reradiation of heat by the atmosphere all affect the amount of heat reaching the earth's surface. Let us examine the ways in which these factors affect incoming solar energy.

At the subsolar point, a "bundle" of solar rays strikes the earth's surface perpendicularly, transferring the maximum possible amount of energy per unit of area to the earth's surface. At higher latitudes, single "bundles" of light energy fall on larger areas (Fig. 4–3), which reduces the amount per unit area. Ignoring for the time being the effect of inclination of the axis of rotation, we see in Figure 4–3 that no energy reaches the earth at the North and South poles. Actually, this situation exists only at winter and summer solstices; at all other times the "no-radiation" point is offset from the poles, just as the subsolar point lies away from the equator at all times except during the solstices.

The inclination of the earth's axis of rotation causes the earth to rock back and forth so that the subsolar point crosses the equator twice yearly. The subsolar point is north of the equator during spring and summer in the northern hemisphere and south of the equator during fall and winter. Thus heat transfer accompanying movement of the subsolar point increases during the summer and spring in the northern hemisphere and decreases during the remainder of the year. This is the mechanism responsible for the earth's seasons.

Inclination of the earth's axis increases energy transferred in one hemisphere and decreases energy transferred in the other hemisphere by equal amounts. In other words, no net increase or decrease in energy transfer occurs as a result of axis inclination. In contrast, variations in distance from the sun due to the ellipticity of the earth's orbit cause differences in net energy transferred to the earth.

The earth approaches the sun from early July to early January, then recedes during the remainder of each year. Differences in distance from the sun cause the greatest amount of radiation to reach the northern hemisphere a few days after the summer solstice (usually June 22) instead of on the summer solstice, which would be the case if the earth orbited the sun along a perfectly circular path.

Once radiation reaches the earth, it is reflected, scattered and absorbed. Atmospheric gases scatter and reflect into space approximately 6 per cent of incoming radiation (Fig. 4–4). Molecules and dust in the atmosphere absorb an additional 14 per cent. Thus, on a very clear day, about 80 per cent of incoming solar radiation reaches the ground.

Cloud cover causes the greatest variation in intensity of solar radiation reaching the earth. Clouds may reflect 30 to 60 per cent of the total incoming solar energy and absorb an additional 5 to 20 per cent. A dense cloud cover can reduce radiation reaching the ground to almost zero, as everyone knows who has been caught in a bad storm.

Having evaded water molecules, dust particles and other obstacles which tend to absorb or reflect it, a bundle of solar energy reaches the earth at last. Some of the energy returns immediately to the atmosphere as land, water and vegetation reflect a percentage of it. Wide variation of albedo is an important factor affect-

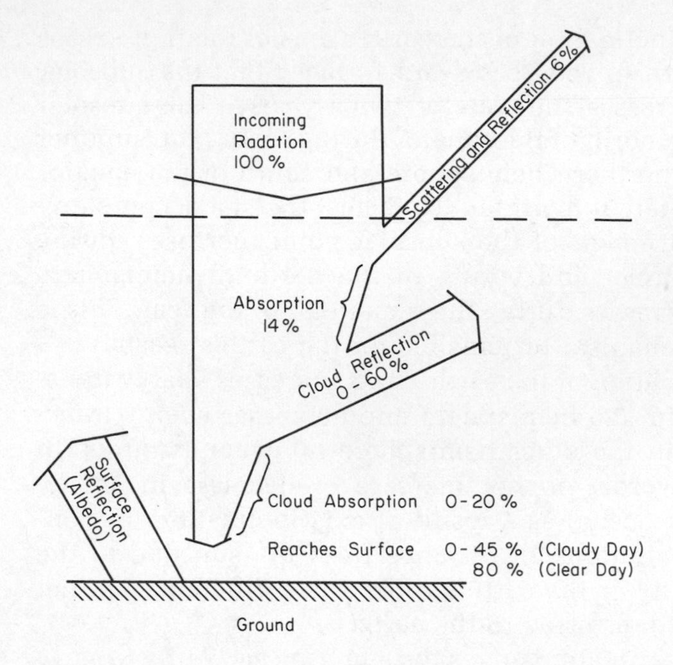

Figure 4–4. Losses of incoming radiation due to scattering and reflection, atmospheric absorption, cloud reflection, cloud absorptions and surface reflection (albedo).

ing climate. Water's albedo ranges from an insignificant 2 per cent for vertical rays to almost 100 per cent for rays striking the surface at low angles. The albedo of the world's oceans averages between 6 and 10 per cent. Albedo of snow and ice is extremely high; 45 to 95 per cent of incident radiation is reflected. Albedo of vegetation and bare ground varies from 5 to 30 per cent.

So far, we have considered the amount of solar energy reaching the earth (our gross income, so to speak), losses due to reflection and scattering (since this part of solar energy is deducted before reaching the earth's surface we can consider it to be the "income tax"), absorption by the atmosphere (since we don't see it at the earth's surface but it provides long-range benefits for us, it is a sort of savings account) and reflection from the earth's surface (outright loss since it reaches the earth's surface but is not utilized). Now let us consider how the earth spends its remaining energy or net income.

The energy absorbed by the ground eventually reradiates into the atmosphere as infrared (or heat) waves. Infrared radiation has longer wavelengths than incoming radiation, a factor which affects subsequent heating. Atmospheric water vapor and carbon dioxide allow short wave radiation to pass but readily absorb much of the longer wave infrared radiation. Glass walls and roofs of greenhouses also allow incoming short wave radiation to pass and block outgoing long wave radiation. This phenomenon gives rise to the term *greenhouse effect,* which is widely applied to the atmosphere's tendency to stop infrared radiation.

The temperature of a body determines the theoretical maximum rate of radiation from a surface. The ratio of the actual rate of radiation to the theoretical maximum rate of radiation for a given material is known as its *emissivity*. Terrestrial vegetation, water, snow, ice and bare ground all emit radiation at rates of 85 to 99 per cent of theoretical maxima; that is, they have emissivities of between 85 and 99.

Energy radiated from the earth's surface adds heat to the atmosphere. The atmosphere radiates this and other heat in all directions. Some of it escapes into space, and some returns to the earth from whence it reradiates again. Heat absorbed by the earth and reradiated as infrared radiation accounts for about 40 per cent of the energy transmitted from the earth's surface to the atmosphere. *Latent heat* and *conduction* transfer the remainder.

Latent heat is defined as the energy absorbed during evaporation and transpiration of water vapor (see Chap. 11). Just as heat transferred to water in a kettle makes the water boil, solar energy absorbed by water on the earth's surface and by plants causes water to evaporate from the surface of the earth and assists plants in giving off water vapor (*transpiration*). Water vapor rises into the atmosphere and changes into liquid droplets, forming a cloud. Energy originally acquired by the water as it changed from liquid to vapor is released. This is the energy that comprises latent heat, and it accounts for about 40 per cent of the energy transmitted from the earth's surface to the atmosphere.

The second transfer mechanism, *conduction*, consists simply of direct transfer of heat from the molecules at the ground surface to *adjacent* air molecules. The air heated by conduction, now lighter than overlying air, rises upward, carrying with it large quantities of heat—and incidentally contributing to the atmospheric motions we call winds. Conduction transfers the remaining 20 per cent of the heat from the earth's surface to the atmosphere.

Thus, the earth's surface "spends" its heat in three ways: by infrared radiation, by latent heat and by conduction. The atmosphere traps much of the heat given off by the earth, later reradiating it.

Circulation

Three factors control the circulation of the earth's atmosphere: the earth's rotation, frictional resistance of the air layers moving over one another (and especially the resistance of air moving over the earth's surface) and pressure gradients (related to the highs and lows of weather maps).

The earth's rotation causes atmospheric air to move generally from west to east. The wind pattern accompanying the west-to-east motion is very complex; only the net vector sum of wind motion is west to east.

Frictional resistance slows winds somewhat, especially over land areas. Friction due to surface irregularities sometimes strongly modifies wind direction and intensity. Friction between air and sea water creates waves ranging in size from gentle ripples to the great waves accompanying a hurricane.

Pressure gradients, or differences in pressure, cause air to move from areas of high pressure to areas of low pressure. Differential heating of land and sea surfaces is one of the principal causes of differences in air pressure. For example, the sun heats the sides of a mountain during the day. Heat from the mountainside warms the overlying air. A pressure gradient develops between warm surface air and overlying cooler air, forcing the warmer air upward and creating a breeze. At night, the mountainside cools more rapidly than the surrounding air, creating another pressure gradient in the opposite direction, which forces the air downward and creates a breeze that blows in opposite direction of the day breeze. Many bright, clear summer mornings in Flagstaff, Arizona, gradually change as warm, relatively moist air flows up the slope of the neighboring San Francisco Peaks. Condensation of the relatively moist air first produces beautiful white cloud pillows over the mountaintops. But the white pillows change to a crumpled gray wet blanket in the late afternoon, from which fall drenching rains.

The mechanism is responsible for the cloud cap in Figure 3–25. On a larger scale, thermally produced pressure gradients control seasonal shifts of wind currents. During the summer, heat radiating from land surface in Siberia and Canada warms the overlying air. The air rises, forming a low pressure area, which draws air inward from surrounding seas. Warm moist air from the Indian Ocean cools as it rises slightly over the Indian continental land mass, and heavy *monsoon* rains result. In winter, the pattern reverses as cooling of the Siberian land surface generates a large high pressure cell. Air fleeing this high pressure cell cools and dries surrounding areas. A similar pattern is found over the eastern United States, where northbound warm, damp air dominates the summer wind pattern and cold dry Canadian air dominates the winter wind pattern.

From a global point of view the most important feature of air motion is the transport of heat from equatorial regions to polar regions. Without the atmosphere's (and the ocean's) ability to transport heat, equatorial regions would be much hotter and polar regions much colder than they are. In this respect, the atmosphere and ocean moderate temperature extremes of the earth's surface.

Precipitation

Chapters 5, 6 and 11 include discussions of weathering, erosion and fluvial processes largely dependent on precipitation. The amount of water available determines the nature of the weathering process and erosion, the shape of the land and, in conjunction with characteristics of near-surface rocks, the role of surface and ground water in the ecology of an area.

Clouds and fog consist of tiny droplets of water or of tiny ice crystals. Each droplet or crystal forms around a nucleus such as a tiny particle of salt (see Chap. 9) or dust. As long as the droplets are small, they remain airborne, but as they bump into each other, coalesce and grow, the likelihood of precipitation increases. The minimum diameter of a raindrop is about 0.02 inch and the maximum roughly 0.2 inch. Larger raindrops are unstable and break up.

The two principal causes of precipitation are *convectional uplift* and *forced uplift* of air. Precipitation may occur at any time after the air cools below its *dew point* (temperature at which water vapor condenses). The dew point varies with humidity; if humidity is high, the dew point is high and vice versa.

Convectional uplift and subsequent precipitation is common during the summer, when uneven heating of

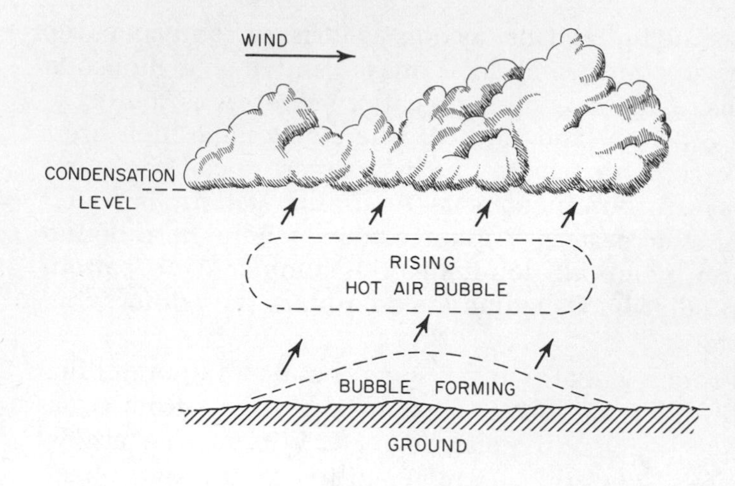

Figure 4–5. A "bubble" of heated air rising into the atmosphere and forming a cloud.

the ground causes a "bubble" of warm air to form over a "hot spot" and begin to rise (Fig. 4–5). The bubble cools as it rises, and water vapor condenses as temperature falls below the dew point. The flat base of a cloud layer dramatically marks the level at which the air temperature coincides with the dew point. Condensation releases latent heat into the bubble, increasing the temperature of the bubble. This may result in instability and cause the bubble to rise rapidly. The result is a *cumulonimbus* cloud, the familiar summer afternoon thunderhead. Precipitation usually follows soon after.

Forced upward movements of air masses occur most commonly when an air mass strikes a mountain range or other topographic barrier, or when collision of two air masses forces one air mass up and over the other. If the upward-moving air mass is warm and wet, and the underlying air mass is cool, heavy rainfall and high winds may result. Tornadoes arise from this type of forced uplift of an air mass.

Mountains blocking the path of prevailing westerly winds in the northwestern United States and northwestern Europe cause *orographic* precipitation (mountain-related rainfall). The Olympic Mountains along the coast of Washington receive more than 100 inches of rainfall annually as a result of eastward flowing moist Pacific air being forced upward by the mountains. The damp nature of the Scottish highlands results from a similar uplift of moist air from the North Atlantic.

Pacific air loses moisture (but gains heat) on the western slopes of the Rocky Mountains. The warmed air drops down the east face of the Rockies and into the Great Plains, sometimes increasing temperatures by 30° to 40°F. within a few hours. Residents of the Great Plains call these winds *chinooks*; in Europe the northward flowing Mediterranean air that descends into the valleys and forelands of the Alps is called a *foehn*.

Rain shadow deserts frequently lie side by side with areas of intense orographic precipitation. The air pushed up and over a mountain range loses its moisture but gains large amounts of latent heat. Dropping down the lee side of the mountain range, the formerly cool and moist air is now hot and dry (Fig. 4–6). The eastern side of the Olympic Mountains, the southwestern slopes of the Hawaiian Islands and the western part of the Great Plains of North America lie in rain shadows.

Mountain ranges paralleling nearby coasts do not everywhere result in extensive rainfall. Along the coasts of Chile, Peru, northern Mexico and southern California, prevailing winds either move parallel to the coast or blow seaward, producing desert or semidesert climates. In these areas, dry air blowing toward the sea often overrides a thin layer of air cooled by cold currents flowing near the coast. A side effect of this process is that pollutants accumulating in the trapped air mass can cause severe smog.

The worldwide precipitation pattern can be divided into seven zones based on the amount and cause of precipitation (Fig. 4–7). These zones are as follows:

1) Near the equator, warm temperatures and abundant moisture favor convective rainfall. Frequent thunderstorms cause precipitation to reach more than 80 inches/year.

2) The eastern coasts of Brazil, Central America, Madagascar, northeastern coast of Australia, and others receive 60 to 80+ inches/year of orographic rainfall. Coastal orographic rainfall belts tend to be narrow.

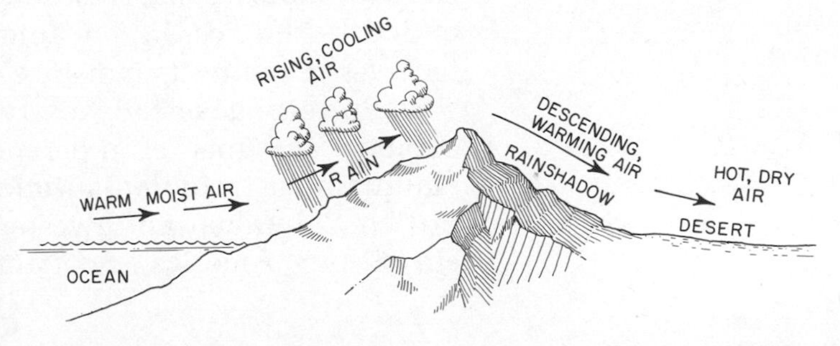

Figure 4–6. Orographic precipitation and a rainshadow desert. Rising moist air cools, loses moisture and releases latent heat. Drier, hotter air dropping on the lee side of the mountain forms a rainshadow desert.

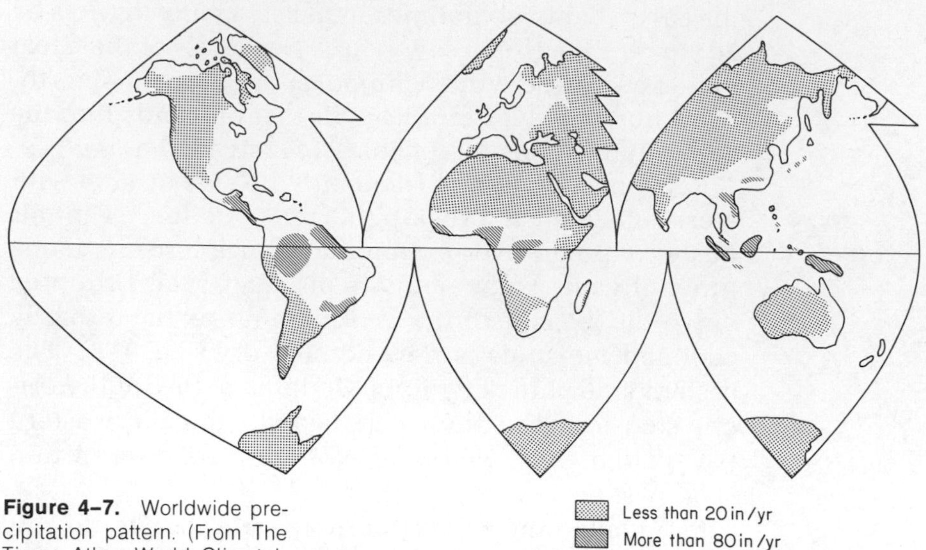

Figure 4–7. Worldwide precipitation pattern. (From The Times Atlas, World Climatology. London, The Times Publishing Co., 1958.)

3) Two belts of tropical deserts straddle the equator between roughly 15 degrees and 30 degrees north and 15 degrees and 30 degrees south. The movement of heat from the equatorial regions to polar regions creates these belts of hot, dry, descending, high-pressure air masses. The deserts of the Sahara, Kalahari, central Australia and the west coast of South America result from these hot air masses.

4) Remoteness from sources of moisture causes the "deserts" of western North America, the plains of Argentina and the Russian *steppes*. Strictly speaking, these "deserts" are all steppes. Virtual lack of vegetation characterizes true deserts like the Sahara. Steppes contain abundant low brush and some grass. The mountains of western North and South America and of Europe and Scandanavia remove moisture from prevailing eastward moving winds, creating vast rain-shadow steppes.

5) Tropical ocean air moving inland provides moisture to subtropical regions on the southeastern sides of North America, Asia, Australia and on the eastern side of South America. In the United States, the Gulf coast receives much of its moisture in this manner.

6) Prevailing westerly (eastward flowing) winds lose copious amounts of orographic precipitation along mountainous middle latitude coastlines of Alaska, British Columbia, northwestern United States, southern South America, Scotland, Norway and South

Island of New Zealand. Cooler temperatures in these regions favor snowfall. Resultant glaciers chisel deep valleys and knife-edge mountain ridges.

7) Cold air north and south of 60° latitude retains only small amounts of moisture. Consequently, these regions form the Arctic and Antarctic deserts. Cold air also impedes evaporation. Therefore, relative humidity and soil moisture are high in these regions where precipitation levels are comparable to levels in the western United States steppes.

Because similar belts probably existed in the geologic past, observation of climatic clues in rocks permits reconstruction of ancient locations of continental masses.

4 THE OCEAN

The oceans comprise the second major part of the fluid outer layer of the earth. Ocean waters and the atmosphere collaborate, the ocean providing moisture with which the atmosphere washes itself and the surface of the land. Ocean waters moving away from the equator join atmospheric currents in carrying heat toward the polar regions. Winds drive waves across the seas, and radiation differences between land and sea drive the winds across the earth. Both air and sea store heat, releasing it slowly and thereby preventing extremes of temperature like those found on waterless, airless planets. Unfortunately, much of man's waste eventually enters the atmosphere and the oceans. Clearly, the ocean is a critical part of our geologic environment. Let us examine some of the characteristics of this vast body of water.

The Blue Planet

More than any other feature, the oceans differentiate the earth from the other planets. Interiors of Venus, Mars and the moon consist of the same silicate rock material as that found in the interior of the earth. Venus and Mars have atmospheres as do Jupiter, Saturn, Neptune and Uranus. But as far as we know, waters cover significant areas of no other planet.

Viewing the earth from Mars, Martians would almost certainly call our 71 per cent water-covered planet, the *Blue Planet*.

Sea water comprises about 87 per cent of the earth's surface water. Water trapped in sediments and rocks comprises 12 per cent, and frozen water in the Antarctic and Greenland ice sheets total a little more than 1 per cent. Rivers and lakes hold a few hundredths of 1 per cent, and about .001 of 1 per cent of the whole makes up clouds and water vapor in the air.

The Pacific Ocean is the largest of the three major ocean basins, and its average depth is greater than that of others. Its immense surface area dwarfs land areas draining into it. Consequently, the Pacific Ocean most closely resembles what might be found in a completely water-covered earth.

Parallel margins of the Americas on the west and Europe and Africa on the east constrict the Atlantic Ocean into a long narrow basin. Shallow marginal seas (Caribbean, Mediterranean, Baltic and others) reduce the average depth of the Atlantic Ocean to below that of the Indian and Pacific Oceans. The area of the Atlantic only slightly exceeds the drainage area of the Congo, Nile, Mississippi and other great rivers. Fresh water from these rivers lowers the average salt content of the Atlantic below that of the Pacific and Indian Oceans.

The Indian Ocean is the smallest of the major ocean basins. Average salt content (*salinity*) and depth are intermediate between those of Atlantic and Pacific.

Several important marginal seas rim the ocean basins. The Mid-America Sea (Caribbean and Gulf of Mexico), the Mediterranean, the Arctic "Ocean," the Japan Sea and others are in this category.

High evaporation rates increase salinity in the Mediterranean and Mid-America Seas. During the Jurassic period (136 to 190 m.y. ago), restricted circulation precipitated salt on the floor of the Gulf of Mexico; during the past few million years, the Straits of Gibraltar seem to have opened and closed several times, causing intermittent salt precipitation in parts of the Mediterranean Sea. At present, warm salty Mediterranean water flows over a ledge between Africa and Europe and enters the Atlantic. This salty underwater "river" can be detected far into the Atlantic. During World War II, Italian submarines caught in "turbulence" in this salty river bobbed to the surface and were destroyed by English forces at Gibraltar.

Very old sedimentary rocks containing pebbles and other features indicative of erosion by water and deposition in a marine environment tell us that the oceans are very ancient. They seem to have formed very early in the earth's history, probably at the time when core, mantle and embryonic crust differentiated from one another.

Water expelled from the earth's interior carries dissolved salts. Geologic processes rapidly remove some elements such as potassium from sea water; other elements, such as calcium, are removed more slowly. Erosion from the land adds these and other elements to sea water. Chapter 9 includes a discussion of the mineral resource potential of the ocean waters.

Thermal Shock Absorbers

Each day, temperatures on the moon's surface soar above the boiling point of water, and each night temperatures plunge far below freezing. On earth the atmosphere, the land surface and the oceans collaborate to reduce the wide swing of temperatures. During the day, air, land surface and sea water absorb heat; at night, each slowly releases heat. These are the earth's thermal shock absorbers, which cushion the violent thermal bumps caused by variations in solar energy input.

The ocean is the most important of the thermal shock absorbers, because water has a somewhat higher heat capacity than soil and a much higher heat capacity than air. In addition, ocean waters reflect a smaller percentage of incident radiation than either atmosphere or land. These factors cause water to absorb more heat during the day and release heat more slowly at night. One result of this is that coastal cities are cooler during the day and warmer at night than inland cities at the same latitude and elevation.

Circulation

It should be no surprise to learn that the same three factors controlling atmospheric circulation (pressure gradients, rotation of the earth and friction) also control oceanic circulation. In the case of friction, the link

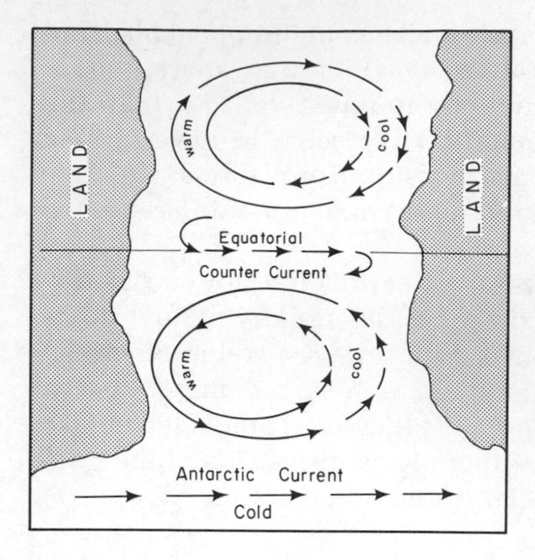

Figure 4–8. Generalized atmospheric and oceanic circulation. Note clockwise circulation of the major air and sea currents in northern hemisphere and counterclockwise circulation in the southern hemisphere. (From Gross, M. G.: Oceanography. Columbus, Ohio, Merrill Publishing Co., 1971.)

between surface winds and ocean waters is especially important.

Surface circulation within the major oceans consists of two big loops or *gyres*, one in each hemisphere. Smaller gyres occur around some margins. Circulation of waters of the major gyres is similar to circulation of atmospheric winds (Fig. 4–8). Forces associated with the rotation of the earth cause the air to rotate clockwise in the northern hemisphere and counterclockwise in the southern hemisphere, and frictional drag of prevailing wind currents pulls oceanic surface waters in the direction of wind flow.

The eastward rotation of the earth compresses the western portion of the ocean currents into narrow, fast-moving streams. In the northern hemisphere, the Gulf Stream races past (relatively speaking) the east coast of the United States, and the Kuroshio Current similarly flows rapidly past Japan. In the southern hemisphere, Peru and Benguela currents flow along the western margins of South America and Africa, respectively. The Indian Ocean has counterrotating gyres similar to Pacific and Atlantic gyres, but they are smaller and less well developed.

Beneath the sea surface, currents carry water in other directions. In 1952, scientists studying water movements in the Pacific were surprised to discover a current flowing beneath and in the opposite direction to the Equatorial Countercurrent. This deeper current, named the *Cromwell Current*, is 250 miles wide, 3500 miles long, lies at depths between 100 and 800 feet and flows much faster eastward than the overlying Equatorial Countercurrent, which carries surface waters westward.

Hydrodynamic theory predicted a deep, oppositely directed current beneath the Gulf Stream and Kuroshio Current. Oceanographers tracking currents by means of weighted floats confirmed the existence of the Gulf Stream Countercurrent (Fig. 4–9) in the mid-1950's.

Cold Arctic and Antarctic waters sink and flow toward the equator. Antarctic water is colder and therefore denser, so it slips under Arctic water in the middle latitudes. As a result, Antarctic water covers much of the floors of Pacific and Atlantic basins. Another body of Antarctic water not as cold as the bottom water slides northward over the Arctic water. The result is an oceanic layer cake in which different water bodies lie on top of one another (Fig. 4–10).

Deep ocean currents move very slowly, requiring up to 4000 years to complete one "revolution." The combined effect of the earth's rotation and prevailing winds blowing parallel to a coast will tend to deflect surface currents either toward or away from coastlines. Along

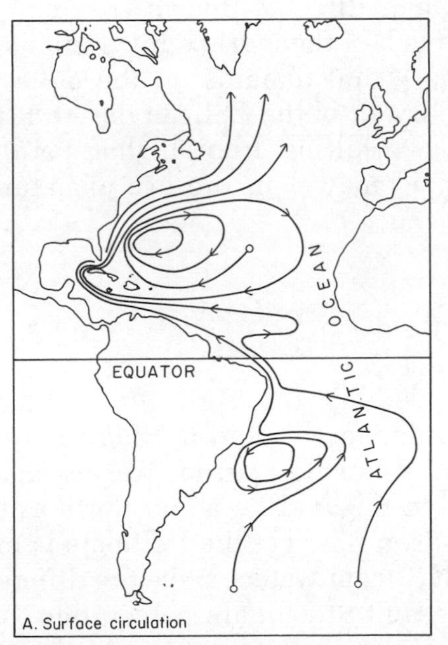

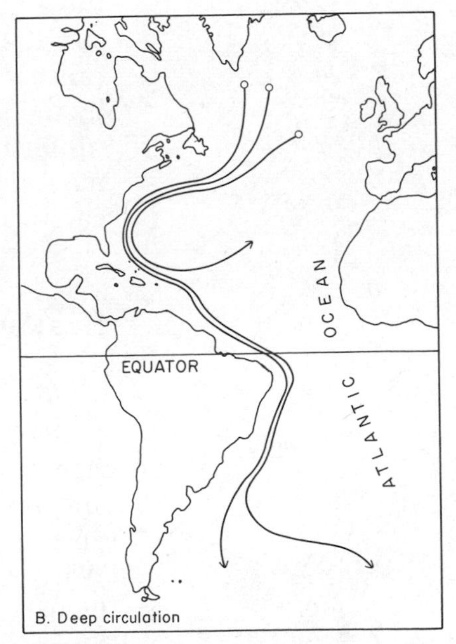

Figure 4–9. Surface circulation (*A*) and deep circulation (*B*) in the North Atlantic. The Gulf Stream flows northward near the surface of the western Atlantic. The Gulf Stream Countercurrent flows beneath, returning water to equatorial regions. (From Stommel, H.: A survey of ocean current theory. Deep Sea Research *4*:149–184, 1957.)

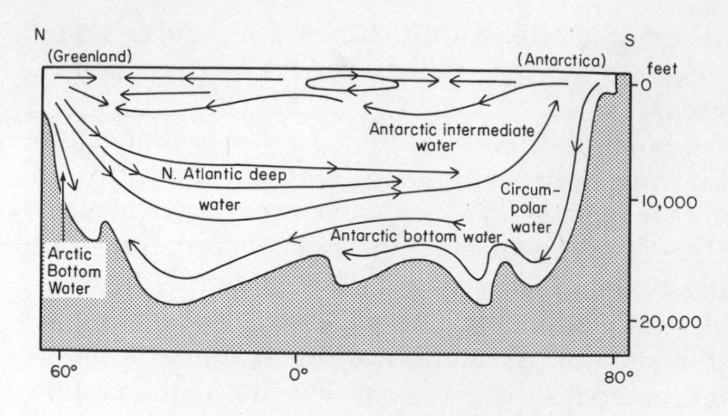

Figure 4–10. Generalized north-south cross-section of the Atlantic Ocean, showing major water masses and their directions of movement. (From Williams, J.: Oceanography. An Introduction to the Marine Sciences. Boston, Little, Brown and Co., 1962.)

the coast of northern Chile, northerly winds and the earth's rotation deflect the Peru Current away from the coastline. Cold Antarctic water rises, or *upwells,* filling the gap between the Peru Current and the shoreline. (Upwellings also occur off the coasts of southern California, Morocco, and southwest Africa.) The upwelling water contains abundant nutrients collected as dead organisms sank into these deep, cold, oxygen-deficient waters as the waters slowly journeyed northward from the Antarctic. Fish thrive in these nutrient-rich waters, and birds feeding on the fish have painted white with excrement the sea cliffs of northern Chile. Before the development of techniques to remove nitrogen from the atmosphere directly, this bird excrement, or *guano,* supplied much of the world's nitrogenous fertilizer. At present the guano remains on the cliffs, mummified by the hot dry air of the Chilean desert. Price increases in fertilizer resulting from higher petroleum prices may spark a renewal in the use of natural fertilizers such as guano.

Heat and Cold

Heat carried by the Gulf Stream current warms eastern North America and northern Europe; heat carried by the Kuroshio Current warms Japan. Scotland and northern England lie as far north as the southern part of Hudson Bay, yet their climate is much milder. Warm Gulf Stream waters make the difference. The effects of the Gulf Stream are perceptible as far north as Murmansk, Russia's ice-free port on the Arctic Ocean. The Viking expeditions and indeed the habitability of the Viking homelands in Scandinavia were made possible by the warming effect of the Gulf Stream. Without the Gulf Stream, Iceland, Great Britain and much of northern Europe would be as inhospitable as central Canada.

In the northern hemisphere, southward moving currents carry cooling waters along the coasts of Spain, North Africa, Southern California and Mexico. In the southern hemisphere, similar cooling currents brush the coastlines of southwest Africa and western South America. The earth's rotation intensifies motion of the Gulf Stream and Kuroshio Current but has the opposite effect in the southern hemisphere, where the cold Peru and Benguela currents flow. Consequently, Peru and Benguela currents *cool* nearby land areas less efficiently than the Gulf Stream and Kuroshio Current *warm* nearby land areas.

Waves

Wind blowing across the sea surface causes most wave motion. Waves form as wind exerts a frictional drag on the sea surface. After formation of small initial ripples, wind action becomes more effective, both pushing from behind and frictionally dragging the wave forward.

The longer the wind continues to blow, the higher the waves rise. The speed of the wind and the *fetch* or open space where the wind can blow unimpeded are also important factors in wave height. Waves continuing after the wind veers or dies are known as *swells*.

Blowing wind transfers to waves energy that may erode sea cliffs or sea floor or may roar onto a beach, destroying man's feeble structures. Wave erosion will be discussed further in Chapter 6.

In addition to wind waves, three other types of waves are important to man. *Solitary waves*, a group including tsunamis (Chap. 2), occur when a sudden jolt disturbs the water. Landslides and earthquakes are the primary causes of solitary waves. *Storm surges* are dramatic rises in water level occurring when the speed of a storm (such as a hurricane) equals the natural speed of waves in shallow water. The deltaic coast of the Ganges and Brahmaputra rivers in Bangladesh is particularly susceptible to storm surges. A storm surge in 1737 supposedly rose 40 feet above normal high tide, swept inland and killed 300,000 people. A storm surge accompanying Hurricane Audrey rolled over the coast of Louisiana in 1957, taking many lives and causing extensive damage.

True *tidal waves* (not tsunamis) occur daily in many river mouths and estuaries. Tidal waves are usually small, but large waves or *tidal bores* move up the Amazon, Colorado, Yangtze and other large rivers. The Amazon tidal bore rises as high as 16 feet as it moves upstream at about 12 knots (14 miles per hour). Tidal waves can also be used as a source of local electric energy (see Chap. 8).

5 ICE

The glacial ice that choked off Viking communication between Iceland and Greenland was by no means the first episode of glaciation recorded in the book of rocks covering the earth's surface. Grooves radiating from southern Africa show where Permian Age (225 to 280 m.y. ago) glaciers gouged underlying rocks (Fig. 4–11), plastering soil and rock onto the ancient

Figure 4–11. Grooves gouged in rock near Durban, South Africa, by overriding glaciers during the Permian period (225–280 m.y. ago). (Photo courtesy of J. L. Snyder.)

Figure 4–12. Geologists examining Permian tillite in South Africa. Boulders, cobbles and pebbles in tillite provide clues about the path of glacier movement.

ground surface. The plaster hardened into a rock called *tillite* (Fig. 4–12), which is widely distributed in India, Australia, South Africa and Antarctica.

Other tillite deposits indicate that glaciers covered parts of the land surface during the Cambrian period, 500 to 570 million years ago, and during earlier geologic periods as well.

Pleistocene Epoch

The *Pleistocene Epoch* is the geologic name for the period of time encompassing the most recent sequence of continental glaciation; that is, the period during which great ice sheets advanced and retreated over large areas of the North American and European mainlands. The brief period of time following the disappearance of the last great ice sheet is called *Holocene* or *Recent Epoch.*

The Pleistocene Epoch is important to man for two reasons. First, it was during this time that the great glaciers scoured vast areas, diverted streams and rivers, chiseled out lake basins and deposited accumulated debris over much of North America and Europe (see Chap. 6). Secondly, man developed from a primitive being to his present status during Pleistocene time. His evolution, his wanderings and his culture are intimately related to Pleistocene climate, glaciers and other phenomena. More importantly, man's future activities also depend to a large extent on future changes in climate and glacial growth or reduction.

Beginning of the Pleistocene

Geologists have found it difficult to establish precisely the onset of the Pleistocene Epoch. Weathering, erosion and advances of subsequent glaciers have disturbed older glacial deposits to the extent that they are of little value in deducing the time of formation of the first ice sheet. Geologists have had to turn elsewhere for evidence of this event. The sea floor has proven to be one of the most rewarding areas of Pleistocene research. Slowly accumulating layers of sea floor sediment record changes in fauna, in atmospheric and oceanic isotope ratios and in sediment types. These phenomena provide clues to variations in water temperature, atmospheric temperature, and extent of glaciation.

Fossil shells of two species of tiny floating marine animals called foraminifera have proven especially useful in estimating the rise and fall of water temperatures. One species, *Globorotalia menardii* (Fig. 4–13), prefers warm water; hence an increasing population indicates warming temperatures. Geologists count the number of shells of *Globorotalia menardii*, count the total number of foraminifera shells and compare the ratio with tables showing variations of ratios with latitude and water temperature.

Another foraminifera species, *Globorotalia truncatulinoides* (Fig. 4–14), coils to the left in cooler water and to the right in warmer water. The ratio of right to left coilers indicates water temperature.

Radiocarbon dating techniques work well for shells deposited during the past 40,000 years, although other radiometric age-dating methods are not useful on these relatively young sea floor sediments. The lack of an effective radiometric age-dating technique posed a serious problem before investigators hit on paleomagnetism (see Chap. 2) as a means of dating sea floor sediments.

Sediments enclosing tiny foraminifera shells include small amounts of magnetic minerals, which tend to align themselves with the earth's field at time of dep-

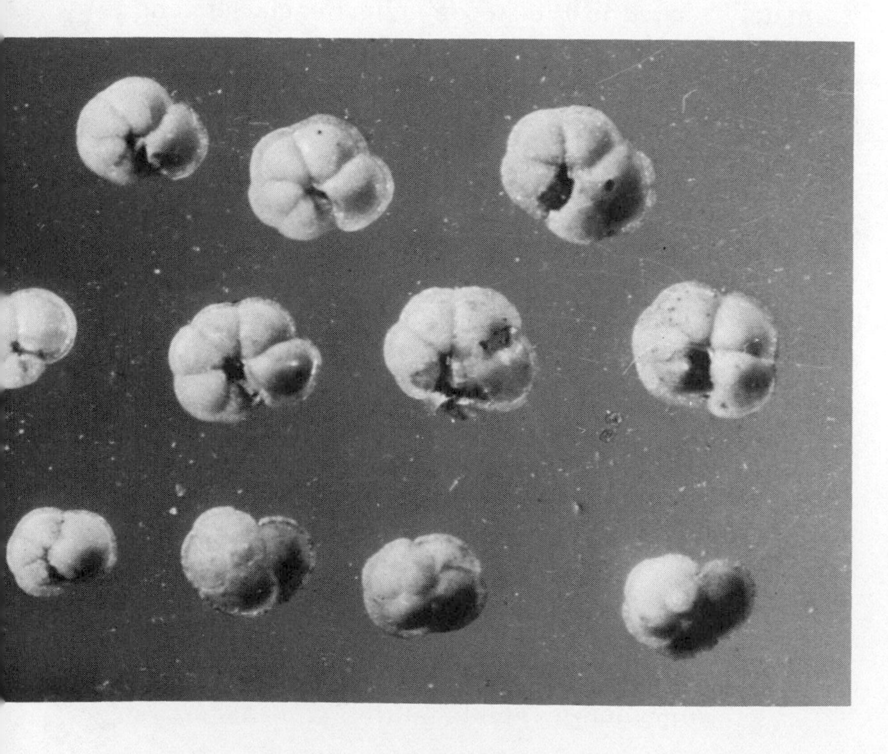

Figure 4–13. Shells of *Globorotalia menardii*, a foraminifera species useful in paleotemperature research. (From Ericson, D. B., and Wollin, G.: Science *162*:1233, 1968.)

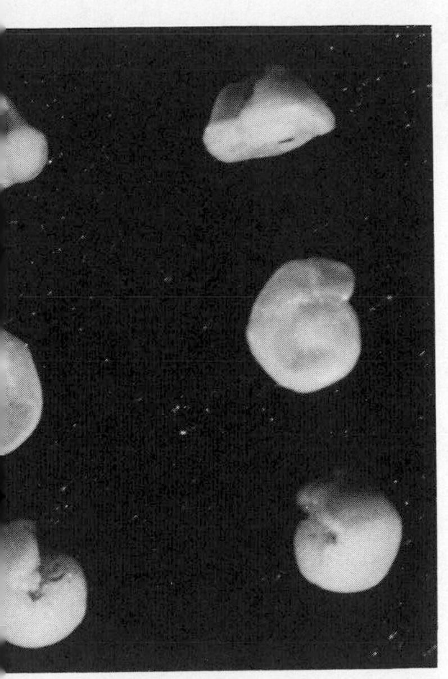

Figure 4–14. Right and left coiling shells of *Globorotalia truncatulinoides*. (From Ericson, D. B., and Wollin, G.: Science *162*:1233, 1968.)

osition. Careful measurements of the magnetism of these rocks showed that Pleistocene glaciation began before the Olduvai normal magnetic epoch (Fig. 4–15). Potassium-argon radiometric dates from basalts elsewhere in the world had shown that the Olduvai Epoch (see Fig. 2–4) began 1.8 million years ago, so it became clear that Pleistocene glaciation began at least that long ago. Before this discovery, geologists had thought glaciation to have begun much later.

Within the upper few dozen feet of the sea, the ratio of oxygen-16 to oxygen-18 depends primarily on water temperature. Because animals use both isotopes indifferently for shell construction, measurements of oxygen isotope ratios in sea floor fossils were used to confirm temperature findings deduced from foraminifera population and coiling data.

Geologists now estimate the beginning of the Pleistocene to be about 2 to 2.5 million years ago. However, ice-rafted glacial material 4 million years old has been found in the Indian Ocean, and ice cores taken from ice island T-3 are more than 6 million years old. During 1973, scientists drilling into the earth's crust were startled to find rocks recovered from the Antarctic Ocean indicating that Antarctic glaciation began as much as 25 to 30 million years ago and that Antarctica had probably been continuously covered by glacial ice since the beginning of this glacial episode. The rocks further indicated that glaciation

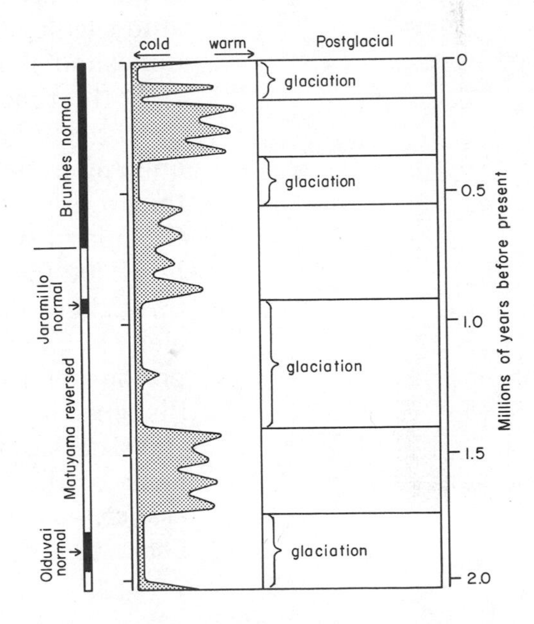

Figure 4–15. Ocean water temperatures deduced from foraminifera shells, and age dates deduced from paleomagnetic measurements. (From Ericson, D. B., and Wollin, G.: Science *162*:1233, 1968.)

began in Antarctica after separation of Antarctica and Australia and development of the Circum-Antarctic current. This discovery may lead to important revisions in theories of the origin and evolution of glaciation.

Temperature

Oxygen isotope measurements made on shell material and ice of the Greenland ice sheet suggest that mean temperatures varied 5 to 9°F. during the later Pleistocene. There is a close correlation between Greenland oxygen isotope temperature data and the geologic record of the last major advance and retreat of ice in the Great Lakes region, which began as temperatures dropped 3 to 5°F. and retreated abruptly as temperatures rose about 5°F.

Sea Level

Ice of the great glaciers consists of water evaporated from sea water, frozen in the atmosphere and deposited as snow. The quantity of water involved is great enough to have caused a perceptible fall in sea level during advances of the great glaciers and a comparable rise during retreat. During the last great advance, sea level fell more than 400 feet below present-day levels, exposing extensive areas of the continental shelves (Fig. 4–16). Marshes, swamps, beach dunes, corals, seashells and fossil terrestrial animal bones lie beneath today's oceans, enabling marine geologists to study geologic history by dredging soil, rock and fossils deposited above or slightly below sea level.

At the time of maximum advance, glaciers covered 27 per cent of the world's land surface, and Greenland and Antarctic ice sheets still cover about 10 per cent of the world's land surface. Melting of glacial ice in the Greenland and Antarctic ice sheets would raise sea level 200 feet or more (Fig. 4–16). Elevated shorelines throughout the world show that alternating periods of glacial growth and retreat caused corresponding periods of rise and fall of sea level throughout the Pleistocene.

Asiatic nomads colonized North America during the peak of the last glacial advance. Lowered sea level associated with glacial advance exposed a land bridge between Asia and North America. Rising waters ac-

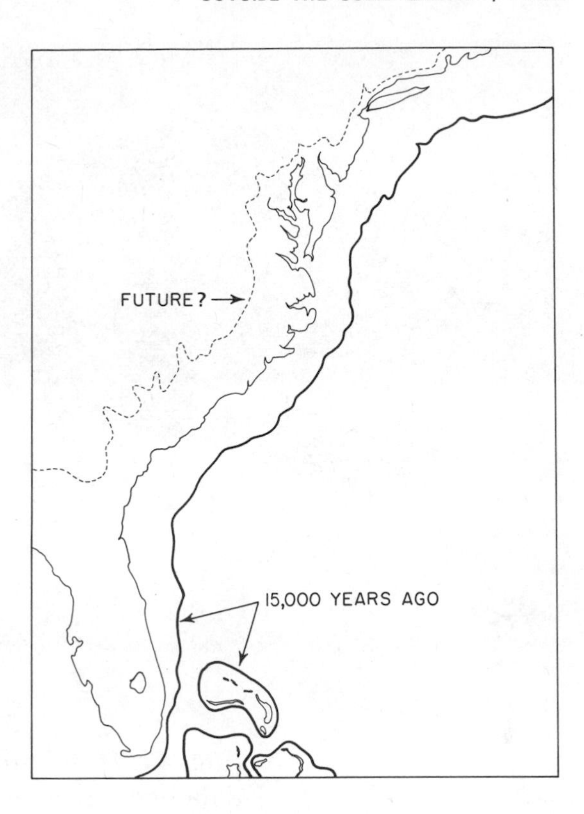

Figure 4–16. Extent of land area in North America during the peak of the glacial advance 15,000 years ago, now and in the future when all glacial ice has melted.

companying the retreat of the glaciers later covered the land bridge and cut off or at least severely restricted immigration.

Rainfall

The great ice sheets modified circulation and rainfall patterns in many areas of the world, especially in the northern hemisphere. Drier climates around the southern margins of North American and European ice sheets resulted in widespread windblown deposits called *loess*. Conversely, wetter climates in the western United States greatly enlarged intermontane lakes

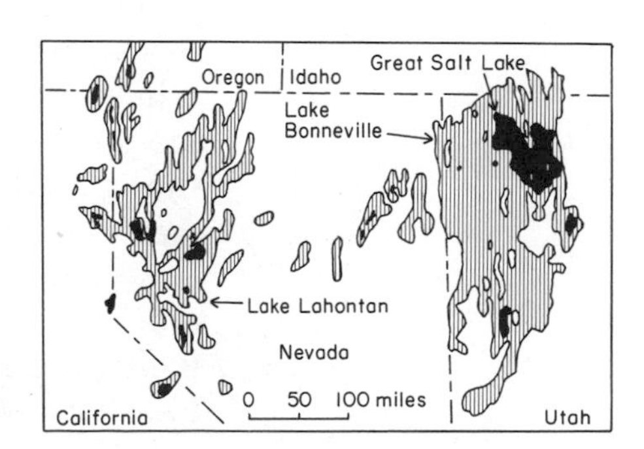

Figure 4–17. Great Pleistocene lake complex in Utah, Nevada and adjacent states. Black areas indicate extent of present-day remnants. (Adapted from Gilluly, J., et al.: Principles of Geology, 3rd ed. San Francisco, W. H. Freeman and Co., 1968.)

Figure 4–18. Lake Bonneville terraces and shorelines on a mountainside far above and to the east of the present-day Great Salt Lake.

(Fig. 4–17), including the Great Salt Lake of Utah. The enlarged Great Salt Lake, Lake Bonneville, was 1000 feet deep and rivaled Lake Michigan in area. Lake Bonneville shorelines and terraces show former lake levels on surrounding mountainsides (Fig. 4–18). At one point, lake waters overflowed a land barrier and reached the sea via the Snake River. None of the other intermontane lakes rose high enough to top surrounding land barriers. Some Pleistocene lake beds in the western United States contain brine shrimp eggs, which will still hatch if placed in warm salty water.

Holocene Epoch

About 15,000 years ago, the sea receded to its lowest level as continental ice accumulation peaked (Fig. 6–1). Three thousand years later an ice advance overwhelmed trees at Two Creeks, Wisconsin, but the great ice sheets, already in retreat, soon disappeared from the United States, leaving great gouges to be filled by waters of the Great Lakes.

By about 5000 BC, the Scandinavian ice sheet was almost gone, and about 2000 years later the last of the meltwater from the great *Laurentide* (North American) ice sheet was flowing into Hudson Bay. The temperature of the hydrosphere began to fall in about 4000 BC (shortly before the Laurentide ice disappeared), marking the beginning of the first of three cycles occurring in the past 6000 years (Fig. 4–19). We call this period the *Neoglacial* (neo = Greek *new*) interval. Ice sheets have not reformed in Eurasia and North America during the Neoglacial interval, but mountain glaciers and ice sheets of Greenland and Antarctica have waxed and waned several times. The most recent glacial surge began in the Middle Ages and culminated about 100 years ago. During this advance the Viking colonists in Greenland were cut off from Iceland and Europe. Alpine glaciers overran villages built in previously safe places. Some of these villages remain frozen beneath glacial ice today. Progressively cooler weather associated with glacial advance reached a peak in Norway in 1741. Crops failed to ripen that year, famine was widespread and the Norwegian death rate far exceeded the birthrate. Glaciers remained at or near maximum lines of advance from about 1600 AD until the mid-nineteenth century.

Pollen preserved in northern Europe during the period AD 50 to 400 (the Roman Iron Age; see also Chap. 9) indicate that the climate was relatively mild during this time. Following this period of warmth, northern Europe began to cool. A brief reversal in the cooling trend warmed the area between 800 and 1000 AD, permitting Vikings to penetrate the North Sea and colonize Iceland, Greenland and North America. But continued cooling and associated advancing sea ice blocked Viking sailing routes to the New World before the colonies became truly viable.

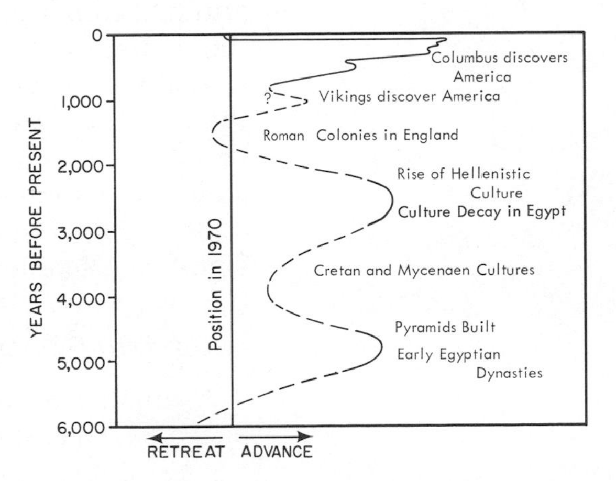

Figure 4–19. Advance and retreat of mountain glaciers during the past 6000 years. (Based in part on Denton, G. H., and Porter, S. C.: Neoglaciation. Sci. Am. *222*(6):107, 1970.)

The cooling trend abruptly reversed in the middle to late nineteenth century, and the earth warmed rapidly. Mountain glaciers retreated all over the world as the average worldwide temperature rose about 2°F. In the 1940's mountain glacier retreat slowed, and in some areas glaciers again began to move forward. As of the late twentieth century, it appears that the warming trend has stopped and the earth is now cooling again.

Possible Causes of Glaciation

The large number of current hypotheses proposed to explain Pleistocene glaciation reflects our continuing ignorance about phenomena affecting global temperatures. Most hypotheses assume that glaciation is caused by one of the following processes:

1) Decreased solar radiation,

2) Astrophysical phenomena reduced the amount of radiation reaching the earth, or

3) Terrestrial phenomena reduced either the heat reaching the earth or heat transport to polar latitudes.

Solar Radiation

Solar radiation hypotheses derive from known short-term variations. It is possible that longer term variations in solar radiation may occur, but at present data adequate to evaluate these hypotheses is lacking.

Astrophysical Phenomena

The simplest hypothesis of this class assumes that our solar system moves through "dust" clouds in space and that from time to time some of these clouds reduce the amount of radiation reaching the earth. Other hypotheses derive mainly from long-term variations in the ellipticity of the earth's orbit and in the inclination of the earth's axis of rotation. Since ellipticity and axial inclination affect the distribution of solar heat on earth, it has been suggested that they could cause significant temperature differences.

Terrestrial Phenomena

A number of terrestrial phenomena have been proposed to account for glaciers. An early and still widely held hypothesis is the *greenhouse effect* hypothesis. It is well known that the amount of carbon dioxide in the atmosphere has risen steadily since the beginning of the Industrial Revolution (see Chap. 9). Atmospheric carbon dioxide absorbs heat reradiated from the earth's surface, slowing its escape into space. It has thus been suggested as the agent responsible for the rise in temperature since the mid-nineteenth century. Atmospheric carbon dioxide cannot be the sole agent, however, since worldwide temperature began to drop after 1940 while the carbon dioxide content of the atmosphere continued to rise.

Some investigators have suggested that volcanic dust has been sufficiently abundant in the atmosphere to reflect enough heat to cause Pleistocene glaciation. Temperatures dipped worldwide following the Krakatoa eruption (see Chap. 3), and the Pleistocene seems to have been a period of unusually great volcanic activity. However, most other Cenozoic (0 to 70 m.y. ago) epochs were also periods of high volcanic activity while glaciation occurred extensively only in the past 25 to 30 million years. In addition, rain rapidly cleans dust from the lower atmosphere. The volcanic dust hypothesis requires frequent mega-eruptions of the Krakatoa type, for which solid evidence is lacking.

Orogenic (mountain-building) hypotheses have also been proposed. According to most of these hypotheses, rising mountains deflect air currents, causing redistribution of terrestrial heat. However, mountain building requires somewhat longer periods of time than glaciation; therefore, these hypotheses appear inadequate.

A very interesting hypothesis dependent in part on orogenic assumptions was proposed by Maurice Ewing and William Donn. Ewing and Donn suggested that continental drift brought the continents into positions favoring continental ice sheet accumulation in the late Cenozoic. They point out that a North American continental ice sheet would require considerably more moisture than now falls on the northern part of North America. In order to obtain the necessary moisture, they suggest that small changes in the Arctic ice pack heat budget could bring on melting of the ice. When free of ice, Arctic Ocean evaporation provides moisture that feeds growing ice sheets on nearby continents. Growth of the ice sheet deflects prevailing wind upward, causing increased precipitation. The cycle ends when the ice sheet expands and cools surrounding waters and the Arctic Ocean refreezes. Freezing of the Arctic Ocean cuts off the principal source of moisture of the ice sheet, which then begins a precipitous retreat. This model explains the more or less periodic advances and retreats of North American and European glaciers, but it contains other features requiring verification.

Problems

It is difficult to decide which of the many hypotheses best explains Pleistocene glaciation. Relatively small changes in solar radiation, terrestrial albedo and terrestrial emissivity cause large global temperature changes. A *one per cent* decrease in solar radiation will decrease global temperature by about 1.5°F., a one per cent increase in the earth's albedo will decrease global temperature by about 3° F., and a one per cent increase in the earth's emissivity will decrease global temperature by about 2.25°F. At present measurements of solar radiation, emissivity and albedo are not sufficiently accurate to be related to observed global temperature variations, but such data may soon be available through the use of satellites. It seems that final determination of the origin of glaciation must wait until the contemporary global heat budget is better understood.

6 CONCLUSION

He was a violent man in a violent age, this Viking of the twenty-first century. Like his ancestors, he had spent his life fighting, as had several generations before him. It began with the famine that had been brought on by worldwide cooling. In some areas such as northern Africa, the Near East and Middle East, cooler weather meant better crops, but in northern Europe, northern Asia and northern North America, cooler weather meant shorter growing seasons and less food.

At first, the northern countries appealed to the United Nations, but few of its members were willing to part with food, especially because their own populations were growing and demanding more food.

Several years separated the First Great Famine and

the Second Great Famine. As the Second Great Famine took its toll, the industrialized countries looked hungrily at the warmer more fertile lands of their southern neighbors. Then, just as the Viking raiders two millenia earlier had lashed out from the cold and icy fiords of Sandinavia, the armies of the northern nations rolled southward.

The actual conquest of the southern lands had taken less than a decade, but guerrilla fighting continued in remote regions for another 30 years. With the taste of blood fresh in their mouths, the northern countries turned on each other, vying for a larger piece of the continually shrinking arable land area.

Of course, the preceding is science fiction.

Or is it?

Drought in the farmlands of the central United States threatens the 1974 grain crop. Are we on the edge of the next Dust Bowl period? Or are we already in it?

What will be the long-range effect of deforestation in the United States? Overgrazing in the area south of the Sahara has apparently converted much of that region into desert with resultant widespread famine and starvation.

The Green Revolution, which was to solve the world's food problems, has fallen short of expectations because of the large amounts of fertilizers and pesticides needed by hybrid grains. These in turn are for the most part petroleum derivatives (see Chap. 8). What can we try next?

Clearly the first step must be toward a better understanding of the sun-powered processes. Because these processes are related, we must study them as a group, using all facilities at our disposal: meteorologic, oceanographic and geologic investigations.

There are many tools we can use. Satellites, computers, deep ocean cores, devices to measure subsea parameters and detailed geologic mapping are but a few.

Satellite observations have contributed immensely to our understanding of meterologic processes and to the short-range prediction of weather patterns. It is said that savings to farmers as a result of better weather prediction are great enough to pay several times over for the costs of the space program. Satellite observations have also been of significant value in studies of water movement on or near the earth's surface. A single photograph, for example, provides detailed flood and ground moisture data virtually unobtainable from ground-based observations. Many scientists hope that satellite data used in conjunction with earth-based data will lead to the solution of the earth's heat budget problem.

Computers have made it possible to construct for the first time elaborate mathematic models of circulation, heat transport and heat transfer in the ocean basins and in the atmosphere. Without such models, it is very difficult to make long-range prediction of rainfall, temperatures and ocean circulation.

Scientists have recently been involved in key experiments to determine the role of air-sea interaction. These investigations suggest that the region encompassing the lowermost atmosphere and uppermost regions of the seas and oceans is especially important insofar as weather is concerned. This area of research is very complex and its investigation requires simultaneous observations by satellite, aircraft, ships and subsea monitors.

Geologic investigations, expecially study of deep-sea cores, provide us with the history of solar-powered processes. Several discoveries made from studies of deep-sea cores have been mentioned earlier in this chapter, including the dramatic discovery in 1973 that Antarctic glaciation had been active for much longer than previously suspected.

Results of geologic (and archaeologic) investigations are also highly significant. In covering relatively long periods of time (several hundred to several billions of years), these data provide us with a view of what has happened in the past regarding global climate and what may well happen again.

The tools necessary to provide the data we need for long-range coexistence of *Homo sapiens* with atmospheric, oceanic and glacial processes seem to be available today. They must be applied intensively over the next few years and resulting information used with wisdom if we are to avoid a state of affairs equal to or worse than the scenario at the beginning of this section.

5

EROSIONAL PROCESSES: MASS MOVEMENTS AND RUNNING WATER

The efficiency of river erosion is shown by the Colorado River at Dead Horse Point, Utah.

1 INTRODUCTION

In 1956 a landslide occurred on the slopes of the Palos Verdes Hills, Los Angeles County, California, destroying many homes in the luxurious suburban development. Sliding continued intermittently for three years, during which time more than $10 million damage was done to the homes in the area. Investigation showed that the housing development had been built on a slope composed of material that had once slid downhill. The slope created by the old slide was stable until the homes and roads were constructed upon it. Then the weight of these reactivated the slide, causing the material to move downslope again.

The courts awarded over $5 million in damage suits against the county that had built a road across the top of the old slide. But who was to blame? Was it the taxpayers who paid the bill for damages? The county engineers who built the road? The developers who constructed the homes? The home owners who bought the

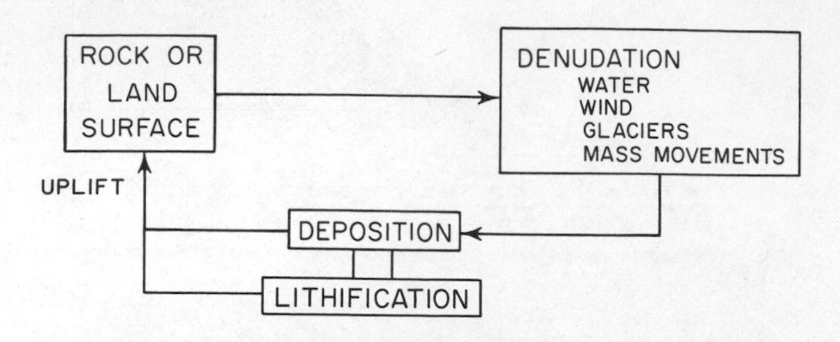

Figure 5–1. The geologic cycle, geomorphic point of view.

houses? Or was it an act of nature, for which no one was to blame? This is the kind of problem we are forced to face more and more often as man urbanizes and builds in places where he does not recognize the hazards of natural forces.

We have seen in previous chapters how dynamic forces work within the earth to uplift continents and build mountains. In this chapter and the next we will examine the natural forces that reduce mountain ranges and alter the surface configuration of the land. Such environmental changes result from the interaction of atmosphere and hydrosphere working with gravity to remove exposed rocks and soil. The types of forces to be considered in this and the succeeding chapters are as follows: movements of earth materials downslope under the pull of gravity, erosion by water running unconfined over the surface, erosion by water confined in a channel, erosion by waves along the coast, movement of loose material by the wind and erosion by water frozen into glaciers. We will examine the principles by which these natural processes function and see how man comes into direct conflict with nature by getting in the way of these forces. These processes, which all work together to wear away the land, detaching and transporting material, are not wholly destructive: it is from the deposition of this material that new rocks are derived for the building of mountains. This is the geologic cycle (Fig. 5–1).

2 RATES OF DENUDATION

Rain falling on the earth, wind blowing over the land, waves beating on the shore and ice slowly moving down a valley all erode and transport the soil or rock. In addition, water moving over the surface or infiltrating through the ground may not only physi-

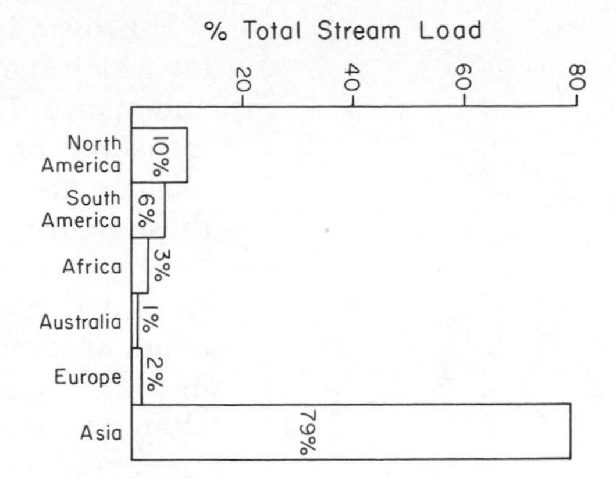

Figure 5–2. Comparative rates of denudation of the continents. (From Holeman, 1968.)

cally move grains but may also dissolve material and carry it away in solution. Such chemical denudation must also be considered as an extremely important natural process, through which many soil nutrients may be lost.

According to Holeman, approximately 18.3×10^9 metric tons of sediment are delivered to the oceans by rivers each year from a world land area of about 100 million square kilometers. Rivers of Asia contribute by far the largest amount each year (Fig. 5–2), because that continent is the largest and highest, with plentiful rainfall (Table 5–1). For example, the Yellow River annually transports 2802 metric tons per square km of drainage, whereas the Mississippi River annually carries only 97 metric tons per square km of drainage. The muddiest river in the United States, the Colorado, has an annual load of 212 metric tons of sediment per square km of area it drains.

TABLE 5–1. RIVERS OF THE WORLD WITH LARGE SEDIMENT LOADS*

River	Basin Area 10^3 km^2	Suspended Load 10^6 metric tons/yr	Suspended Load metric tons/km^2
Yellow, China	673	1886	2802
Ganges, India	956	1451	1518
Red, North Vietnam	119	130	1092
Brahmaputra, Bangladesh	666	726	1090
Irrawaddy, Burma	430	299	695
Indus, W. Pakistan	969	435	449
Mekong, Thailand	795	170	214
Colorado, USA	637	135	212
Mississippi, USA	3222	312	97
Amazon, Brazil	5776	363	63

*From Holeman, J.: The sediment yield of major rivers of the world. Water Resources Research *4*(4):737–747, 1968.

However, it must be remembered that measurements on which the figures in Table 5–1 are based are scanty and spotty. The load of many large rivers has not been measured or may be represented by only one or two measurements which may not be typical. Secondly, these figures represent material carried in suspension, hence only the finer particles. Larger, heavier debris carried along the bed of the rivers or carried in solution is not accounted for in this table. Thirdly, many rivers in the United States and elsewhere have recently been altered by man. Dams have been built which trap sediment and thus reduce the load brought to the sea. Material eroded from the land may also be deposited naturally in ponds, lakes, river channels or flood plains. However, the data in Table 5–1 provide a rough estimate of the magnitude of sediment removed from the land.

These figures represent only the denudation caused by rivers and carried off in suspension. Livingston has calculated that the total dissolved load of all rivers annually amounts to 3.9×10^9 metric tons. Judson gives an estimate of 0.1×10^9 metric tons per year for erosion by ice and 0.06 to 0.36×10^9 metric tons annually for wind erosion. These figures indicate that rivers account for most of the sediment brought to the sea.

Controls on Rates of Denudation

What are some of the controls on rates of denudation? The primary control is climate and its effect on vegetation. The highest denudation rates generally occur in regions of approximately 10 to 14 inches of rainfall annually. Less than 10 inches of rainfall does not provide enough run-off to cause erosion, and with more than 14 inches of rain there is enough vegetative cover to protect the ground. Vegetation protects the ground in three ways: by preventing splash when raindrops impact the ground, by intercepting rain and by transpiring moisture. In addition, vegetation aids infiltration into the ground, both by slowing up the surface flow and by providing a root network to increase soil permeability. The network of plant roots also tends to hold the soil against erosive forces. Density and type of vegetation determines the amount of soil protection.

Temperature also affects erosion as in Greenland,

where the denudation rate is estimated at 9 to 37 m/1000 years. This rapid rate of erosion occurs because annual cycles of freeze and thaw affect downslope movement of large amounts of debris.

Another control of denudation rates is topography, particularly the effect of elevation on rainfall. As air rises, it cools, and moisture in the air condenses. Thus high mountain areas have more rainfall than nearby lowlands. Elevation, or relief, also contributes to the energy of rivers, which depends upon the rate of fall of the river water. Rivers that fall quickly have more erosive energy than rivers flowing over gentle slopes. Hence, areas with high relief and steep slopes generally have high sediment production. The Appalachians have, at the present time, a mean denudation rate of 0.062 m/1000 years, whereas the Himalaya Mountains, which are the highest in the world, have an erosion rate of 0.21 m/1000 years. Studies in Japan, a region of steep high mountains, indicate that denudation rates tend to increase with uplift; the higher the mountains, the faster the rate of erosion.

Using a starting maximum figure of 1 m of surface erosion per 1000 years with erosion slowing with time, Schumm calculated that it would take 15 to 110 million years to erode a region at 5000 m elevation down to sea level. Why, then, are there any mountains on the earth? If erosion can occur so rapidly, the earth should be low-lying or flat topographically. The answer is threefold. Firstly, the earth is not stable, as you have seen in Chapters 2 and 3 of this book. Tectonic forces within the earth cause its surface to be uplifted and high mountains to form. For example, at the present time various ranges in California are being uplifted at rates of 4 to 10 m/1000 years. The rate at which the Japanese Islands are rising is approximately 5 m/1000 years. Plate tectonics, then, may cause mountains to rise faster than they are eroded.

Secondly, isostatic adjustment of the continental masses may cause uplift as the tops of the mountains are eroded away. Many regions once covered by ice are now rising in response to the removal of the load. Their uplift has been measured at rates of .03 to 1 m/1000 years. Although isostatic adjustment takes place at rates that may be lower than those of erosion, such uplift does decrease the net erosion.

Finally, the rate of erosion becomes less as relief

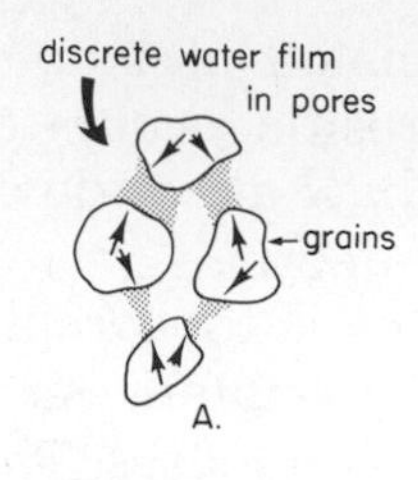

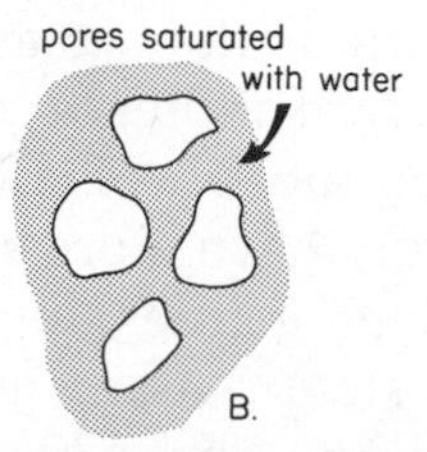

Figure 5–3. Role of soil moisture in binding grains. *A*, Small, discontinuous films of moisture hold grains by tension of the water film. *B*, Pores saturated with water relieve the tension so grains are loosened.

decreases. Hence only the very high mountains are eroded quickly. As the mountains are worn down, rate of erosion decreases because the amount of energy available is less; the lower the land, the slower denudation proceeds, and the earth never becomes totally flat or completely reaches base (sea) level.

The geology of a region is also an important factor in determining rate of erosion. Hard, well indurated rocks such as quartzites or granites tend to be more resistant to erosion than softer, less well consolidated ones such as shales. Sandstone and other porous rocks are often resistant to erosion, because water sinks into them rather than running off and carrying grains with it. Limestone, which is soluble, is easily corroded in humid regions but is more resistant in arid climates.

Characteristics such as size and composition of the mineral grains composing the soil are important determinants of erosion. Soils composed of silt are often more cohesive and resistant to erosive forces than coarser or finer materials, because silt grains can be held together by moisture in the soil. Discrete films of water exert a surface tension that binds the grains (Fig. 5–3). However, when the silty soil is saturated and the water in the pores continuous, the capillary tension is destroyed and the soil particles are no longer bound together. Under such conditions of saturation the soil is not only erodible but may even flow under gravity.

Finally, man is a controlling factor in denudation of the earth's surface. Judson calculated that the total natural transport of sediment to the sea is about 9.3×10^9 metric tons per year and that with man's help, erosion rates have risen to 24×10^9 metric tons per year. Man's devegetation and urbanization of large areas have stripped the earth of its protective cover. Urbanization generally increases run-off because regions become less permeable when covered with cement. Increased amounts of water running off the land provide more energy to rivers for erosion. Man's impact as an erosive force will be discussed in detail in later sections.

3 MASS MOVEMENTS

Material loose on the surface of the earth moves downslope under the pull of gravity. Evidence of this miles long and 1 to 3 miles wide, now known as the Valley of Ten Thousand Smokes, as well as covering the original fissure. Loss of gas and magma from beneath the summit of Mt. Katmai during the eruption caused the summit to collapse, reducing the height of the mountain by 1150 feet.

Perhaps the least dramatic but by far the most important eruptions in terms of volume of lava are the fissure eruptions that form lava plains and plateaus. The Deccan Plateau of India (Fig. 3–26), which now covers about one million square miles, may have occupied twice as much territory before removal of some of the basalts by erosion. Basaltic lava covers extensive areas in Africa, particularly near the Rift Zone (Figs. 3–27 and 3–28).

In the northwestern United States, plateau basalts underlie the Columbia Plateau and the Snake River Plain (Fig. 3–29). Plateau basalts must have covered much of the eastern United States about 150 million years ago. Exposed remnants of this vast plain now consist mainly of fissures filled with solidified magma, which were originally beneath the ground but have been exposed by erosion. The volume of lava comprising the great plateaus staggers the imagination. The Columbia Plateau consists of an estimated 100,000 cubic miles of lava, and the volume of the Paraná basalts of Brazil and Uruguay may be twice as great.

Figure 3–27. Blue Nile in Ethiopia flowing over flat plateau surface created by basalt flows. A fault gashing the plateau created Tissisat Falls in lower right.

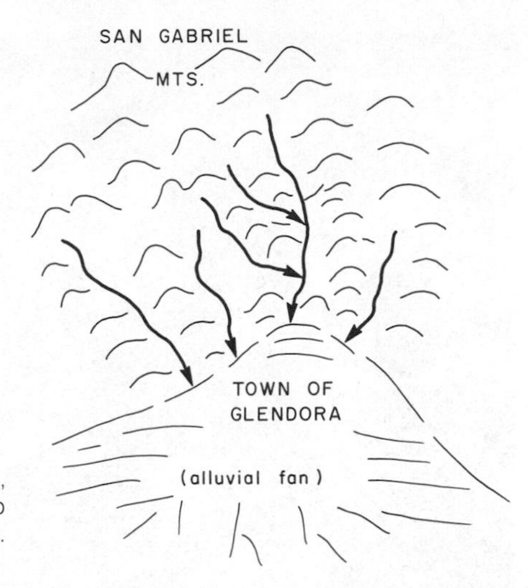

Figure 5–4. Setting of the disastrous debris flow at Glendora, California, in 1969. Debris flows occurred in the two valleys to the right. (Adapted from National Geographic, Oct., 1969, p. 566.)

principle is all around us but is most easily visible when disaster occurs. In 1969, winter storms deluged Southern California with a rainfall 800 per cent above the normal rate, soaking the slopes of mountains that had previously been denuded by forest fires. The water-saturated slopes gave way as a huge mass called a *debris avalanche*—a mixture of mud, boulders and water—poured out of the steep canyon mouths, bringing death and disaster to many. Figure 5–4 illustrates the path of two debris avalanches, which flowed out of two canyons and onto Glendora, a suburban center lying at the foot of the San Gabriel Mountains in a setting typical of many towns in California. The scope of the disaster is also shown by the figures of Table 5–2.

Another instance of destructive mass movement was the Nevados Huascaran avalanche of 1970, which was

TABLE 5–2. LANDSLIDE COSTS, WINTER 1968–1969, NINE BAY AREA, CALIFORNIA*

Public Costs		$10,184,948
state highways	$4,995,800	
county costs	5,177,148	
tax revenue lost	12,000	
Private Costs		9,088,808
depreciated property	7,105,546	
other	1,983,262	
Miscellaneous		6,120,200
Total		$25,393,956

*From Taylor, F. A., and Brabb, E. E.: Map showing distribution and cost by counties of structurally damaging landslides in the San Francisco Bay Region, California, winter of 1968–69. Washington, D.C., U.S. Department of Interior, Geological Survey MF-327, 1972.

Figure 5–5. The scene of the Nevados Huascaran debris flow. (From Ericksen, Plafker, and Concha, USGS Circular 639, 1970.)

triggered by an earthquake. Nevados Huascaran is the highest mountain of the Cordillera Blanca range of Peru (Fig. 5–5). It is separated from another range, the Cordillera Negra, by the Rio Santa valley. An earthquake of magnitude M = 7.7 shook the area on May 31, 1970, causing rock and glacial ice to be shaken loose from the face of Nevados Huascaran. The huge mass traveled down the valley at tremendous speed, reaching Rio Santa, some 15 km away, in less than four minutes. The momentum of the slide carried debris across the valley and 45 m up the other side. Downstream, the debris became more fluid, incorporating river water and melting ice into its mass. Roaring on down the valley it overran a ridge in its path and flowed into the village of Yungay, where 15,000 of 19,000 inhabitants were killed, most of them buried under 5 m of mud and debris. Another village, Ranrahira, was also partially destroyed by the same avalanche. In 1974 another earthquake shook the Andes, loosening another avalanche in a nearby area.

Besides such spectacular disasters, smaller downslope movements of material occur often and can be very costly. It is estimated that slope failures in the Appalachian region of the United States entail an annual highway repair bill of more than $100 million. The United States Department of Agriculture stated that damages from debris avalanches as a result of a torrential rainfall in Virginia on August 19–20, 1969, amounted to almost $24 million. Table 5–2 presents approximate expenses to Californians in the Bay area from landslides that occurred over a period of only three months.

4 TYPES OF EARTH FAILURES

There are several ways to classify downslope movements that take place when debris lies loose on a hillside. We will classify them in terms of the way the material moves (Table 5–3). If the debris falls in such a way that movement is distributed throughout the mass, we will call it a *flow* (Fig. 5–6); if it moves as one solid mass along a definite slip surface, we will call it a *slide*. In general, flows move more slowly than slides and hence may be less dangerous to life, although they may represent as much of an economic cost as slides. We will examine the types of slope failures, their causes and some possible remedies or precautions.

Flows

Flows may be subdivided into types of movement depending on rate of flow, material involved and amount of water present or needed. Consistency of material varies with amount of water present. The

TABLE 5–3. TYPES OF MASS MOVEMENT

	Rate of Flow	Amount of Water Present
Flow (movement distributed throughout material)		
creep	slow	water not necessary
rock glacier, rock stream	slow	water not necessary
solifluction	fast	water-saturated
mudflow, debris flow	slow or fast	much water
earthflow	slow or fast	much water
Slide (movement as one mass on a slip surface)		
debris avalanche	fast	wet or dry
slump	fast	wet or dry
landslide, rockslide	fast	wet or dry
Fall (free fall of rock or soil)		

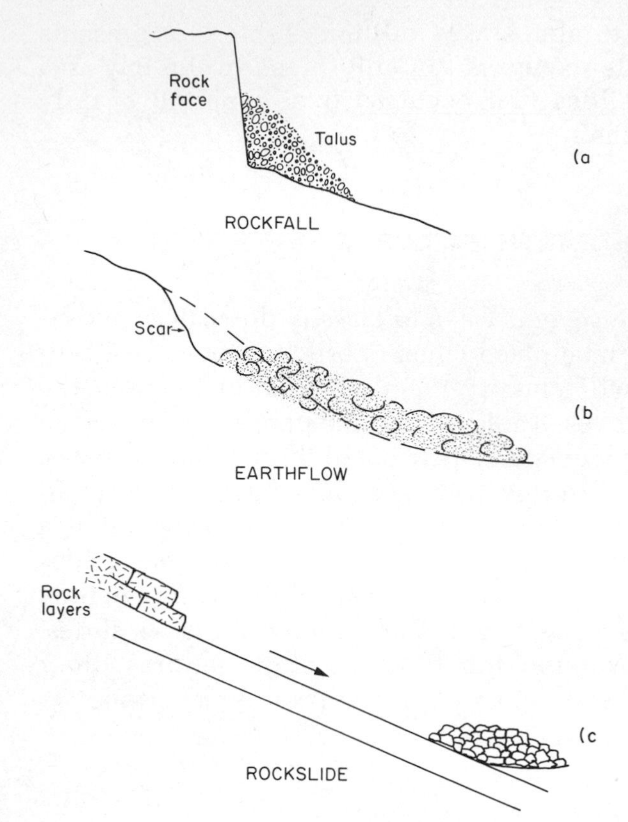

Figure 5–6. Some types of earth failures or mass movements.

more water, the faster the velocity of flow downslope and the less consistent and coherent the material.

Creep is the slow downslope movement of individual particles under the influence of gravity. The movement is imperceptibly slow, generally less than 5 to 10 mm/year. Creep is taking place on many hillslopes all around us but is not itself dangerous to man. It may, however, serve as a warning of an unstable slope. Creep is evidenced by tilting of poles and fences and by curved tree trunks.

Rock glaciers and *rock streams* are usually found in mountainous areas such as the Colorado Rockies, or at high latitudes such as Alaska. Both forms require a plentiful supply of bare rock and a rigorous climate. They consist of large masses of rock slowly moving downslope in a valley. Rock glaciers are larger in overall bulk than rock streams. These two also differ in map view: rock glaciers are lobate or tongue-shaped, whereas rock streams are more linear in form. Because of their geographic locale and slow movement, at present neither represents a serious hazard to mankind.

Solifluction is the downslope movement of a mass of water-saturated debris over an impermeable layer which water cannot penetrate. The result is an almost fluid mass which is highly unstable and which will flow on an extremely low slope. This type of movement occurs most often in regions such as Greenland, where there is permanently frozen ground (*permafrost*) beneath an upper unfrozen mantle. Solifluction is one of the dangers that must be forestalled in building the Alaskan pipeline.

Mudflows, which also take place under conditions of water saturation of fine soil or weathered material, consist of about 60 to 70 per cent solids, ranging from clay and silt to large boulders, which may be rafted along in the flow. Because the flows have a consistency somewhat like thick pudding and a high density, they may act as a natural bulldozer, pushing anything along that is in their way. Conditions conducive to mudflows include a thick mantle of loose weathered material, steep and barren hillslopes and heavy rain.

Mudflows quite often debouch from narrow steep mountain canyons and build up extensive deposits called *fans* at canyon mouths. These fans are highly desirable locations for development because they provide a nice view, they are close to the mountains, they are cool (owing to their elevation) and they are not hard to build upon, because they are composed of loose material. Many such fans have been intensively developed as suburban communities, especially in California. Because the mudflows occur sporadically, the general public may be unaware of their existence. The shortsightedness of both developer and buyer has led to many disasters.

One such calamity occurred in the town of Wrightwood, California. The city lies at the foot of the San Gabriel Mountains (Fig. 5–7), which rise abruptly some 400 to 1000 feet above the surrounding plains. In May, 1941, a sudden warm period brought about melting of a thick snowpack in the mountains. The meltwater saturated the clayey weathered bedrock, which moved down into Heath Canyon, reaching out 15 miles from the head. The eastern edge of the town was inundated by the flows, which were renewed each day for more than a week as the snow melted. Investigators determined that the canyon was the site of previous mudflows and predicted future ones.

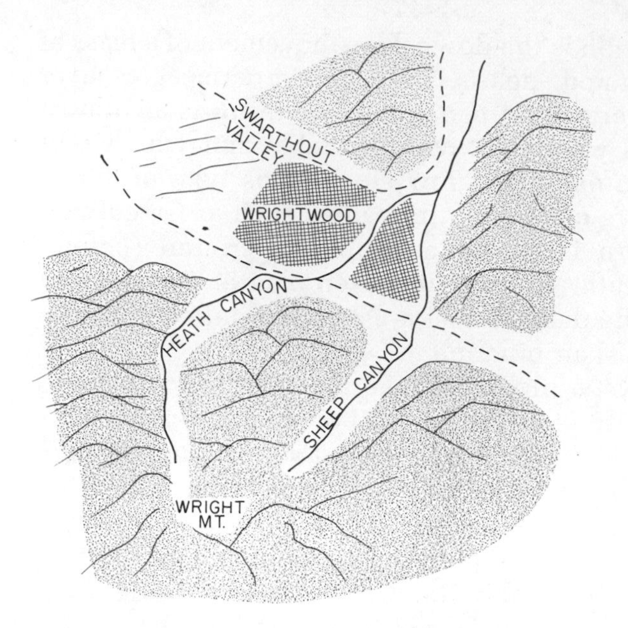

Figure 5–7. Sketch map of the Wrightwood mudflow area.

Earthflows are movements of large masses of debris down a slope (Fig. 5–8). They are not confined to a channel as are mudflows but spread out laterally over the surface. Although they require a supply of water to lessen the cohesiveness of the soil, they are generally more consistent than mudflows and thus flow more slowly. The Gros Ventre earthflow, which took place over a four-year period (1908–11), is a typical example. The Gros Ventre River lies in the Gros Ventre

Figure 5–8. The Slate Creek earthflow, Teton National Forest. (From Bailey, R.: Landslide hazards related to land use planning. US Forest Service, 324 25 St., Ogden, Utah. P. 30, Fig. 12. 1971.)

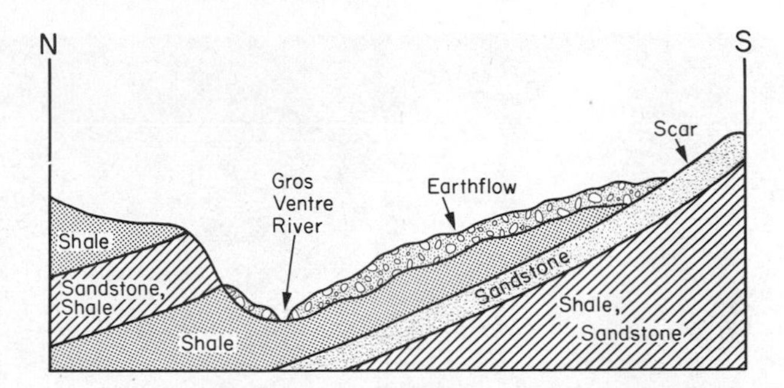

Figure 5–9. Geologic conditions along the Gros Ventre River, at the upper Gros Ventre earthflow. (Adapted from Bailey, 1971.)

Mountains of Wyoming and has cut its valley into layers of soft shales and clays (Fig. 5–9). The unstable slopes on the south side of the valley began to move slowly and intermittently downward and are thought to be still moving. Movement was slow but definite, as shown from destruction of a road traversing the valley and the overturning of telephone poles. Speed of the flow varied, not only from place to place but seasonally, moving faster in the spring and more slowly in the autumn. The flow, which eventually dammed the river forming Upper Slide Lake, is a tongue of heterogeneous debris some 4.8 km long. The mass is narrow towards the head but has spread out at the base. The scar at the head is steep, and the earthflow material there is broken up into blocks by cracks parallel to the scarred slope. The surface of the flow is very hummocky with low mounds and hollows.

Slides

Slides, although composed of discrete grains or blocks of soil or rock, move as a rigid mass, rapidly and generally with definite boundaries. The amount of water present varies, but slides generally occur under conditions of unusually heavy precipitation or snowmelt. Little or no vegetation, a thick mantle of loose debris and steep slopes also contribute to sliding. Slide material moves on a slip surface, which may be a bedding plane, a rock surface or a curved surface of failure. Slides are the most spectacular mass movements, primarily because of the speed with which they move. They are common and worldwide and are responsible for much damage and loss of life.

A *debris avalanche* is a rapid downslope movement of rock and/or soil mixed with water. These avalanches

Figure 5–10. Lower Gros Ventre slide, oblique air view. (From Bailey, R.: Landslide hazards related to land use planning. US Forest Service. Fig. 41, p. 63, 1971.)

have a high consistency; that is, they contain much more solid material than water. Morphologically they can be recognized by the long narrow track they leave on a hillside, by the scar or open area at the head and by the irregular or hummocky mass of debris at the bottom of the slope (Fig. 5–10). Although they generally occur in mountainous areas after heavy rains, avalanches are also common along the bluffs of the Illinois River, for example, where the slopes at the edge of the valley are steep, the soils are clay-rich shales and the region is susceptible to heavy rainfall.

An avalanche occurred on the lower Gros Ventre River valley in 1925 after a heavy storm. The rain saturated the porous sandstone layers lying above a clayey bed. A huge mass of rock, soil, water and vegetation with an estimated volume of 38 million cubic meters cascaded down the mountainside and across the valley. The momentum was so great that the debris reached almost 120 meters up the north side of the valley, damming the river and forming Lower Slide Lake. The slide moved downward about 700 km in minutes. The scar at the head (Fig. 5–10) is so large that it can be seen from Highway 89 through Jackson Hole. Evidence that it moved as a unit is shown by the many trees still growing on the slide material, survivors of the movement.

Slides in which the slip surface is curved are called

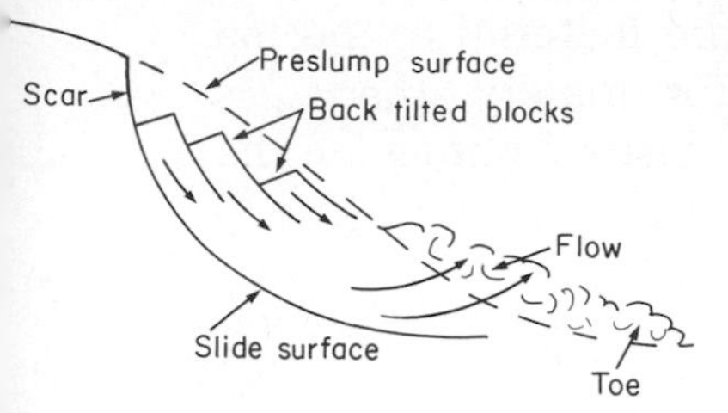

Figure 5–11. A slump. Side view showing slump blocks and a flow at the toe.

slumps (Fig. 5–11). From above, a slump looks somewhat spoon-shaped. The broken blocks move along the curved surface in a rotational slip, which tilts the blocks backward (Fig. 5–11). A slump is characterized by a steep scarp at the head, slip blocks tilted back into the hillside and the usual irregular hummocky topography at the toe (Fig. 5–12). Slumps often occur in cohesive soils that have been oversteepened, such as those along eroded river banks, cliffed seashore or man-excavated slopes. Slumping may also take place where overloading of a slope causes internal stress of the underlying material. Such slumps are numerous and are usually of the type where a weak clay-rich rock or soil underlies a stronger more resistant rock such as sandstone or limestone. Seepage of water through the upper layers reduces the strength of the clay material beneath, and failure follows.

The large slump that occurred in 1964 in the vicinity of Matsushiro, Japan, was due to a steep volcanic hillside saturated with water from a typhoon. The head of the slump was a crack that had formed as a result of earthquake shocks. Water had infiltrated the crack and the soil to the weathered rock beneath, reducing its strength and causing it to fail. Numerous large blocks at the head were rotated backward and tilted, forming hollows, which filled with water. An apple orchard growing on the slope moved downslope, with trees still standing. However, a shed on the toe of the slump was

Figure 5–12. A slump along the roadside in Colorado. Note the broken blocks and circular form of the movement. (Photo by Hansen, USGS Circular 689, 1973.)

destroyed as the lower part of the material at the toe moved out as an earthflow. Fortunately, the mass stopped at the road and did not destroy homes on the other side.

Rockslides

Rockslides or *landslides* are movements of masses of rock or large blocks on a planar slip surface, usually a bedding surface. They are commonly drier than debris avalanches or slumps. Rockslides are frequent in mountains formed on folded rock layers such as the Alps, Appalachians or some of the Rocky Mountain ranges. Figure 5–13 illustrates a rockslide along the front of the Bighorn Mountains in Wyoming, where the rock strata are tilted sharply upward over the range, with the layers dipping eastward. Porous *dolomite* (a magnesia-rich sedimentary rock, resembling limestone) overlies a layer of softer shale that forms a lubrication plane along which the upper material can slide.

Rockfall

Free fall of rock or soil fragments occurs on slopes or cliffs that are so steep that loose material cannot remain on the surface. On such slopes, weathered material falls as rapidly as it is loosened and no soil mantle can be maintained.

Imminent *rockfalls* are visible along bedrock cuts on highways. They are dangerous, and usually signs are posted to warn motorists of them (Fig. 5–14). Rockfalls

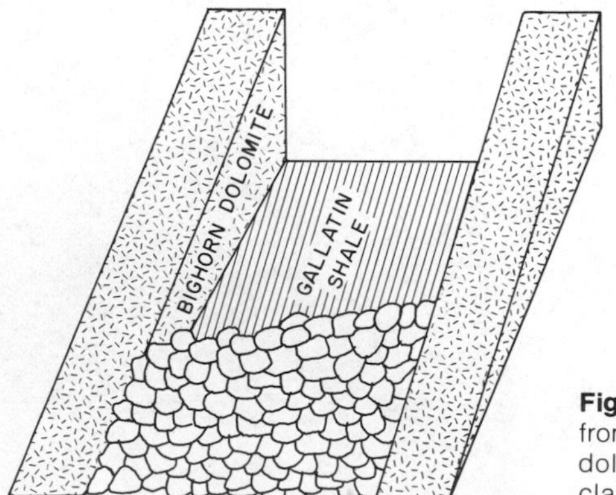

Figure 5–13. Diagram illustrating slides along the eastern front of the Bighorn Mountains, where steeply dipping Bighorn dolomite slides down the bedding surface of the underlying clayey Gallatin shale.

Figure 5–14. Road sign warning motorists of an imminent rockfall.

occur seasonally, when freeze and thaw cycles enlarge cracks and may even shove rocks away from the cliff face.

Rocks or soil fallen from a cliff face build up piles of debris called *talus* (Fig. 5–15). Talus may collect in a number of ways. If there is an equal wasting away all along a cliff face (Fig. 5–16), the talus will form a somewhat even slope between the cliff and the ground surface below. When the rockwall weathers or is eroded unevenly along joints, for example, chutes or funnels

Figure 5–15. Talus formed below a cliff on the glacially oversteepened slopes of Buckskin Creek, Colorado.

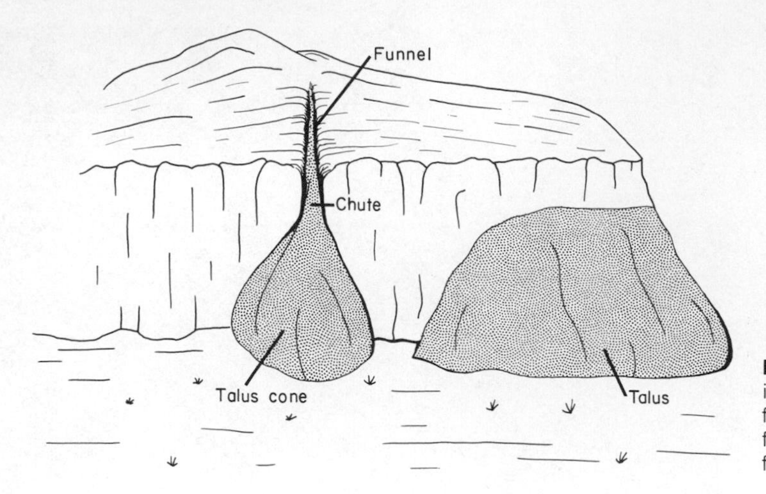

Figure 5–16. Diagram illustrating talus that has fallen evenly from a cliff, and a talus cone formed by debris fallen through a funnel.

will form. Debris is channeled through the chutes and funnels and shed on the surface below, building up a cone-shaped talus (Fig. 5–15).

The rocks of the talus assume a stable slope, called the *angle of repose,* which varies from 25 to 40 degrees. Factors such as size, shape and density of the fragments, surface roughness of the rocks, slope of the ground beneath and height of fall interact to determine the exact angle of repose. Any forces acting on any of these factors may disturb the talus and cause the debris to move. For example, if the toe is steepened, the rock debris, which had assumed a stable angle of repose, will adjust to the new toe angle by moving. Or, again, additional weight on a stable talus may cause the rocks to shift and assume a new angle of repose. The stability of natural talus is demonstrated by the fact that many taluses are covered by trees or brush in humid regions.

5 CAUSES OF SLOPE FAILURE

In discussing the types of downslope movements, we have referred often to causes of failure. These will now be treated in more systematic detail. The stability of the mass of loose debris on a slope depends on the equilibrium between the forces tending to pull (or push) the material down the slope and those forces resisting such movement. We can express the equilibrium as follows:

$$\text{downslope forces} = \text{resisting forces}$$

Anything which upsets this equilibrium toward the left of the equation will cause slope failure. Factors determining the strength of the downslope forces are the angle of slope and the mass (weight) of the material. Resistance to downslope movement depends upon friction with the slope, internal cohesion of the material and pressure of water in the pores. If the angle of slope is increased by cutting into it, the material may fail. If the weight on a slope is increased by building a house, highway or other structure on it, this may be sufficient to displace the equilibrium and cause slope failure. Accelerated mass movement along a logging road in a forested region of California was caused by the steepening of slopes for the road. During the winter of 1967–68 there were 19 slides along a 0.75 km stretch of road.

The exact angle and weight that will cause a slope to fail depends upon the internal strength of the material comprising the slope. In general, although other factors play a part, the strength depends upon the reaction of the debris to moisture. Water penetrating the material not only adds to the mass but also produces a buoyant effect, which will float the material downslope. If the material is rich in clay, water may cause swelling and increased pressure, which may force downslope movement. Clays also become plastic when wet, losing their strength and deforming in a downslope direction. Silts and sands may be strong when slightly wet because of the capillary tension of water (see Fig. 5–3), which tightly binds the grains. But when all the voids are filled with water, capillary tension is relieved, and the material loses coherence.

Water, then, is the culprit in many cases of slope failure. For instance, on November 15, 1972, an intense cloudburst poured down on the Big Sur region of California, soaking slopes that had been denuded by a brush fire the previous summer. The saturated bare rock and soil gave way, and the resulting debris avalanche flowed down on River Village below, virtually destroying it. Ironically, the town was just recovering from a landslide that had taken place in October after a similar storm. Several hundred avalanches were triggered by heavy rains in August, 1969, in Virginia. A storm in May, 1973, in the vicinity of Denver, Colorado, caused hundreds of landslides and flows. Most of the slope failures were along highways (Fig. 5–17) and

Figure 5–17. Rockfall along Interstate 70, west of Denver, Colorado. Failure resulted from a storm on May 5–6, 1973. (Photo by Hansen. USGS Circular 689, Fig. 18.)

caused considerable inconvenience, hazard to motorists and great expense to taxpayers.

Liquefaction of ground results in a sudden decrease in the strength of slope material. This may occur not because of the addition of water but rather as a result of a shock that causes the material to recompose itself into a new structure. During this restructuring, the water already present in the pores is resituated, and the material flows. Such materials are called *quick clays.* During the Alaskan earthquake of 1964, many slope failures occurred as a result of liquefaction of quick clays. At Anchorage, for example, the weight of buildings became too great when the clays beneath them were liquefied by the earth's shaking. As the liquefied clay moved, the block of earth above it floated along. As the blocks moved, cracks at the head gaped open, removing support from walls of the cracks, which collapsed (Fig. 5–18). There were five major slides of this type in Anchorage, and they caused a great deal of destruction. The largest and most destructive was the Turnagain Heights slide, which involved 12.5 million cubic yards of material extending over

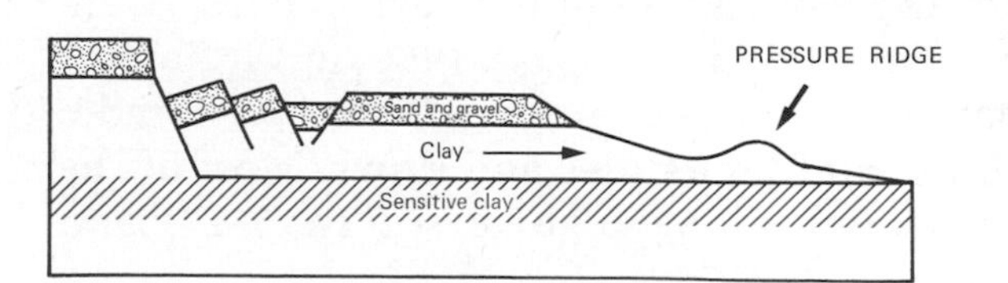

Figure 5–18. Diagram of slippage along liquefied clay planes, the cause of many slope failures during the Alaskan quake of 1964. (After Hansen, USGS Prof. Paper 542.)

130 acres. The area was a jumble of large and small blocks (see Fig. 2–27) that had moved along the clayey slip surface of Bootlegger Cove clay.

What Can Be Done?

Since mass movements are so common, variable and destructive, what can be done to prevent them? One solution is to recognize regions of unstable slopes and avoid them. This has been done in many areas of California, where zoning laws prohibit building on slopes prone to failure. There is presently a commendable effort being made in California and elsewhere to map unstable slopes so they can be avoided by builders.

In cases where unstable slopes have already been built upon, what can be done for future stabilization? The first measures to be taken should be in reference to drainage. Capture and diversion of water falling on or flowing into the slope is necessary. This can be accomplished by drainage ditches or subsurface pipes. Compaction of the soil to increase bearing strength and reduce pore space can also be accomplished and will help to stabilize the slope. Vegetating the slope serves two purposes: it increases the absorption of moisture, and the root network makes the soil more cohesive.

Changing the angle of slope or its shape is also a viable solution. Regrading the slope to a lower angle or

Figure 5–19. Steel wall put along a highway to control slope failure. Note how the debris pressing on the wall has broken it in several places. Along Route 11, New York.

terracing it in order to change the distribution of weight is often effective in stabilization. Structural measures such as retaining walls, barriers, wire mesh, rock bolts and pilings can be used to restrain slope material. However, debris has been known to break through or overtop such man-made barriers (Fig. 5–19). Hardening the soil by packing or by *grouting* (filling pores and cracks with cement) has been successful in some areas.

However, many of these measures are costly, and it is best to avoid building on an unstable slope if possible. The old adage, "An ounce of prevention is worth a pound of cure," is very apropos in the case of mass movements.

6 SHEET EROSION

Sheet erosion refers to the detachment and transport of particles by sheets of water flowing over ground surfaces. Such sheets form when rain falls faster than the soil can absorb it or when the ground is already saturated. The excess water then flows along the surface, detaching and carrying off the soil particles as it moves downhill. Sheet erosion can be a very effective erosional agent. Sheet erosion is also called *rainwash, slope wash, sheetwash* or *sheetflood*.

Erosion by overland flow of water as it moves toward a stream channel is difficult to assess. Certainly, much load is brought into a river in this way. The rate and severity of such erosion are determined by the type and density of vegetative cover; by characteristics of the soil that influence porosity, permeability and cohesion; by the intensity and duration of the rain; and by the steepness and length of the slope over which the water is flowing.

Construction sites, parking areas, highways and cultivated fields are all favorable sites for sheetfloods to occur. A number of studies have shown that there is a sharp rise—as much as a hundredfold—in soil erosion during development and construction in suburban areas and that most of this increase is probably due to rainwash. Sediment production resulting solely from surface erosion by sheetwash on logging roads (during and after construction) gave an increased yield 220 times that previous to roadbuilding. Rainwash erosion jumped tremendously during construction and gradually tapered off as the more erodible particles were removed and as vegetation started to grow over the area after construction.

The first phase of sheetwash erosion is loosening of grains by raindrop impact. The splash of falling raindrops on bare soil effectively loosens it and detaches particles. The amount of splash depends on the size of the drops and intensity of the rain, which determine the force of impact. Duration of the rain is also a controlling factor. Vegetative cover is important, since rainsplash occurs more readily on bare surfaces. Characteristics of the soil such as density, grain size and mineral content also affect splash. Fine sands seem to be very susceptible to raindrop splash, and fossil raindrop imprints are commonly found in fine-grained sediments.

Severity of erosion also depends on steepness and length of the slope. The energy of flow varies with the velocity, which in turn varies with the slope. A long slope allows more concentration of water, so the mass increases as the length of overland flow increases. Both velocity and mass provide energy, so a steep and long sloping surface will have the greatest amount of erosion by rainwash.

Sheets of water pouring down a slope may move loosened particles by rolling them along or causing them to be lifted momentarily and carried downslope in a jumping process called *saltation*. Although sheet

Figure 5–20. Shoestring rills on a denuded Oregon slope. This type of denudation is common along roadside slopes. (Picture from USDA, SCS Ore 75,165.)

flow usually is not very deep, fine particles can be moved in suspension in the water, and other material can be removed by solution.

Because of irregularities in ground surface, sheet flow becomes interrupted and easily concentrated into *shoestring rills* (Fig. 5–20), which widen and deepen as rainwash from the areas between them pour into them. In time, they may become gullies and later be integrated into stream systems.

7 STREAM EROSION

Water flowing off the surface of the ground as sheetwash eventually enters a stream channel. This overland flow, combined with some of the rainfall that infiltrates the ground and moves laterally toward a channel as *ground water*, comprises the discharge (volume of water past a given point on a river per unit time, measured in cubic feet per second) of rivers. Over the earth as a whole, approximately one third of the rainfall flows off the land in rivers. Most of this is overland flow, especially during intense storms. Only a small percentage of the streamflow is derived from ground water. However, this effluent is very important as it is the steady groundwater flow which provides the year-round discharge of rivers.

Figures given at the beginning of this chapter offer vivid proof that water channelized in rivers is probably the most important erosional agent denuding the earth. Yet it took man a long time to realize the efficacy of rivers in eroding the land. Perhaps because he has an eye for the spectacular and the grandiose, man attribuuted the valley of a river not to the erosion by water flowing in it but to an earthquake or single great flood. It was hard to believe that a deep canyon could be slowly carved over a period of millions of years by running water. It took astute observers a long time to persuade fellow scientists that rivers do indeed erode their own valleys. Hutton in 1795 clearly demonstrated this in his writings, but his ideas were not accepted until the mid-nineteenth century. Since that time numerous quantitative measurements, such as those presented in Table 5–1, have well established the fact that rivers erode material and transport it to the sea, where it is deposited.

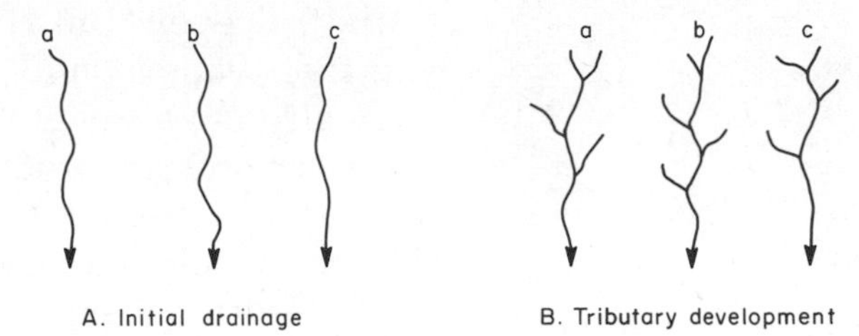

Figure 5–21. Development of an integrated drainage system. *A*, Consequent streams flowing down the slope of a land surface. *B*, Development of tributaries to each stream. *C*, Stream b has captured the drainage of stream c. *D*, Stream b has captured the drainage of stream a. The whole system is now integrated with stream b as the main river.

Let us examine erosion as it proceeds on land surface newly exposed to denudational forces. We have previously described how sheet flow occurs and is concentrated into *rills* and then into *gullies*. As rills and gullies are lengthened and widened, one may cut into the channel of another and divert its flow (Fig. 5–21). Thus one stream has *pirated* another. It is in this way that a stream system becomes integrated and a drainage basin grows. Drainage basins are separated from each other by *drainage divides*. A divide is a line between adjacent basins; rain falling on different sides of the divide will flow into different basins.

The streams that flow down the original slope of the land surface are called *consequent* rivers. As erosion proceeds, other streams develop in the system by erosion at the head (Fig. 5–21). A stream that erodes along some weakness in the underlying rock is called a *subsequent* river. It may develop along a very weak, easily eroded layer or along a fault or joint in the rocks. In any event, it finds a weakness in the surface and exploits it in order to erode its channel. Other tributaries (called *obsequent* rivers) may develop along valley walls, flowing in a direction opposite to that of the consequent river. Thus the *drainage system* grows.

As drainage systems grow and rivers cut into the land, divides between basins become narrow and valleys deepen. But rivers not only downcut, deepening their valleys, and headcut, lengthening them, but they also cut away at their banks, widening the valley. As river erosion proceeds, working along with sheetwash and mass movements, divides become lowered and the whole landscape more subdued.

The erosional aspect of rivers provides us with some of the most pleasing scenery in the environment. These include such wonders as the Grand Canyon and Niagara Falls. To understand how these are formed, let us look more closely at the work of rivers. There are three steps in the fluvial processes: *detachment* or *entrainment* of grains, *transportation* of the material and *deposition* of particles.

Entrainment

Particles, loose on the bottom or held on the banks of rivers, can be detached and entrained by hydraulic lift or drag. Entrainment results from chaotic or turbulent stream flow; that is, the flow is composed of many individual water masses, called *eddies* or *vortices,* which flow with different velocities and in different directions (Fig. 5–22) and which exist in a variety of forms. Turbulence can be generated by logs, rocks or other obstructions in a channel. Bends in a river, changes in slope, entrance of a tributary and many other factors affect the velocity or direction of water movement and thus cause turbulence.

One important type of disturbance is called *separation of flow* (Fig. 5–23). When an obstacle deflects the flow of water, a separation of the main flow from the obstacle occurs, and eddies are formed on each side of the obstruction. Separation of flow results in a localized area of low pressure with movement of water back toward the main flow, enabling the eddies to lift a particle from the stream bed or remove it from the bank.

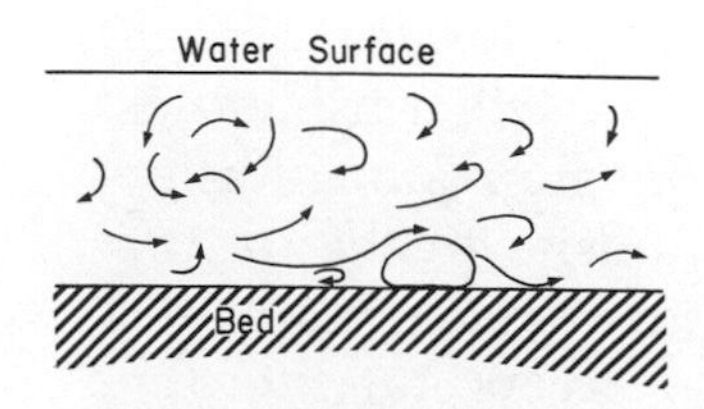

Figure 5–22. Turbulent flow of water masses.

Drag is the result of differential flow velocity at varying depths in the stream. Where the upper flow velocity, V_1 is much greater than velocity of flow lower in the water, V_2, the difference in velocity force above and below the particle causes an overturning of the grain and entrainment (Fig. 5–24).

Grains may also be entrained by sheer *hydraulic force*. When you hose down a sidewalk or driveway,

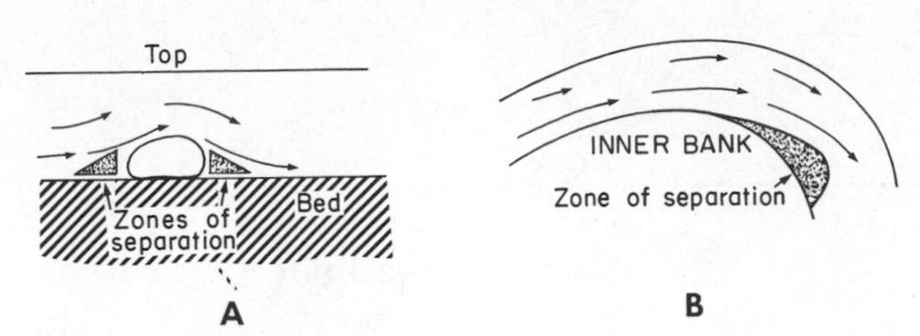

Figure 5–23. Separation of flow. *A*, Around obstacles. *B*, At a bend.

you are using hydraulic force of the water coming from the hose to remove debris; the same principle is applied in hydraulic mining. Natural hydraulic force causes stream erosion.

Transportation

When particles have been entrained, they may be transported as either *bedload* or *suspended load*. Bedload consists of grains that roll or slide along the river bottom or that "jump" along by saltation (Fig. 5–25).

The suspended load consists of particles that are generally smaller than bedload particles and are carried above the bed, totally supported by the turbulence of the water. The amount of suspended load and the size of particles carried varies with depth in the river. Most of the suspended load is transported near the bottom, decreasing toward the surface of the water. The larger particles are transported near the bottom, and only very fine grains are present near the water surface.

The amount and size of particles that can be transported depends upon the energy of the river, which is derived from the mass of water flowing and from its velocity of flow. The mass of water, essentially the *discharge* or volume of water flowing per unit time, depends on rainfall and in particular the amount of rainfall that flows into the river channel and downstream. Velocity of flow depends upon a number of variables, including the slope or gradient of the river, the volume of water, the roughness of the bed and banks, the width and depth of the channel and the sinuosity or "bendiness" of the river. These can be expressed as

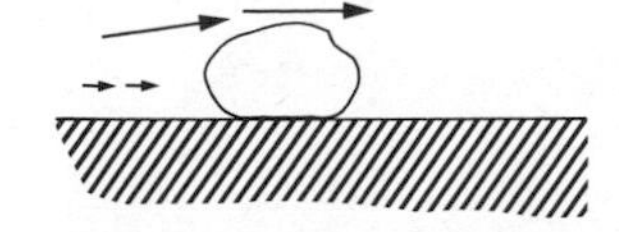

Figure 5–24. Hydraulic drag on a pebble lying on a stream bed. Size of arrows represents velocity. Difference in velocity causes differential fluid pressure on the pebble.

$$E_k = \frac{MV^2}{2}$$

$$V = f\,(s, Q, n, w, d, P)$$

where E_k is kinetic energy, M is the mass of water, V is velocity, f represents some mathematical function, s is

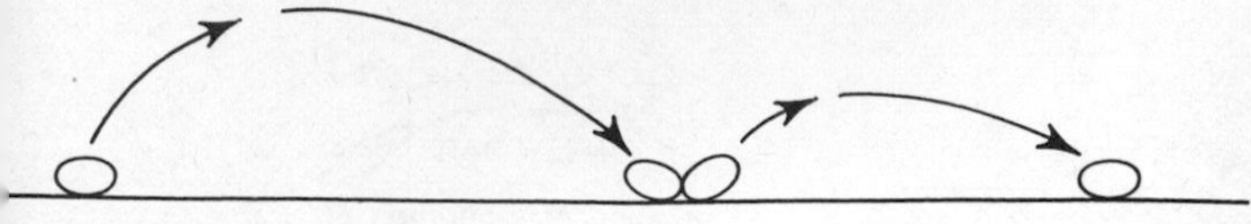

Figure 5–25. Saltation or jumping of a grain. It is lifted up into the current and is carried forward as it falls. The impact may loosen another grain into saltation.

gradient, Q is discharge, n is roughness, d is depth, w is width and P is sinuosity.

The effects of these variables on velocity can be described as follows, when all other variables are held constant: increase in gradient or discharge increases velocity; increase in width, depth, roughness or sinuosity decreases velocity. If the mass flowing remains constant, increase in velocity will increase the energy and thus the erosional capacities of a river. However, it cannot be emphasized too strongly that these relationships in nature are complex, and combinations of changes may cause a change in velocity and load-carrying ability different from the preceding principles, in which variables were taken singly. For example, an increase in width should decrease velocity, and a decrease in depth should increase velocity. However, if both changes occur at once, a wide shallow channel is formed, and more bedload movement results, partly because of the increased roughness of the bottom, which increases turbulence. Thus in a natural state all the variables interact with each other to achieve a delicate balance.

The exact effect of changes in the aforementioned variables on velocity and load carried also depends on the size of the material comprising the load and the relative erodibility of the bed and banks. The most easily eroded materials lie in the size range of .01 to 4 mm. Fine silts and clays are cohesive and tend to resist removal, whereas coarse sands and gravels are difficult to move because of their size and weight. It takes more energy to entrain a particle than it does to transport it once it is entrained.

It is important to remember that a river has energy and that energy has the capacity to do work. Any factor that increases the energy or decreases workload will cause the river to erode either its bed or banks.

Deposition

Material entrained and transported by a river is ultimately deposited. It may be deposited in the channel or valley of the tributary or main stem, at the mouth of the river or in the ocean. Deposition results when the energy of a river is lowered or when the load is greater than the river's capacity to transport material.

Lowering of energy takes place when variables in the velocity equation change so that velocity is decreased or when discharge is decreased. A decrease in discharge may result from a climatic change (such as less rainfall) or a change in the basin so that less run-off flows into the stream. A velocity decrease may follow from a decrease in gradient, channel widening or deepening or an increase in roughness or sinuosity. However, again it is not easy to predict precise events in a river system if a single factor is altered. Rather, there will be an interacting chain of events, which may be determined by positive and negative feedbacks.

In a river system with constantly increasing discharge in a downstream direction, deposits tend to be laid down at points where water velocity is decelerated. *Fans* may be deposited at tributary junctions, where the flow of tributary and main stream intercept and interfere with each other. Deposition is also caused by deceleration in areas where there is a separation of flow (Fig. 5–26), at the inside of bends (*meanders*) or around obstacles. Deceleration as a stream flows into a pond, lake or the ocean results in the building of *deltaic* deposits.

Change in size of debris or supply of load may also

Figure 5–26. Meander bend of the South Platte River near Fairplay, Colorado. Note deposits on the inside of the meander.

Figure 5–27. The Green River meandering over a broad flood plain near Green River, Wyoming.

cause the river to lay down material. If additional load is supplied or a large-sized load added to a river, it may be unable to carry the material. This often happens at tributary junctions, where a more steeply sloping, turbulently flowing tributary carries boulders larger than the main stream is capable of moving. The huge rocks are left in the main channel, providing exciting rapids for adventurous river-runners.

Deposits are also laid down during floods. When water overflows its channel and spreads out over the land, it tends to drop the load it is carrying, partly because of decreased velocity and partly because of decrease in discharge as water infiltrates the ground. This material deposited over the banks, as well as that deposited in the channel as bars, forms the river *flood plain.* A river flowing on a broad flood plain may wind its way back and forth across the plain in meander bends (Fig. 5–27). These flood plains are among the most fertile lands in the world, because each time the water overflows, it brings nutrients to replenish those taken away by crops. Often, as the river overflows its banks it will deposit material immediately at the river edge. If the overflow is not great and this occurs a number of times a mound of deposits will be built up all along the edge of the river bank. This is a natural *levee.* Man sometimes tries to protect the flood plain away from the stream channel by increasing the height of these natural levees or by building artificial ones.

The effect of the Aswan Dam on the Nile River is an example of the far-reaching results of changing the sediment load of a river. The Nile valley was long known for its fertility, renewed each year by a flood that not only deposited some 100 to 130 million tons of silt on its flood plain but also washed away salts that had accumulated from evaporation. In addition, an equilibrium had been established at the extensive delta at the mouth, whereby the river replenished deposits eroded away by currents of the Mediterranean Sea. Since 1964 the Aswan Dam has been storing water of the Nile to produce power and to irrigate the desert.

Let us consider the impact of the dam on the transport regime. The muddy flood waters of the Nile are caught behind the dam, and their nutrient-rich load is deposited in Lake Nasser. Below the dam, clear water flows downstream to the Mediterranean, no longer refertilizing the flood plain nor replenishing the delta. As a result the flood plain now requires artificial fertilization at enormous expense and labor. The soil, no longer cleansed of evaporite salts by the flood waters, is fast becoming too saline for cultivation. The equilibrium of the delta coast has been upset. Because there is no more deposition, marine currents and waves are eroding the coastline. In addition, the clear water flowing below the dam has been deprived of its load and now has sufficient energy to pick up new material from the channel. Riverbed scour since the dam was built has undermined hundreds of bridges. These are but a few of the problems that may make the Aswan Dam a tragedy rather than a blessing to Egypt.

8 EQUILIBRIUM CONCEPTS

The overriding concept in considering the effects of any change in the environment of a watershed is that of river equilibrium. A river tends to establish a steady yet dynamic condition of balance within a given hydrologic and geologic environment, so that discharge and load are adjusted to river morphology and hydraulics. This process involves the energy concepts already discussed and illustrated in Figure 5–28. If the environment is changed in such a way that the energy becomes greater than the load, erosion results. If, on the other hand, conditions are altered so that the load

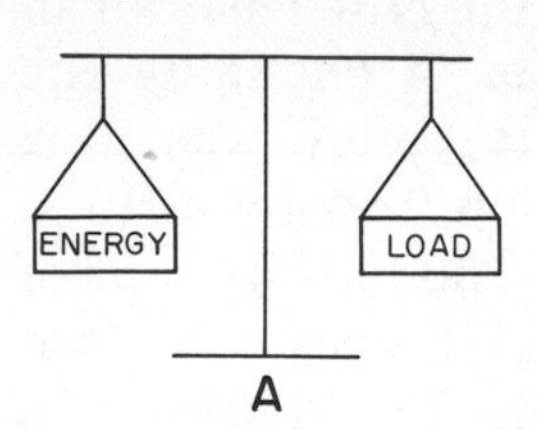

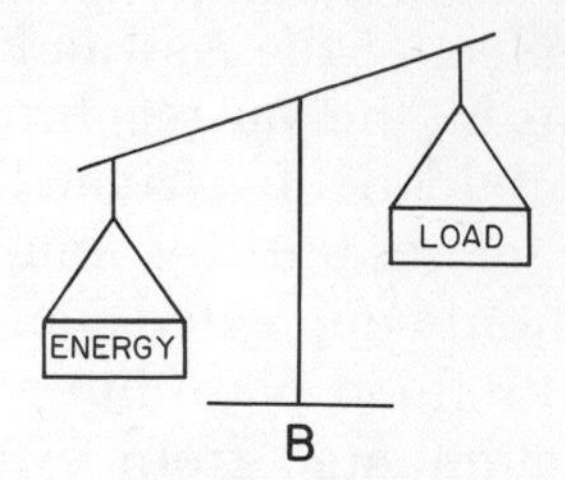

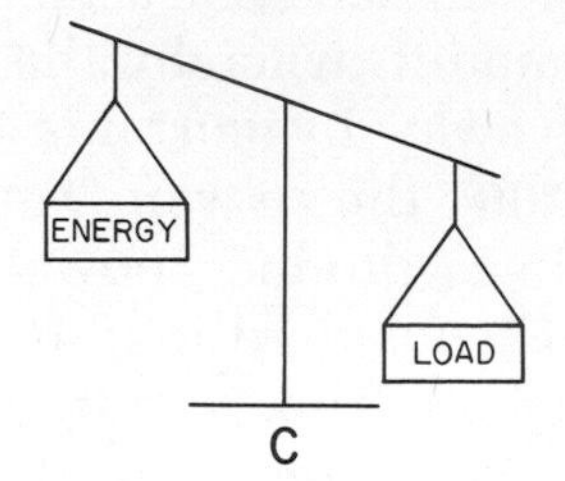

Figure 5–28. Stream equilibrium and disequilibrium expressed as a balance of energy and load. *A*, Energy and load are balanced; the river is in equilibrium. *B*, Energy is greater than load. The river is in disequilibrium. *C*, Load is greater than the energy. The river is in disequilibrium.

exceeds the energy of the river, deposition results. The change is always in the direction that will enable the river to reach an equilibrium again. Man must heed this fundamental law of river reaction.

Another principle which man has disregarded in his efforts to control rivers is that a river is a system that includes the complete watershed, land surface and tributary channels. Whatever happens to any part of the watershed or to any member of the system affects the entire basin. Sometimes the resultant readjustment is slow or faint, but often it is quick and catastrophic.

Let us examine some examples of how man has changed a river and how the river effected its readjustment and new balance. In 1935, Hoover Dam was built on the Colorado River to prevent flooding of towns and lands on the lower reaches. The river channel, however, was already in equilibrium with these hydrologic events which man considered detrimental. The Colorado had adjusted its morphology to the high discharge and heavy silt load that it carried to the sea. Construction of the dam resulted in deposition of great quantities of sediment in Lake Mead, changing the equilibrium conditions of energy and load both above and below the dam. Immediately below the dam, the clear water eroded the channel, increasing its load to balance the energy available. Further downstream, at Needles (Fig. 5–29), the equilibrium channel had been established with a low gradient, which was sufficient to carry the load under the previous hydrologic environment. However, after the dam was built, the gradient was no longer steep enough for the river to transport the new load with the reduced discharge. Bars

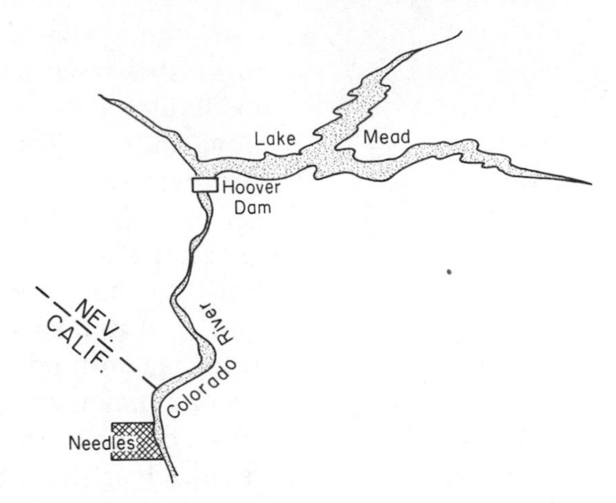

Figure 5–29. Map of the Colorado River from Lake Mead to Needles, California.

were deposited in the channel at Needles, building up islands that caused channelways to be divided and small. After 1941, when the reservoir behind the dam was filled to capacity, outflow was increased. When the increased discharge reached the channel at Needles, the river readily overflowed the silt-choked channel, inundating an area previously flood free. By 1944 there was a 125-acre swamp along the river at Needles, and approximately 100 homes had to be relocated. Later amelioration cost taxpayers nearly a million dollars for drainage, levees and dredging.

The second example concerns the Mississippi River, a river that drains 41 per cent of the area of the United States (Fig. 5–30). The Mississippi presents many problems to man, the most important of which involve navigation and flooding. With such a large drainage area, huge discharges move down the valley,

Figure 5–30. The Mississippi River drainage basin. (After Army Corps of Engineers slide.)

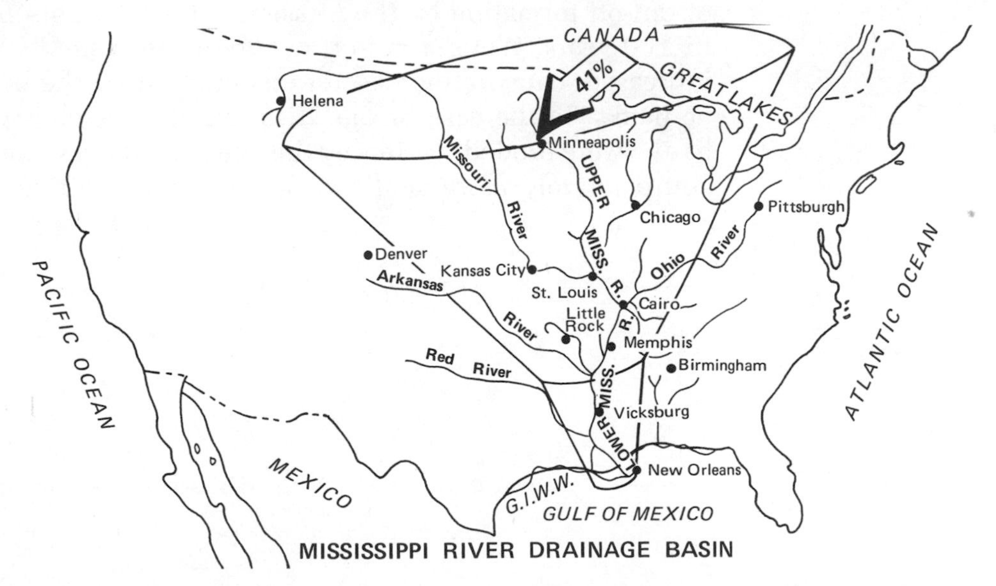

inundating the vast flood plain. Because the river was (and still is) a major transportation route, large cities exist along its banks. These must be protected from devastation. Two main methods of attack on the flood problem were construction of levees and artificial *cut-off* of meander bends.

What are the effects of artificial levees on a river in equilibrium? During floods, as a river overflows its banks, it deposits the material it is transporting. If the river is confined by levees, it will deposit the debris in its channel as the discharge lessens. So, with each high flow, deposits are left on the bed of the river, raising its level. Over time, the bed is built up (Fig. 5–31). With increased elevation of the bottom in relation to the levee, less water may be required to overflow the original levee height. Thus, to protect the area behind it, the levee has to be raised – and the cycle is repeated. Thus the Mississippi levees had to be built higher and higher to cope with an ever increasing water height.

Now let us look at the results of an artificial cut-off. In a naturally meandering river, erosion occurs on the outside of a bend and slightly downstream of it, according to the path of maximum velocity and energy (Fig. 5–32). Deposition takes place on the inner side of the meander bend because of deceleration of flow at the inner bank (Fig. 5–32). This results in a downstream migration of the bend and formation of a neck. With continued erosion the neck may be cut through at high flow and a natural cut-off formed (Fig. 5–33). Sixteen artificially induced cut-offs on the Mississippi River were created during a 10-year period. The natural rate of cut-off formation by the Mississippi is 13.5 cut-offs in 100 years. When a cut-off occurs, water velocity is increased, causing increased erosion of either the bed or banks. In the case of the Mississippi, noncohesive banks were eroded, widening the channel. But a wider, comparatively more shallow channel is less efficient

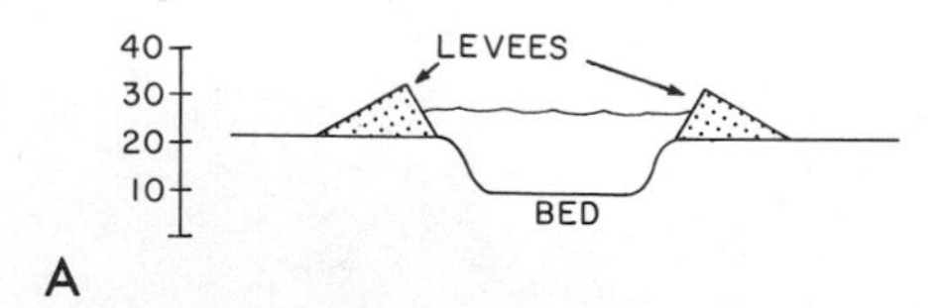

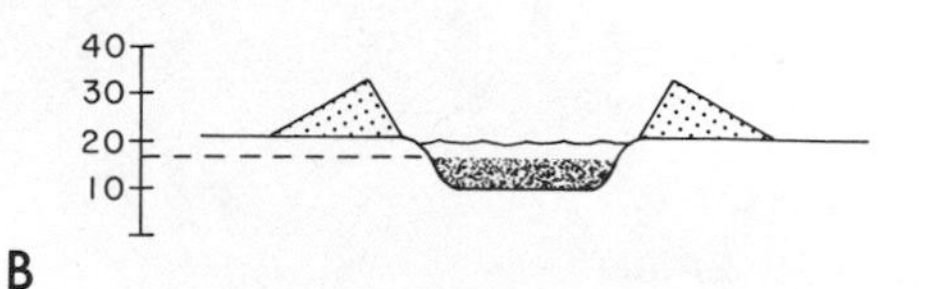

Figure 5–31. Deposition on the bed of a channel confined by levees. *A*, Confinement of a river at high flow. *B*, After flow has receded and the river has deposited material on its bed.

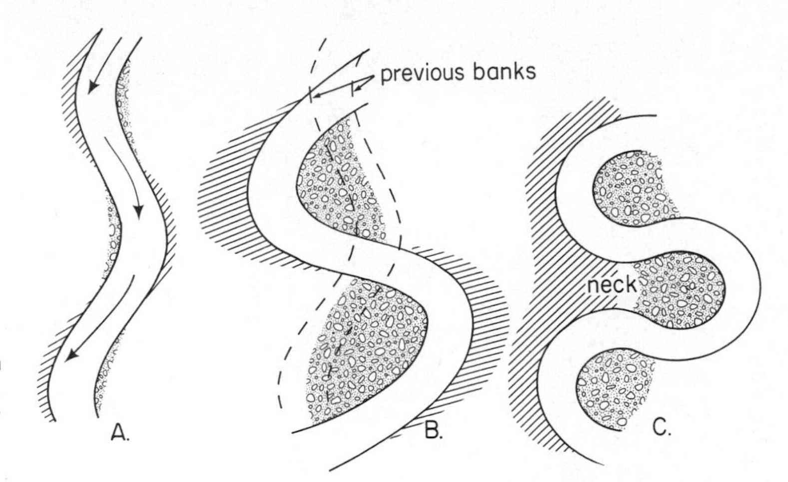

Figure 5–32. Downstream migration of meander bends. Arrows on *A* indicate mainstream of flow.

in transporting material, so in order to steepen the gradient deposition occurs after widening. Thus, along the stretch of river at Vicksburg, Mississippi, the number of divided flows around bars deposited in the channel increased from 41 to 78 between 1936 and 1973. Such an aggrading channel, choked with debris, presents navigation problems. Deposition has also raised the height of the river bed and made the flood plain more susceptible to flooding.

Admittedly, we have presented a one-sided picture of events. Dams, cut-offs, and other structural controls of rivers are important in many ways. Our concern here is to illustrate how rivers react to man-induced changes, and to urge the application of current knowledge about concepts of energy/workload equilibrium in a river system to guide our actions. We should learn to make use of and work with natural forces.

Figure 5–33. Meander cut-offs. *A* and *B*, Development of a natural cut-off. *C*, Artificial cut-offs on the Mississippi River.

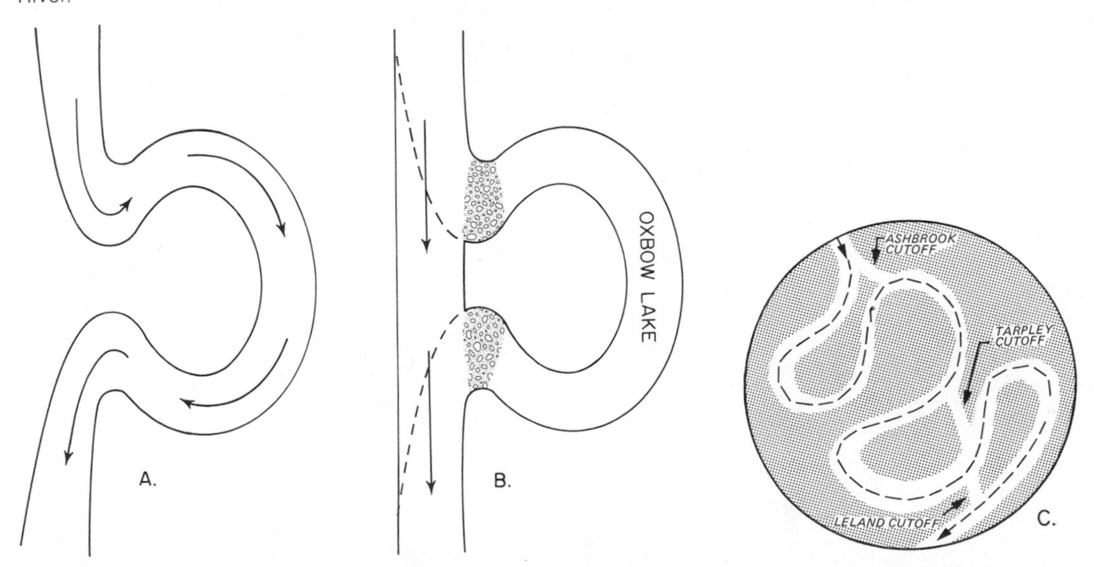

6

DENUDATIONAL PROCESSES: GLACIERS, WAVES AND WIND

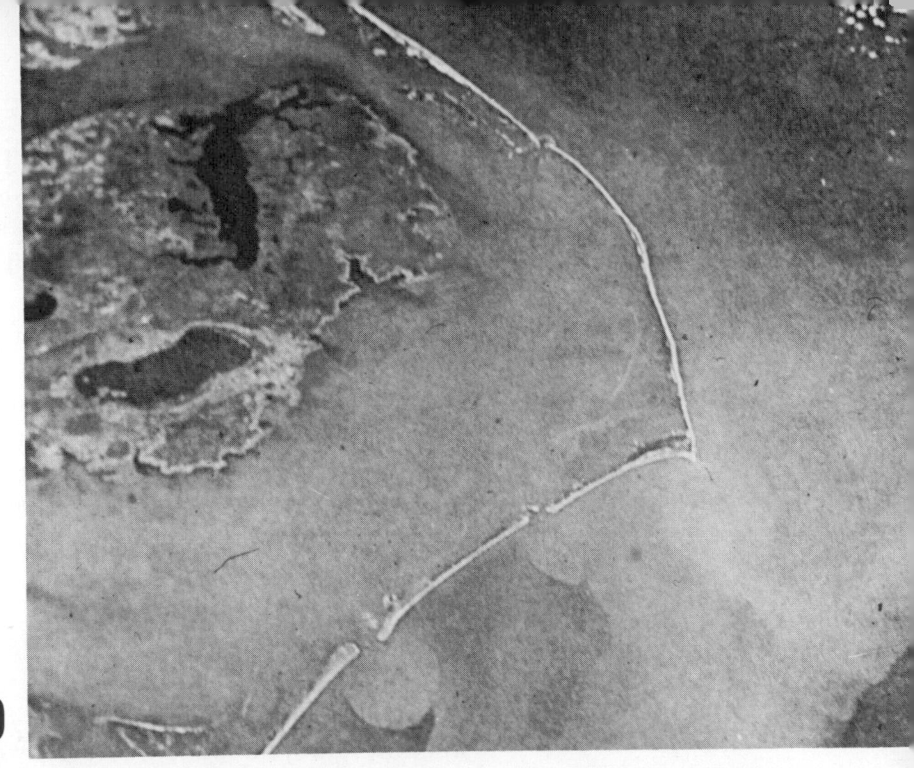

Satellite photograph of the North Carolina outer banks, which consist entirely of sand and dunes formed by the breaking of waves on the North Carolina coast. (Photo courtesy of NASA.)

Mass movements and running water, unconfined and in channels, were discussed in the previous chapter. Because they are powered by gravity, they take place everywhere and are the most important processes denuding the earth's surface. The denudational processes to be discussed in this chapter are those limited either spatially or by energy source, and they are not as effective in eroding the land.

Although glaciers are gravity-powered, they occur only where (or when) the climate allows formation of extensive ice fields. Waves are effective only along a shore where their energy, which is derived from wind and tides, can work. Both of these agents of denudation are tremendously powerful in action but limited in locale. Perhaps the least effective agent of erosion, transportation and deposition is the wind, although there are certain regions where wind depositional features dominate the landscape.

1 GLACIAL DENUDATION

Although only 10 per cent of the earth's land surface is now covered by ice, glaciers covered almost one-third of the land in the recent past. This Ice Age left its mark upon the earth in many ways.

Glaciers are defined as moving bodies of ice derived from snowfall. They contain slightly more than 1 per cent of the water in the hydrosphere. The Antarctic and Greenland ice sheets and the glaciers in many high mountain ranges are actually shrunken remnants of great continental glaciers that once covered large parts of North and South America and Eurasia. Although they melted away from much of the earth about 10,000 to 15,000 years ago, the glaciers left their imprint on the present-day landscape.

Glacial Ice

Glacial ice forms from perennial snowfields that accumulate year after year. The old snow is melted and refrozen, growing larger and denser each time. This granular old ice, called *firn* or *nevé*, becomes tightly packed as layers accumulate. When the specific gravity of the firn reaches about 0.8 to 0.9, it is defined as glacial ice.

The area in which ice is formed is called the *zone of accumulation* (Fig. 6–1). The glacier moves downward or outward toward a *zone of ablation*, or loss. If there is a net gain of accumulation over ablation, the glacier advances; if the ablation rate is faster than the rate of accumulation, the glacier retreats. Ice is said to be stagnant if it is no longer moving.

There are three kinds of glaciers: *alpine glaciers*, which occupy mountain valleys; *piedmont glaciers*, which lie at the foot of mountains; and *continental ice*

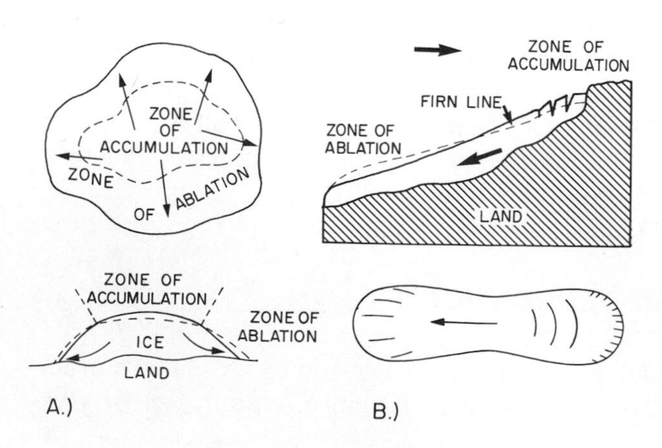

Figure 6–1. *A*, Map and section through a continental ice sheet, showing zones of accumulation and ablation. *B*, Map and section through an alpine glacier, showing accumulation and ablation zones.

sheets, which move out radially from a center. Another classification denotes glacial types by their thermal properties. A *polar*, or cold, *glacier* is one that is well below freezing and is solid ice throughout. A *temperate*, or warm, *glacier* is one that is at the melting point throughout and hence contains water between the grains. However, because the melting point of ice decreases with confining pressure, warm glaciers become increasingly colder toward the bottom, even though there are still meltwater films present.

Flow of Glacial Ice

Ice crystals deform easily, flowing plastically under stress without melting. The flow is a response to gravity or the weight of ice above. This is probably the mode of movement of polar ice, and it is slow. In temperate glaciers, water films are present and facilitate slipping of the ice, both internally and along the bottom. In addition, because of its sensitivity to pressure-melting, a temperate glacier also moves by melting and refreezing in a downward or outward direction. Warm glaciers, thus, generally move faster than cold ones.

Although rates vary, typical movement of a glacier is approximately several cm per day but may be as much as a meter. Sometimes a glacier will surge for-

Figure 6–2. Crevassed surface of a glacier. The upper brittle ice is broken as it is rafted along on the lower plastically moving layers.

ward with a speed of tens of meters per day. Such an unusual velocity is caused by an increase in accumulation of snow or debris on the ice. The sudden weight increase sets up a wave that moves rapidly downglacier.

The upper, more brittle ice at the surface of a glacier is rafted along on the lower plastically moving material. This results in a surface which is broken and crevassed (Fig. 6–2).

Glacial Erosion

Glacial ice is an extremely competent erosional agent. As the water freezes, large boulders can be incorporated into the glacier at the head and removed in a "quarrying" action. Flow and subsequent movement of ice around blocks plucks them loose as part of the frozen load. Debris also reaches the glacier from water flowing down valley sides or from avalanches and other mass movements. All of this material becomes part of the load and moves downslope with the ice. Held firmly as if in a vise, large boulders may gouge and otherwise wear away valley walls and bedrock. The ability of glaciers to erode the land is evident from the landforms created, such as those in Figure 6–3.

Figure 6–3. Topography resulting from mountain glaciation, Front Range, Colorado. The view shows cirques, arêtes, and horns.

Figure 6–4. Glacially deepened U-shaped valley, Cottonwood Canyon, Utah.

Mountains that have been glaciated are rugged and precipitous in aspect. Frost action and erosion at the head, where the glacier forms, create large steep-walled, rounded basins (Fig. 6–3). As glaciers erode both sides of a mountain, the top is eaten away until only a sharp, craggy *horn* is left, and the divide is sharpened to a knife-edged ridge. Hills overridden by ice are cleared of soil, and the bedrock is smoothed and rounded. A glacier moving down a valley deepens and widens it into an easily recognized broad, **U**-shaped valley (Fig. 6–4). Glacially eroded basins are often occupied by lakes.

Glacial erosion of mountain regions has created some of the most spectacular scenery on the earth. The Sierra Nevada Mountains (in Yosemite National Park), Glacier National Park and Banff, Canada, are three of the many glaciated mountain areas of the United States and Canada which millions of tourists visit yearly to gaze on the effects of glaciation.

The Great Lakes, the Finger Lakes of New York and many other such bodies of water are the legacy of glacial erosion. For example, before glaciation, river valleys occupied the site of the present Great Lakes. Glacial erosion deepened and widened the valleys, and the scoured-out basins became marginal ice lakes as the glaciers melted back. The present Great Lakes were created by tilting from uplift on the south as the earth rebounded on release from the weight of the ice which had pressed it down.

Glacial Deposition

The vast amount of debris carried by a glacier is deposited when it melts, either directly from the glacier or from meltwater (Fig. 6–5). Meltwater (or *glacio-fluvial*) deposits can be recognized by their stratification; deposits directly from the glacier (*till*) are unstratified. The various types of glacial and glacio-fluvial deposits are listed in Table 6–1.

Moraines are composed of unstratified material and are subdivided on the basis of location with respect to the glacier and on the basis of form. *Lateral moraines* are ridges of till deposited along the sides of a glacier. Mounds of debris deposited at the end of a glacier are called *terminal moraine*. These are often lobate in map plan (Fig. 6–6) and represent fingers of ice at the front of a glacier. *Ground moraine* is level, undulating till, located away from the edges of the glacier. *Drumlins* are rounded, spoon-shaped hills of till that have been remolded into a streamlined form as the glacier overrode its previous deposits. Debris that falls into a crevasse may be left as a ridge when the ice melts away.

Figure 6–5. Debris-covered terminus of the Nisqually Glacier, Mt. Rainier, Washington. This view is typical of glacial snouts; the debris in the foreground has been washed from the glacier.

TABLE 6-1. TYPES OF GLACIAL DEPOSITS

Till	Glacio-fluvial
Moraine	*Ice contact forms*
Lateral	Kame
Terminal	Kame terrace
Recessional	Kame delta
Ground	Esker
Drumlins	*Outwash*
Crevasse fillings	Valley train
Miscellaneous	Outwash plain
Lake beds	

Meltwater flowing alongside, beneath and from the front of a glacier carries debris with it. *Kames* are hills of debris deposited in depressions in the ice. *Kame terraces* are deposits laid down between the ice and

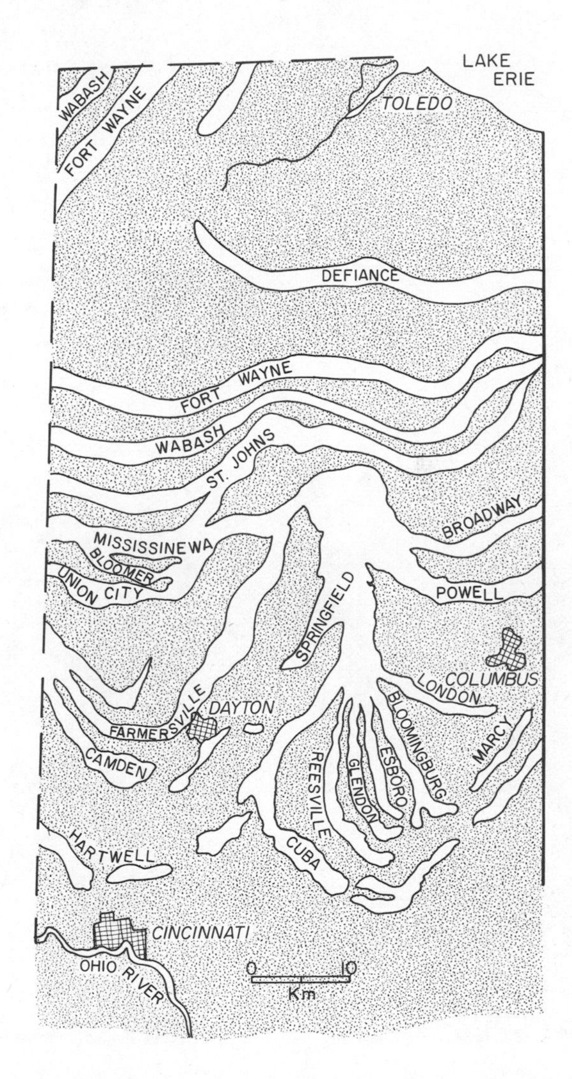

Figure 6–6. Map of lobate terminal and recessional moraines in Ohio. (After Mis. Invest. Map I-316, Ohio Geological Survey.)

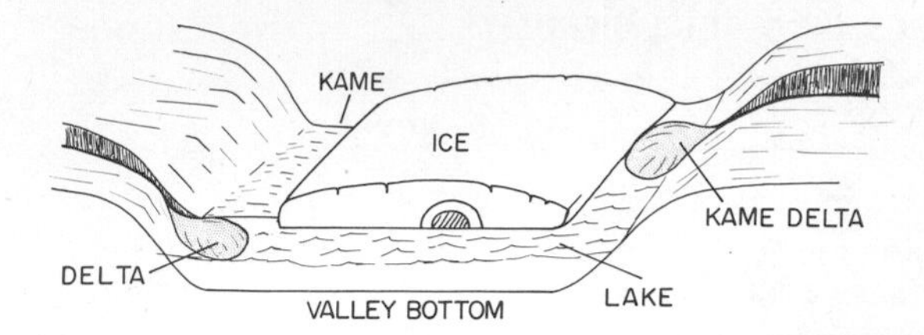

Figure 6–7. Sketch illustrating types of glacial deposits.

the adjoining valley wall, or between two ice lobes. As their name implies, they are level on top. *Kame deltas* are deltas that were deposited against the ice in glacial lakes. *Eskers* are sinuous ridges, deposited underneath the ice or at the terminus in tunnels. Flat plains of material deposited as meltwater flowed away from the terminus of the glacier are called *outwash*. Outwash constricted within valley walls is called a *valley train*.

Kame, esker and outwash sands and gravels are a highly important economic legacy from the glacier. Not only are they sources of road-building material but because of their permeability, they form great reservoirs of valuable ground water.

As ancient glacial ice melted away, vast amounts of water as well as debris were released. In many cases this water became ponded, and vast lakes were formed, such as Lake Agassiz, which existed during the Pleistocene Epoch in present-day North Dakota, Minnesota and part of Canada. Sediment laid down in these lakes

Figure 6–8. An underfit stream. The Nanticoke River in New York, flowing in a valley widened by glacial meltwater. The present-day stream, shrunken in volume, fills only a small part of the valley.

now forms fertile farms. For example, many of the truck farms of New Jersey are on old glacial lake beds, and the bottom of Lake Agassiz in North Dakota and Canada is cultivated and supplies great stores of grain to both the United States and Canada.

Glacial meltwater eventually flowed away, enlarging and carving huge valleys in which we now see only a small stream flowing (Fig. 6–8). Such *underfit rivers* (rivers which are too small for their valleys) are evidence that during glacial times the valley was a spillway for a much larger discharge. The Mississippi, Susquehanna, Ohio and many other rivers owe their wide valleys and thick floodplain deposits to meltwater from the glacier.

Glaciers and the Environment

In many places the results of glaciation have not been beneficial to man. For example, glaciers removed the soil cover from much of the northern part of North America and Europe. In place of the thick soil, the glacier left barren rock or swamps and lakes. Such regions are not suitable for human habitation or cultivation, although they do serve as recreational sites and wildlife sanctuaries.

However, in other areas glaciers deposited fertile lake beds and farmlands. Fluvio-glacial sands and gravels supply ground water for many communities, and the mining of these sands and gravels provides a valuable contribution to the economy of the region. The Great Lakes, which owe their existence to glaciation, are a major factor in the economic development of the adjacent states.

Perhaps the most intriguing aspect of glaciers and the environment is the answer to the question "Are we still in a glacial age?" At present there are two large continental glaciers, as well as many small mountain glaciers, still existing on the earth's surface. These glaciers are remnants of the Pleistocene ice sheets which began to disappear about 15,000 to 10,000 years ago. Scientists have determined that at the time the glaciers were melting away, it was actually warmer than it is now, with a maximum warm climate about 8000 to 5000 years ago. After that very warm period the climate cooled again and the glaciers advanced. During this "Little Ice Age" (from 1300 to 1800 AD), several towns in the Alps were overrun by glaciers, and Norsemen who had settled in Greenland and Iceland were forced from those areas. Then, after 1850 the climate ameliorated and again the glaciers retreated. At the present time, it is hard to determine whether glaciers are advancing or retreating.

If glaciers advance over the land again, the human race will have to make adjustments to cooler and wetter climates. The ice will cover much productive agricultural land and "real estate." On the other hand, if the ice retreats still farther, the melting will release water to the oceans. Measurements show that sea level has been rising at the rate of 1.2 mm/year since 1900. It is estimated that if all the glacial ice were to melt, sea level would rise 60 m above its present level. Such an event would destroy most of the large cities of the world and cause great disaster and suffering.

2 THE SHORELINE

Although only 15 per cent of the land of the United States is touched by the ocean, one-third of the population lives in the coastal area. This interface of land and sea is therefore very important. Large cities are located along the coast, where transportation for overseas commerce is readily available. Because of its esthetic value much of the coastline has been usurped for suburban development. Industrial and utility companies seek sites near the ocean for their plants since it provides a plentiful supply of water for cooling and waste disposal. At the present time there are two nuclear power plants along the California coast, another is in process of being constructed and seven more are planned. The near- and off-shore regions are sources of oil, gas, gravel and mineral resources. Tidal lagoons and river mouths (*estuaries*) represent a region of high productivity. These areas provide twice as much food per acre as inland farms.

Amidst all of this competition for space along the limited coastal zone, there are a few sanctuaries reserved for human recreation and wildlife preservation. But very little land is being set aside for these two necessary purposes. For example, along the 1707 km stretch of California coast there are only 245 km of public beach. There are only seven National Seashore areas set aside for preservation along the total shoreline of the United States.

Man's activities are interfering with natural coastal processes, polluting coastal waters and changing the ecology. It is imperative that we learn to understand the processes that are taking place in this vital corridor between land and sea.

Types of Coasts

In some areas, such as the Gulf and Atlantic coasts of the southern United States, the land surface slopes gently down to the sea. Here the waves have built up long *offshore bars* or *barrier beaches.* A *lagoon* lies behind the bars (Fig. 6–9). The ultimate fate of both the barrier islands and the lagoons is that they will disappear. The bars are washed landward, and the lagoons are filled in with material from sea and land.

A barrier bar is a natural protective device for the shoreline behind it. However, the bar itself is subject to erosion by the waves which break on it. Waves wash up onto the beach and (at times of very strong waves or high water) wash material from the ocean side of the barrier to the lagoon behind. Thus the barrier bar

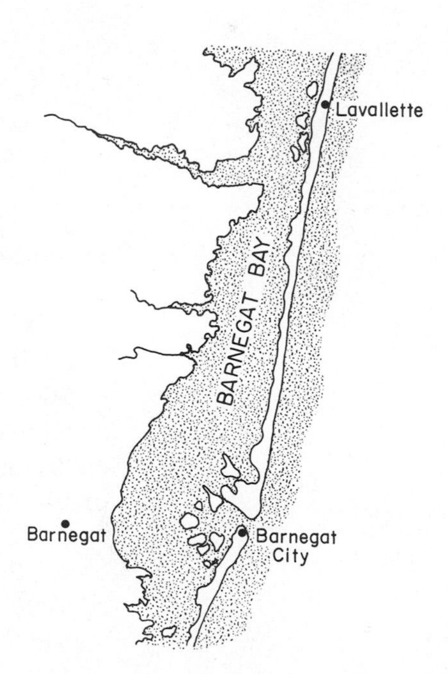

Figure 6–9. A barrier island along the coast of New Jersey. (From the Wilmington topographic quadrangle, scale: 1:250,-000.)

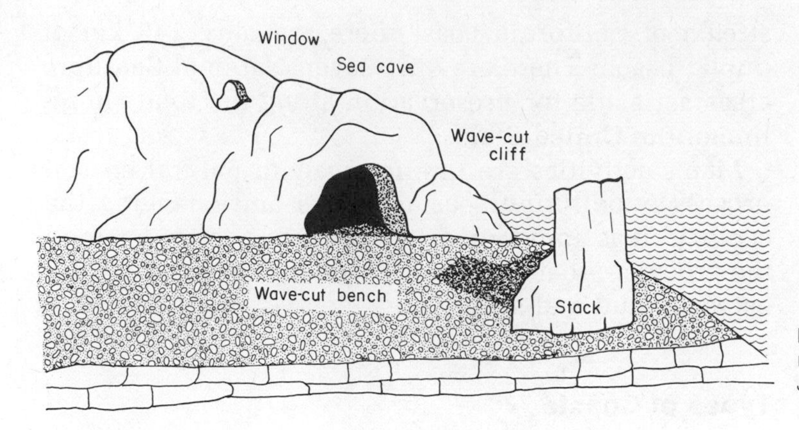

Figure 6–10. Sketch of the rocky coast of Jogoshima, Japan, at low tide.

gradually moves landward. For example, the ocean side of the barrier beaches on the south shore of Long Island is being eroded at a rate of 3.5 to as much as 10 ft per year. All of this material is not being washed into the lagoon behind, however; some is moved alongshore by currents and carried off to be deposited on some other beach.

Along other areas the coastline is irregular, cliffed and rocky. Waves beating on the bare rock have carved steep cliffs and a level bench on which one may find caves, windows and sea stacks (Fig. 6–10). Such coasts may be found in many parts of the world, including New England, Oregon, Washington, Japan and England.

Denudation by Waves

The seacoast represents a region where erosional-depositional processes are limited in space but exceptional in force. Estimates of the amount of coastal erosion by waves vary widely. Beaches along the northeast coast of the United States are retreating at a rate of about 1 to 1.5 m per year, those on the Yorkshire coast of England at approximately 1.8 to 5 m per year. Short-term erosion by great storm waves may be quite spectacular, as evidenced by a hurricane that moved a sandy New Jersey beach back by 25 m. It has been reported that a shingle beach (gravel) in England retreated 1.5 m in a three-hour period during a storm.

A hurricane along the east coast of the United States in 1938 killed nearly 500 people on 12 miles of the Rhode Island coast alone. Hundreds of beach homes were destroyed, and damage was estimated at more than $250 million. The hurricane, which struck at high tide, increased the water level by more than 4 m above mean high tide, and the high waves eroded cliffs and beach dunes as much as 10 m. The coast in this region was a low *barrier island* (Fig. 6–11). The storm caused many inlets to be cut through the island, washing material into the bay behind it. Actually, although much erosion occurred on the ocean side of the barrier, beaches on the mainland side were widened by deposition. (This was, of course, no comfort to those who had their homes close to the edge of the ocean side.)

So the sea not only takes away, it also gives. Although the seacoast of England lost 1900 hectares over a 35-year period, it gained more than 14,000 hectares from the sea during that time. Material eroded from the front of a beach may be washed over onto the backside. Sand removed from a cliff or beach in one area may be transported and deposited elsewhere along the shore. Or material may be exchanged between beach and offshore bar (Fig. 6–12). Winter storm waves erode the beach, carving a *berm* or bench at the edge of the water and carrying the material offshore to form a bar under the water. As the bar is built up, waves break over it instead of on the shore. During the summer, as the waves break on the bar, they remove the sand from the bar and move it up onto the beach, causing a seasonal change in the appearance of beaches (Fig. 6–13). The winter profile is steep with a narrow high beach, whereas the summer profile is wider and lower.

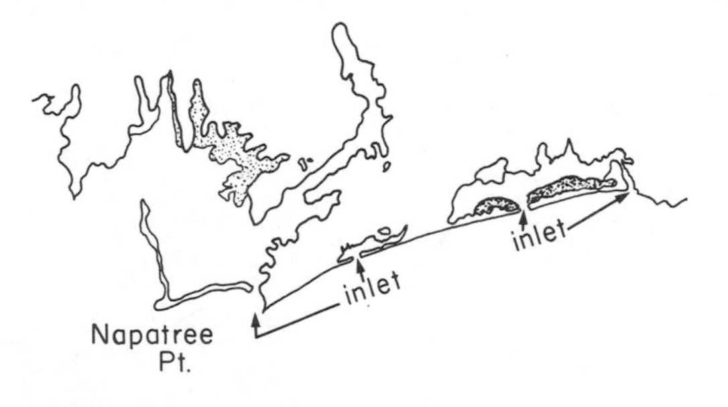

Figure 6–11. Map of the Rhode Island coast where the hurricane of 1938 struck. Beaches on the barrier island were eroded and inlets cut through by the high waves.

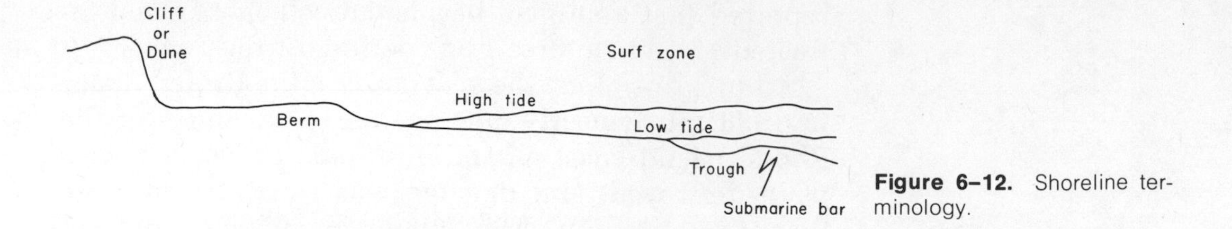

Figure 6–12. Shoreline terminology.

Coastal areas may also be gained or lost by a fall or rise in sea level.

Coastal erosion and deposition and rates at which these processes occur have far-reaching impact on man. The coast is used—and misused—by man primarily as an essential link in transportation systems and as a recreational resource.

Wave Erosion

Along the shore, waves are the most important agent of erosion and deposition, although in some areas winds and tidal currents may play a significant part in shore processes.

Wave erosion along a coast is similar in principle to stream erosion. Beach erosion takes place by chemical solution, mechanical detachment, transport as suspended load or bedload and by hydraulic action. Chemical solution is especially important on coasts composed of limestone or other soluble rock material. However, we will discuss in detail only the mechanical removal and deposition of particles.

The hydraulic action of waves is a tremendously powerful force. Measurements have indicated that shock pressures of as much as 0.593 kg/cm^2 (1210 lbs/ft^2) result from the hydraulic impact of waves 3.3

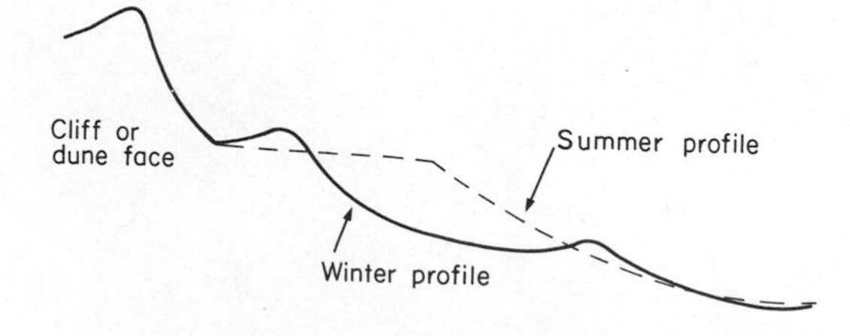

Figure 6–13. Winter and summer beach profiles. The sand beach is high and narrow during the winter and wide and lower in the summer.

m high. These pressures are enough to move extremely large blocks from cliffs or man-made structures.

Normal natural rates of erosion depend upon the form of the beach and its exposure to wind, the type of coast (whether low and duned or steep and rocky), offshore relief, sea-level changes, currents along the shore and the availability of material. Man-made structures may interfere with natural rates of erosion and deposition.

Capitola Beach is an ocean resort along the California coast south of Santa Cruz. Many people enjoyed the wide beach at Capitola as a recreational spot, which was built up every summer by the southward drift of sand. In the early 1960's, the U.S. Army Corps of Engineers built a yacht harbor for the city of Santa Cruz. The structures of the yacht harbor trapped the sand being transported down the shore, cutting off the supply of material to Capitola Beach. While the harbor at Santa Cruz was being filled with sand, the beach at Capitola started to be carried away by the waves. To save the beach, tons of sand were trucked in and dumped on the shore, but that sand too was quickly eroded away by the waves. So *groins* (walls built out into the water, perpendicular to the beach) were built, and new sand was trucked in to fill the beach behind the groins. Although the new sand remained on the beach between the groins, waves began to erode the beaches further to the south. Man had solved the problem at Capitola Beach, but only by moving erosion downdrift. Let us see why this happened.

Wave Energy

Erosion and deposition along the shore may be related to the energy/workload balance of the waves and currents. Wave energy, E, depends upon the wave length, L, wave height, H, and the weight of the water, w:

$$E = \frac{wLH^2}{8}$$

Wave height (Fig. 6–14) depends upon the prevailing wind strength and duration patterns, the *fetch* or distance over which the wind blows and the configura-

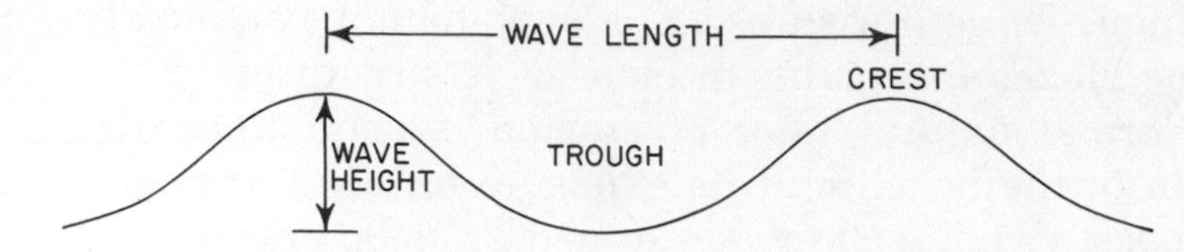

Figure 6–14. Wave terminology. Wave length is distance from crest to crest, wave height is distance from crest to trough in a vertical dimension.

tion of the shore bottom. Waves are high seasonally and during storms, when winds are stronger than usual.

Wave steepness, or the relation of height to length (*H/L*), is the critical factor in determining wave energy, as can be seen by the fact that height is much more important in the wave energy equations. High-energy destructive waves are steep; that is, they are high relative to their length. Constructive waves, which tend to build up a beach, are long and low. Wave height and length are modified as the wave nears shore. As a wave enters shallow water its length decreases. Wave height increases as the wave "feels the bottom" (Fig. 6–15) until it becomes too steep to hold its form under the pull of gravity, and it breaks. As the wave breaks, its energy is dissipated on the shore, detaching particles and rolling them along the bottom or lifting them into suspension.

The difference between destructional and constructional waves has been documented by Fox and Davis (1970) on the Lake Michigan shores and by Hayes and Boothroyd (1969) for a region on the Massachusetts coast. As storm systems move into the coastal area, strong winds blowing onshore increase wave height. These steep waves are destructive and erode the beach, moving the sediment offshore to form an underwater bar. As the storm passes and the wind shifts, blowing offshore, there is a post-storm recovery. Waves moving towards the shore are of low steepness with low energy. They break over the offshore bar and gradually move it landward. Although some of the beach is recovered in this way after a storm, there is a net loss of material moved along the shore.

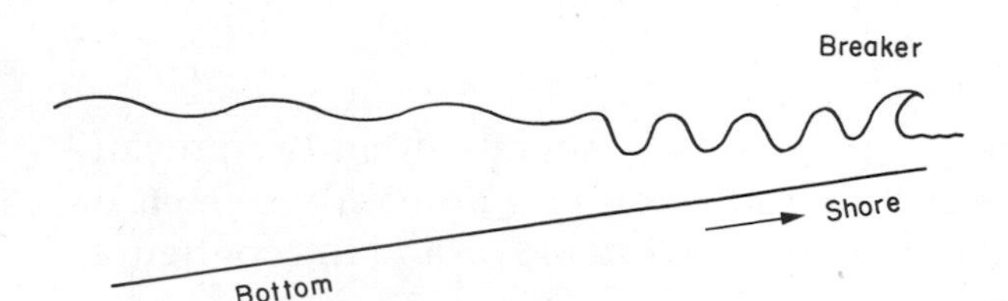

Figure 6–15. A deep-water wave changes as it touches bottom. Length decreases and height increases.

Wave Refraction

Modification of wave form and drag on the bottom as the wave moves into shallow water result in a decrease in wave velocity. If the bottom is irregular, velocity may be decreased in one place and not in another, or may be decreased differentially along the wave front. Differential changes in velocity cause one part of the wave to outrun another, and the wave front will be bent or *refracted* (Fig. 6–16). Thus an irregular bottom configuration results in variations in the direction in which the wave approaches the shore. In turn, these variations cause zones of high and low energy concentration and create currents flowing from the high energy zone to the low energy zone. These currents are the main process by which transport takes place.

Transport Along the Shore

Currents may be generated by winds, tides or waves. Although those induced by winds and tides are often effective transporting or eroding agents, wave-induced currents, formed when waves approach the shore obliquely, are more common. Along a curved or embayed coast, currents move from zones of high energy to zones of low energy. On a straight shore, the cumulative effect of waves breaking in a direction oblique to the shore generates an alongshore or *littoral* current, which is effective in moving material along the shore.

Littoral transport along the New Jersey coast has been estimated as 376,000 m^3 of sand annually. The littoral current can be looked upon as a river flowing

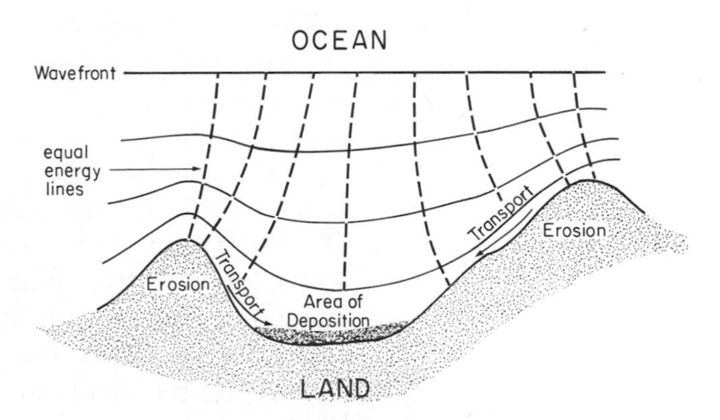

Figure 6–16. Wave refraction on an irregular coast where the wave feels the bottom at different places along the wave front, causing it to bend in conformity with the map plan.

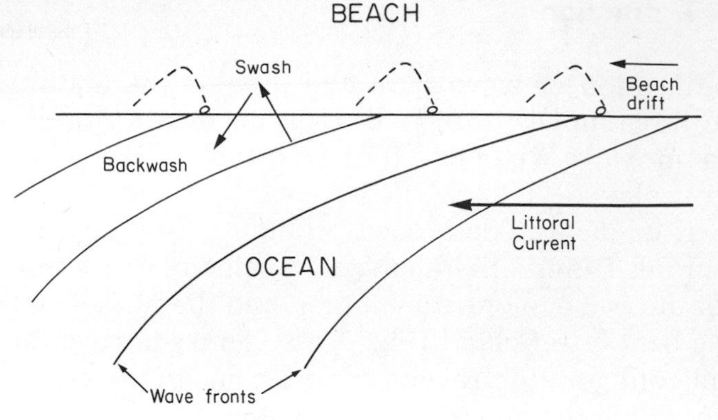

Figure 6–17. Beach drift. Particles on the beach are moved in with the swash and back with the backwash. The result is an overall down-beach movement of grains.

along the shore, with the beach as one bank and the ocean as the other. Like a river it has energy dependent upon its mass and velocity and thus works by carrying a load. If the current is interrupted, for example by an inlet, the load may be deposited. The waves and littoral current downbeach from the inlet, deprived of the load, will pick up new material and move it on. Thus an inlet (as well as the rest of the shore) is a dynamic feature by its very nature and is an integral part of the beach system.

Another process by which material is moved alongshore is beach drift. As a wave approaches shore obliquely, it breaks and runs up on the beach as *swash*, carrying particles along with the swash (Fig. 6–17). As the water flows back oceanward in a *backwash*, it moves perpendicularly down the beach slope, again carrying particles with it. In this way grains move in a zigzag pattern with an overall net movement in a downbeach direction as indicated in the figure. Beach drift can be a most effective transporting agent where waves have a small steepness ratio (H/L) and break near shore.

These shoreline processes of erosion, transportation and deposition by waves and currents are analogous in many ways to fluvial processes (see Chap. 5). The shore-ocean interface represents a system that seeks and attains a natural equilibrium. Once the steady-state beach profile is established, there is a delicate balance of energy and work. The alongshore current system acts as a sand transport conveyor belt, whereby sand arrives at one section of a beach and other sand is carried away from the same area. If the balance of arrival and departure of sand at any one point is maintained at a steady pace, no erosion takes place. If the transport system is disturbed so that more sand arrives than is removed, the beach is built up; if more sand is removed than arrives, the beach is eroded. Unfortunately, tectonic forces within the earth and driving energy in the atmosphere continually act to upset this balance. But perhaps the most disturbing element in the shoreline is man.

Man's Manipulation of Coasts

As with many other natural processes, the work of waves along the shore creates problems for man. As he develops harbors and uses the shore for recreational purposes, man in his effort to stabilize coastal areas comes into direct conflict with the natural forces seeking to change the shore.

On rivers man builds dams to control the waters; on the shore he constructs groins, jetties and seawalls to control the waves. Groins are walls constructed on a beach and extending into the water perpendicular to the shoreline (Fig. 6–18). They are built expressly to trap the sediment being carried along the shore by beach drift and littoral transport.

On the south shore of Long Island, groins built at Riis Park and Westhampton Beach caused such erosion on beaches downdrift that additional groins had to be constructed at great expense. These, in turn, shifted the focus of erosion further downbeach.

Figure 6–18. A groin built along the shore in order to trap sand and build up the beach.

A jetty is similar to a groin but is placed at the sides of an inlet. Its purpose is to stop the transport of material into the inlet, where it would naturally be deposited. Inlets are breaks made in a barrier bar during storms. The inlet is a transient feature that is soon closed by the longshore transport of material. Artificially maintained inlets create a break in the sediment transport system and cause beaches downdrift to be eroded as the energetic waves pick up material to replace that lost at the jetty.

For example, McCormick (1973) showed that there was an increase in the rate of erosion of the beaches between Shinnecok and Moriches inlets on Long Island. The increase was related to the stabilization of the two inlets by jetties.

With both groins and jetties, removal of the load in

transport has deleterious effects downbeach. However, it is possible that by careful study of the waves, littoral currents and the transport system, groins and jetties can be constructed that will work in harmony with natural processes.

Another method used to stabilize the shoreline is artificial nourishment. Sand is fed into the littoral current to supply it with a load so that it will expend its energy in moving the load provided rather than picking up material from the beach. Again, this method effectively stabilizes the beach only if a study has been made of the waves, currents and sediment load already being moved. This method is being used along the New Jersey coast, where it is estimated that 1.3 million cubic meters of sand will be needed annually for nourishment, after an initial 22 million cubic yards are filled in.

The main problem in artificial nourishment is finding a suitable supply of sand to match the wave and current energy needs. Because of loss of beach by erosion, Gilgo Beach on the south shore of Long Island was nourished with 1.8 million m^3 of sand in 1959. In 1961 the same beach required 1.98 million m^3 replenishment and in 1964 more than 1 million m^3 were added to the beach. This is a very popular beach for the metropolitan New York City area and thus these expensive measures are used to maintain the recreational resource. This problem can be ameliorated by by-passing, a method presently used at inlets. Sand is removed on the updrift side of the inlet and piped to the other side, where it is fed back into the system. Adapting this method to beach nourishment would entail catching transported material at the end of the system (perhaps where the material is dropped into an offshore canyon), piping it back upshore and feeding it into the system. Thus the same sand is transported through the cycle repeatedly.

All of these beach stabilization measures are costly, and many states, as well as national governments everywhere, will have to deal with the issue of spending vast amounts of money to protect the shore. This problem is compounded by the fact that much of the shore is privately owned, and the money for the stabilization measures would be taxpayers' money. The fundamental concept behind coastal conservation programs must be that the shore-ocean contact is a dynamic system.

Estuaries

As the flow of fresh water of a river nears its mouth, it encounters salt water moving upstream from the ocean in response to tidal forces. Because the salt water is dense it moves upstream along the bottom of the riverbed and the lighter fresh water moves seaward above the salty layer. Gradually the two layers of water mix along the interface until at the mouth the result is the brackish water of harbors, lagoons and salt marshes.

This region where fresh water flowing to the sea meets and mixes with the salty water of the ocean defines an estuary, an ecologic zone rich in life forms and important as a food provider. Many large cities (New York, London and Tokyo, for example) are located on estuaries, deriving their water supply from them and using the river-sea zone for transportation, discharge of wastes, recreation and food supply. Some of these uses are incompatible with each other.

The Delaware River is an estuary below Trenton, New Jersey. Its tidal waters supply municipal and industrial water to the cities along the banks and carry wastes to the sea. A combination of upstream dams and water pollution is destroying the productivity of the estuary. The fishery catch in shad alone decreased from a high of 20 million pounds (worth $4.5 million) in 1900 to 80,000 pounds in 1970 (worth $14,000), according to Thomann (1970).

The Thames estuary is another river-sea interface where pollution destroyed the fish population. Wastes poured into the river had completely closed all commercial fishing in the estuary near London by 1850 (Mann, 1970). The river became increasingly anaerobic until by 1953 it was putrid, giving off hydrogen sulfide fumes. By 1960 the British had started a clean-up program which is so successful that today the fish are returning to the estuary.

Modification of river discharge by dams and diversions of water diminishes the flow of fresh water and increases the intrusion of saline water upstream and decreases the dilution of ocean water in the estuarine environment. Stabilization of stream flow causes stabilization of salinity and reduces fisheries and shellfish populations that depend on annual freshwater flooding for propagation. Oyster reefs and clam fisheries are disappearing in some estuaries because of the stabilization of salinity.

Reclamation of coastal areas has also resulted in a significant loss of estuarine habitat. Rivers and shoreline systems have been changed by dredging and filling in order to increase "real estate." More than 7 per cent of the estuarine areas of the United States have been lost by the filling of estuaries and lagoons. Most of this loss has occurred in California (see the story of San Francisco Bay, Chap. 13). Reclamation of land from the sea has been most significant and successful in Belgium and the Netherlands. These two countries have reclaimed more than a million acres from the sea by diking and filling. However, in view of the fact that estuarine and ocean habitats provide twice as much food per acre as land cultivation, one wonders about the wisdom of this kind of reclamation.

3 DENUDATION BY WIND

The wind is probably the least effective of all the geomorphic processes acting on the surface of the earth. It is limited in its ability to erode and transport by its lack of competence, see p. 191. Moreover, it is important as a process only in specific areas such as deserts, steppes and along duned seacoasts.

Wind Erosion and Transportation

Over water, energy is transferred from the wind to waves; over land, wind energy is expended in moving fine particles. Grains are entrained by the shear stress of the wind blowing over the surface and by fluid turbulence, which lifts grains into the air. This phenomenon can be seen in the familiar "dust devils," turbulent eddies of wind which swirl and rise, carrying fine particles with them (Fig. 6–19). As the air moves rapidly in toward a center, it must rise. The hydraulic principles described in relation to fluvial processes (see p. 191) also apply to wind erosion. The difference in capacity and competence of the two processes is related to the difference in mass of the two media. Although winds can attain much higher velocities than water, wind can entrain only small particles (less than 1 mm in size).

Wind erosion is accomplished by *deflation* and *abra-*

Figure 6–19. A dust devil on the plains of Arizona, rotating turbulent wind, carrying silt.

sion. Deflation is the removal of material (as bedload or suspended load) by the wind; abrasion is the grinding away of rock as sand is blown against it. Bedload movement is concentrated in an area of one meter or less above the ground. Some of the bedload is transported by *creep*, a slow rolling or sliding of grains along the ground, but most bedload movement is by *saltation*. As in streams, saltation is the "jumping" of grains momentarily lifted into the air (Fig. 6–20). Saltation is initiated by processes of changing velocity and lift and may be propagated by the bumping of grains in saltation on one another (Fig. 6–20). It is thought that creep results from this impact of grains in saltation.

While the wind can move grains as large as one mm in diameter as bedload, only silt or clay can be carried in suspension. However, a dust cloud can contain a large amount of material. In the 1930's the Great Plains became a dust bowl. The land had been "farmed out" and then deserted, leaving it devegetated and ready for wind action. Strong winds blowing over the empty plains removed volumes of topsoil. It is claimed that one dust storm in 1935 created a cloud a mile thick as it moved over Kansas and contained 40,000 metric tons/km^3 of material.

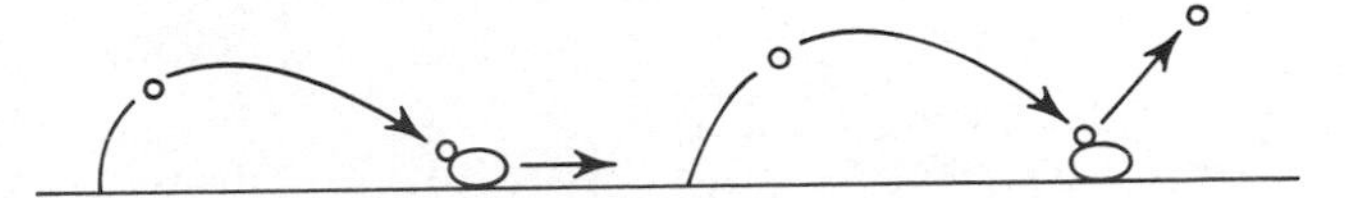

Figure 6–20. Diagram illustrating creep and saltation by wind.

The rate of wind erosion and transport depends on the duration, strength and fetch of the wind and on the material available. Vast open spaces, constant winds and lack of any vegetal cover are necessary for the wind to be effective. Bare ground, open fields and loose fine particles on the ground are conducive to loss of soil by wind, and dry conditions are helpful because fine material is less cohesive when dry.

Erosional Landforms

One of the main erosional landforms created by the wind is a *deflation hollow*. These are common in the Great Plains and semi-arid regions of the West. The hollows are extensive in area but shallow in depth. Because they are depressions on the surface, rainwater may collect in them for short periods of time, during which weathering may take place. The hollows are thus maintained as the wind removes the weathering products. Hollows may also be increased in size by animals which churn the ground, breaking it up so that particles can be removed by the wind. Many such buffalo wallows are found on the Great Plains.

Another feature resulting from wind erosion is *desert pavement*. This term refers to the larger material left behind as a lag deposit when the wind removes the finer particles. Such coarse material may "pave" the surface of arid regions.

Wind Deposition

When wind velocity is decreased or turbulence calmed, the particles in transit are dropped. There are two major types of wind deposits; *loess* and *dunes*. Loess, which is composed of silt with some clay and fine sand, is coherent and tightly bound and is thus able to maintain vertical slopes. It is porous, slightly calcareous and fertile. Approximately 10 m of loess covers the region around the northern and central Mississippi River. It is thought to have blown off the deposits left by glacial meltwaters along the wide

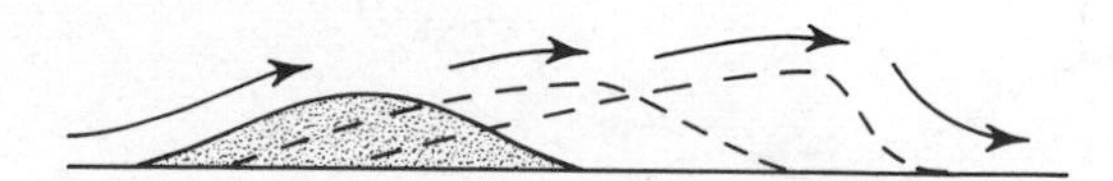

Figure 6–21. Successive profiles of dune movement, showing development of steep lee slope.

Figure 6–22. Transverse dunes, Great Sand Dune National Monument, Colorado.

Mississippi River flood plain. Silt blown from the Gobi Desert was deposited on China, covering a vast area with a blanket of loess more than 50 m thick.

Sand deposits are often blown into assemblages called dunes. Whenever wind velocity decreases, either from gustiness or an obstruction, sand in transit is dropped. As soon as a number of grains aggregate, a wind shadow is created, which results in even more deposition. The size of the sand piles increases by continuous deposition. A profile is created on the sand pile as the gentle upwind slope allows sand to be moved to the crest. When the crest is high enough it loses its stability and the sand slides down the front, creating a steep lee slope (Fig. 6–21). Sand and dune form move as grains are blown up the gentle upwind slope and cascade down the lee slope (Fig. 6–22). Sand dunes are found in desert areas and along shorelines.

Wind Erosion and Man

Wind blowing over the land can be a most effective denudational agent if the land is devoid of vegetation. In many areas man has increased the ability of wind to do its work by removing plant cover for cultivation or by overgrazing. On beaches man has aided wind

Figure 6–23. Partially eroded, windblown sand dunes at Fire Island, New York.

erosion by trampling vegetation so it can no longer live and act against the wind. The same type of destruction is now taking place in many fragile semi-arid regions where sparse vegetation is being crushed by vehicles. Thus, in many ways, man is increasing the desertization of the land and laying it open for wind action.

However, wind erosion can be controlled. Wind breaks, fences and revegetation all can be used to check the velocity of the wind or hold the soil. These measures not only may decrease wind erosion but may also aid wind deposition.

Perhaps the most important area of wind action and wind-created landforms are duned seacoasts. Duned coasts appear where sand accumulates on a beach faster than the waves can move the material alongshore. Quite often this occurs in areas where rivers pour large amounts of sediment into the ocean or where cliffs upshore are being eroded. When the beach is built above high sea level, the winds can move and mold the sand into dunes (Fig. 6–23).

By absorbing wave energy, these dunes provide natural protection for the beach during storms. Many duned coasts are being eroded as a result of man's actions. The outer banks of North Carolina were once covered by dunes and vegetation. Devegetation of the dunes resulted from use by man and his domestic animals, leaving the dunes unprotected and open to erosion by both wind and waves. Then the settlements were deserted, and the outer banks were put under the protection of the National Park Service as part of the Cape Hatteras and Cape Lookout National Seashores. A program of revegetation has been started, with the hope of stabilizing the dunes. The dunes, in turn, will aid in protecting the islands from erosion. It is a wise plan, which will allow these barrier beaches and dune systems to return to a natural equilibrium.

In Chapters 5 and 6 we have discussed how the geomorphic agents of wind, waves, ice, gravity and running water are at work reducing the land to the level of the oceans. Their work is involuntary, the result of energy in the system. The processes are dynamic and function within the system, struggling toward an equilibrium. Man must understand and accept this and learn to work with the system rather than against it. He must realize that he can live safely within the natural environment only if he is aware of the principles by which forces in the environment function. Otherwise natural events become catastrophes.

Unit II

EARTH RESOURCES

7

RENEWABLE AND NONRENEWABLE RESOURCES

This ancient source of water and others in the vicinity were probably responsible for the initial settlement in and around present-day Nazareth, Israel.

1 INTRODUCTION

The subject of renewable and nonrenewable resources is complex. The terms themselves can be clearly defined, but application of those definitions is in many cases closely related to the economics involved, especially the energy cost.

A renewable resource is one which can be used repeatedly. Water is renewable: it can be used, cleaned and reused many times along a single river before returning to sea. On the other hand, fossil fuels such as petroleum and coal are nonrenewable because they are altered completely when they are used and thus cannot be reused. Many substances are partially renewable. Copper can be *recycled* many times because it is generally used in its unaltered form of copper metal. However, the economic feasibility of its reuse depends on the cost of mining new copper versus the cost of reclaiming the used copper.

2 RENEWABLE RESOURCES

Water is a resource which is renewable both by man and by nature. The hydrologic cycle (Fig. 7–1) is the process by which water is recycled in nature. A detailed discussion of the hydrologic cycle is found in Chapter 11. When water evaporates and then condenses as rain it is purified naturally. The process of evaporation entails a low-temperature distillation, which leaves behind most elements. This means that if we evaporate water that is polluted with lead or other metals, the evaporation leaves these pollutants behind, and the rain that condenses from this water vapor is pure. Another part of the cycle, *groundwater* (water that fills the voids and openings below the water table; see also Chap. 11) *movement*, purifies water of bacteria

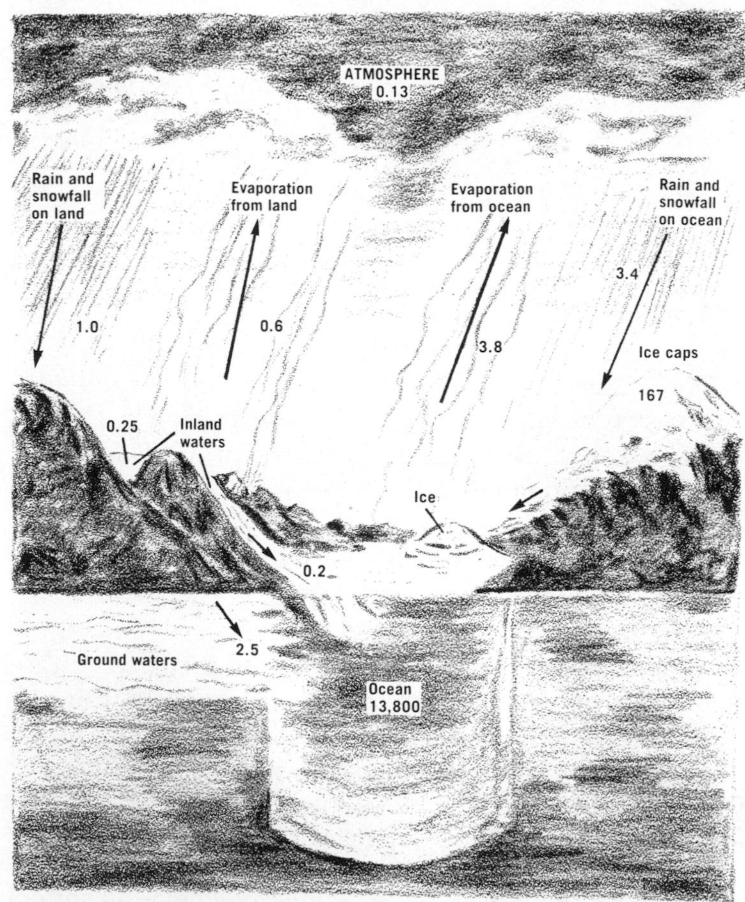

Figure 7–1. The hydrologic cycle. (From Turk, A., et al.: Environmental Science, Philadelphia, W. B. Saunders Co., 1974.)

3600
2400
1200
0
BILLION GALLONS PER DAY
Surface water
(2,800)
1950 1955 1960 1965 1970
HYDROELECTRIC POWER
A

Figure 7–2. *A*, Graphs showing trends in use of water for hydroelectric power and in all other withdrawal uses combined, 1950–1970. *B*, Changes in water withdrawals and consumption in the United States in billion gallons per day, 1950–1970. (From Department of the Interior News Release, August 30, 1972.)

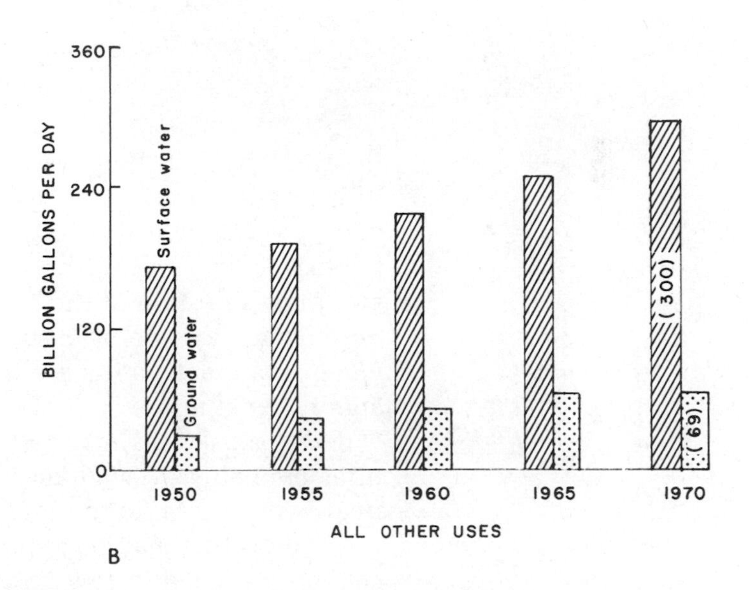

—so long as the system is not overloaded. However, such overloading is becoming more and more frequent as use of water increases.

Rising demands on the nation's fixed supply of water require it to be recycled at a faster rate.

"About half of those who take their drinking water from public water-supply systems in the United States," says Daniel K. Okun, University of North Carolina environmental-engineering professor, "use waters parts of which only hours before had been discharged from some industrial or municipal sewer."*

*From Stuart, P. C.: A drink of water—but is it fit to drink? Christian Science Monitor, April 21, 1972.

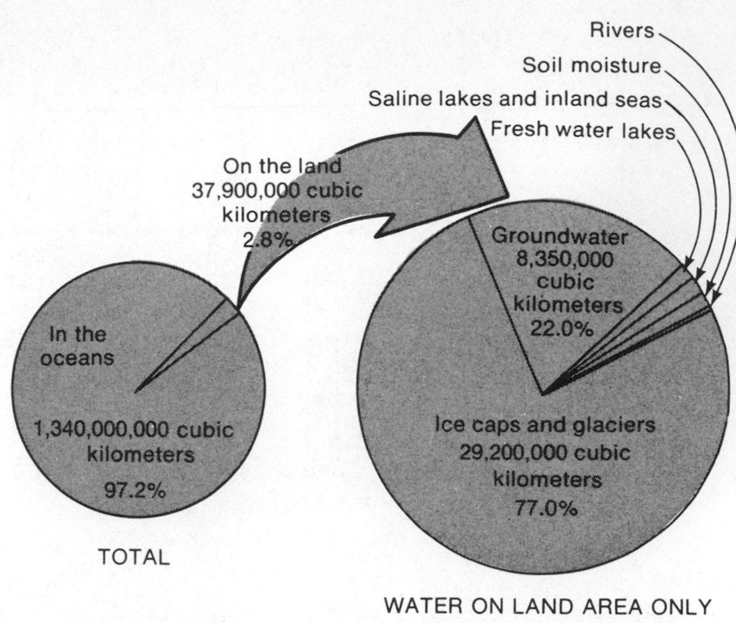

Figure 7–3. Distribution of water on the earth. (From National Academy of Science Committee on Geological Sciences, National Research Council: The Earth and Human Affairs. San Francisco. Canfield Press, 1972.)

It is no longer sufficient to use water, dump it and allow nature to purify it (Figs. 7–2 and 7–3). Water treatment plants must be built. These have the double advantage of providing the needed recycled water and stopping pollution of the body of water into which the used water was being dumped. The disadvantages of water treatment plants are that they are expensive and energy consuming.

If water use reaches the point where available supplies are being overworked even with the use of water treatment plants, other less conventional methods of water recycling may be applied. One rather exotic suggestion has been to tow large sections of glacial ice to areas that need water, such as southern California. A more practical alternative, which is now being done on a small scale, is *desalination*. Desalination is the removal of salts from sea water or saline ground water, generally through the process of distillation. The technology is simple, but the cost of energy is presently too high to permit large-scale desalination. The fact remains that desalination can be accomplished; we literally cannot run out of water so long as energy for desalination is available.

Another example of a renewable resource is soil, which is discussed in detail in Chapter 10. Soils are renewable resources only in the sense that we can use them over and over again for crop and forest lands by rotating crops, employing chemical additives such as fertilizer and lime and irrigating. However, intensive farming does not necessarily destroy soil, and forestation prevents erosion by providing a protective cover and a root system that helps to retain soil.

Generally, however, when soil is lost through erosion it cannot be replaced. The weathering of rocks and minerals to form soils (see Chap. 10) is a process that normally involves a minimum of thousands of years.

3 NONRENEWABLE RESOURCES

Simply stated, mineral resources are nonrenewable because once they are removed from the ground all that remains is a hole. Deposits in which minerals are sufficiently concentrated to be economically extracted are called *ore deposits.* Ore deposits are neither widespread nor uniformly distributed. In general, a combination of geologic processes must occur in order for an ore deposit to form. Such combinations are relatively rare. Ore deposits are limited in size and contain a finite supply of ore, which eventually must be depleted. Some deposits will last for hundreds of years; others will be exhausted within the next few years.

Depletion rate depends on a variety of factors. In general, if the element is a major element in the earth's crust, large ore deposits tend to form. The major elements of the crust are oxygen, silicon, aluminum, iron, calcium, sodium, potassium and magnesium. Of these elements, which comprise 99 per cent by weight of the crust, only aluminum and iron are used as industrial metals. Iron mineral deposits, which are widespread and large, will last for a long time. The iron deposits in Minnesota have been mined for more than 100 years. Table 7–1 gives data on the abundance of certain metals in the crust, the percentage necessary to form an ore and the *enrichment factor.* The enrichment factor is defined as the per cent metal in ore divided by the per cent metal in the crust and is used as a measure of how much the metal must be concentrated in order to form an ore.

TABLE 7-1. ENRICHMENT FACTOR FOR SOME COMMON METALS*

Metal	Per cent in crust	Per cent in ore	Enrichment factor
Mercury	0.000008	0.2	25,000
Gold	0.0000002	0.0008	4000
Lead	0.0013	5.0	3840
Silver	0.00007	0.01	1450
Nickel	0.008	1.0	125
Copper	0.006	0.6	100
Iron	5.2	30.0	6
Aluminum	8.2	38.0	4

*From National Academy of Science Department of Geological Sciences: The Earth and Human Affairs. San Francisco, Canfield Press, 1972.

Some of the earth processes that cause mineral deposits to form are weathering, volcanism and formation of hot solutions at depth (see Chap. 9). Whereas these processes are prevalent, the combination necessary to produce a mineral deposit (or more importantly, an ore deposit) is rare. For example, the process of weathering is common; the formation of an aluminum deposit is not. It is a rare combination of 1) tropical weathering conditions, 2) operating for extended periods of time, 3) where erosion is slow enough to allow *in situ* weathering, 4) of a rock high in aluminum and low in iron which taken together can produce an aluminum ore deposit (see Chap. 9).

Petroleum (a naturally occurring complex liquid hydrocarbon), natural gas (composed principally of CH_4, methane) and coal (a combustible rock) are collectively known as the fossil fuels. The origin of these (see Chap. 8) is related to the alteration of organisms in a non-oxidizing environment; that is, the dead organism must be quickly buried or in some way removed from the presence of excess oxygen or it will decompose to form gases such as CO_2 and H_2O rather than be reduced to carbon or retained as hydrocarbons. Coal and some natural gas are related to alteration of land plants, while petroleum and other natural gas come mainly from the alteration of marine animals and plant organisms. The vast oil shale deposits of the north central United States (see Fig. 7–5) contain a substance, kerogen, which can be distilled to yield a combustible hydrocarbon.

When a fossil fuel is burned it is altered by oxygen,

and the products of this combustion are CO, CO_2, and H_2O—the same major products that occur in the surface decomposition of an organism. In essence the carbon is being returned to the atmosphere in the form of CO_2 at a very rapid rate. The accumulation of hundreds of millions of years' worth of coal, oil and gas is being returned to the natural cycle in a few hundred years (see Chap. 4).

Fossil fuels are presently being formed at an unknown rate. Traces of hydrocarbons were recently found in relatively young deep-sea sediments, indicating that the formation of petroleum can occur at lower temperatures than was previously believed and within less than a million years rather than several million years. However, even this shorter time scale for natural formation of petroleum is still much too long to be of any use to man. Even if the rate of production of fossil fuels in nature were similar to our rate of consumption, the newly formed deposits are dispersed and unavailable. In any event, human consumption is calculated in terms of hundreds or thousands of years, while production of *extractable* fossil fuels requires millions of years (Fig. 7–4).

Thus fossil fuels are nonrenewable resources not only because their supply is limited but also because when they decompose they lose their ability to provide energy. However, new technologies may soon provide ways to extend fossil fuel supplies.

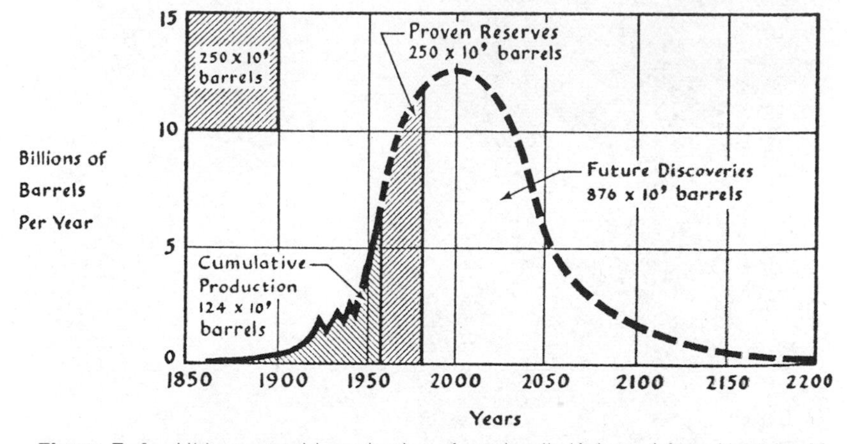

Figure 7–4. Ultimate world production of crude oil. (Adapted from Hubbert, M. King: Energy Resources. Washington, D.C., Committee on Natural Resources, National Academy of Sciences–National Research Council, Publication 1000D, 1962.)

Extension of Nonrenewable Resources

As the preceding discussion indicates, the term "nonrenewable" does not apply equally to all resources. Other factors such as exhaustibility, recyclability and energy cost must be considered. Certain nonrenewable resources can be *extended* by recycling, and application of technologic advances can lower energy costs, making other reserves essentially inexhaustible. For example, the use of oil shales (Fig. 7–5) as a source of energy could become economically feasible by 1985.

The continental shelf areas of the world probably contain large amounts of petroleum and gas. In the more distant future, the possibility of petroleum and gas from deep-sea areas cannot be discounted. Finally,

Figure 7–5. Shale oil pilot plant in Colorado. (Science *183*:642, 1974.)

development of alternate energy sources could and should lead to a situation in which fossil fuels are not used extensively as fuels but rather in production of petrochemicals (chemicals derived from petroleum or natural gas). Such a program would help to extend fossil fuel reserves, but the extension is limited, and eventually these resources will be exhausted (see Fig. 7–4).

Discoveries regarding nitrates provide a classic example of new technology changing the availability of a raw material. In the nineteenth and early twentieth centuries, the principal world supply of nitrates came from the nitrate deposits of Chile. In the early part of the twentieth century, the Haber-Bosch process, which converts the nitrogen of the air to ammonia, was invented. Thus nitrates could be manufactured directly from an inexhaustible reservoir, and we are no longer dependent on a nonrenewable resource. When natural gas became cheap and plentiful in the 1950's, ammonia was produced from natural gas because this method was cheaper than the Haber process. Thus at least in part we retrogressed by needlessly consuming a nonrenewable resource. This is a clear case of the abuse of natural resources owing to a lack of regulatory general policy.

Recycling can extend the availability of a nonrenewable resource. Gold has been recycled since ancient times because its value always exceeded the recycling cost. Most other metals are marginal in their recycling cost relative to production cost. As new sources of metals become scarcer and the search for them more costly, the idea of recycling becomes more attractive. Table 7–2 provides some recycling statistics.

Copper is a metal that can be and is recycled. Approximately 90 per cent of copper is used in ways that do not alter its original form (Fig. 7–6). The rest, however, is difficult (and expensive) to recycle. For example, copper used in fungicides is too dispersed to be reclaimed, and the cost of recovering copper put into landfills is prohibitive.

A significant amount of used copper can be recovered from various sources to help meet the evergrowing demand, but newly mined copper will be needed to replace the unreclaimable supplies and the supplies in current use and to satisfy new demand. Thus recycling is helpful but does not indefinitely postpone the time when copper supplies are exhausted.

TABLE 7-2. TOTAL ANNUAL U.S. MINERAL SUPPLIES AND USES*

	Domestic Primary Production	Old Scrap**	Total Amount Used†	Projected Primary Demand For 2000 AD
Aluminum	450,000	160,000	4,947,000	23,800,000
Copper	1,380,000	422,000	2,122,000	4,860,000
Iron	49,000,000	35,400,000	110,200,000	138,000,000
Lead	517,000	450,000	1,207,000	1,390,000
Mercury	597	378	1995	2730
Bituminous coal & lignite	504,000,000	0	504,050,000	900,000,000
Natural gas (dry)	446,000,000	0	465,080,000	1,030,000,000
Petroleum (including natural gas liquids)	508,190,000	0	668,190,000	1,490,000,000
Uranium	9515	0	9515	55,800

*From National Academy of Science Committee on Geological Sciences: The Earth and Human Affairs, San Francisco, Canfield Press, 1972.

**Preliminary data for 1971, given in metric tons. (From the First Annual Report of the Secretary of the Interior under Mining and Minerals Policy Act of 1970 (P. L. 91-631); March, 1972.)

†Including government stockpiling, industry stocks and exports. The difference between domestic supply and demand is met by foreign imports.

It is possible that nonrenewable resources can be extracted from reservoirs of practically inexhaustible materials. Energy cost and availability will largely determine implementation of this prospect. Consider the major elements iron and aluminum. These metals occur in reasonable abundance in common rocks and minerals. For example, any common mud contains ap-

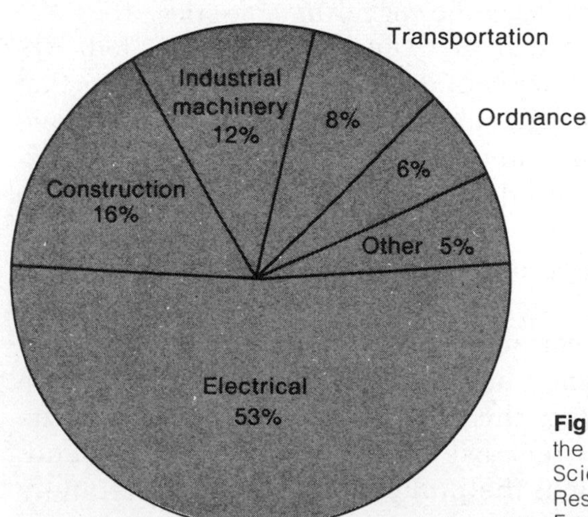

Figure 7–6. Principal uses of copper by industry in the United States in 1970. (From National Academy of Science Committee on Geological Sciences, National Research Council: The Earth and Human Affairs. San Francisco, Canfield Press, 1972.)

proximately 10 per cent aluminum. The average crustal abundance of aluminum is greater than 8 per cent. Clearly, then, aluminum is practically inexhaustible if it can be extracted at reasonable economic and environmental energy cost. The nonrenewable but recyclable resources are energy-dependent in the sense that given very cheap sources of energy, these resources could be almost indefinitely extended.

4 THE NO-GROWTH SOCIETY

The belief that unlimited, unregulated growth—whether of industry, population or consumption of natural resources—is beneficial has become dangerously prevalent and is to a great extent responsible for the current energy crisis. Recently, however, there has been a heartening movement to examine alternative policies that would establish a zero-growth or equilibrium society.

The Limits to Growth is one of the more dramatic studies that argues for a no-growth society. This report predicts a catastrophic result for mankind if unchecked growth continues:

> Although we have many reservations about the approximations and simplifications in the present world model, it has led us to one conclusion that appears to be justified under all the assumptions we have tested so far. The basic behavior mode of the world system is exponential growth of population and capital, followed by collapse. As we have shown in the model runs presented here, this behavior mode occurs if we assume no change in the present system or if we assume any number of technological changes in the system.

There is no definitive answer to the long-range question regarding growth versus no-growth societies, but it is generally conceded that an immediate and vast effort to face these problems must be made.

8

ENERGY RESOURCES

This stone lion in Nara, Japan, was quarried and sculpted largely by hand labor. Quarrying and crushing of stone and subsequent construction with stone products today uses large quantities of fossil fuel.

1 ENERGY AND SOCIETY

Energy and iron are the twin pillars supporting today's technology. Of the two, energy is by far the more important. Without abundant iron, technology would progress more slowly, and common features of our life such as automobiles and high-rise buildings would be less common, but technology would progress nevertheless.

Without abundant energy, civilization would permanently mark time at the level of the previous century. Iron and other materials, which require large quantities of energy to produce, would be in much shorter supply. Additional energy is used to roll steel and shape it into automobile bodies, pipes and so forth. Energy is required to transport iron from factory to consumer.

Energy use is pervasive. Some of these uses, such as gasoline for automobiles, are obvious; others are indirect and complex. Let us examine a few examples of how energy affects our lives.

Energy and Food Production

The relationship of energy to food production provides a good example of an indirect use of energy. We usually think of food as a renewable resource—which it is—but the energy used in food production is nonrenewable. Thus the *rate* at which food can be produced is extremely energy dependent.

In the United States and most other developed countries, the availability of energy determines the rate of food production, which, in turn, determines the cost of food. Thus the price of food goes up when energy is in short supply. The beef shortage in 1973 and resultant increases in meat prices demonstrate this principle.

In developing countries, the situation is far more serious. Shortages of food often mean starvation. Expensive energy means that tractors cannot be built and fields cannot be efficiently tilled, irrigated or fertilized.

To see how food production and energy are related, let us examine corn production in the United States. Corn is the most important grain crop in the United States. Worldwide it is the third largest food crop, and it is second to wheat among cereal grains. In the United States, the yield of corn per acre increased from 26 bushels in 1909 to 87 bushels in 1971 (Fig. 8–1). The sharpest rise in yield occurred in the 1950's and was

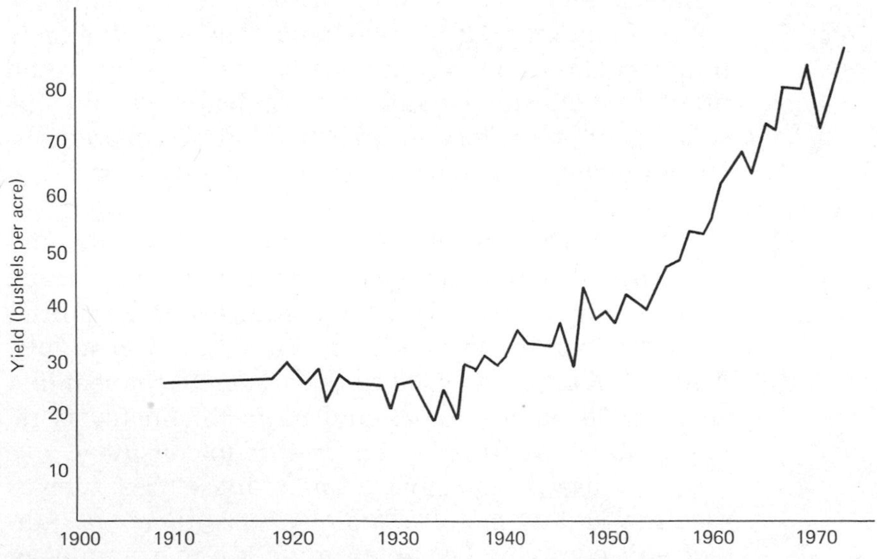

Figure 8–1. Corn production in the U.S., 1909–1971. (Adapted from Pimental, D., et al. Food production and the Energy Crisis. Science *182*:443–449, 1973.)

due to planting of hybrid corn and increased fertilization. Optimum yields from hybrid grain require relatively large amounts of fertilizer. The cost of fertilizer depends to a great degree on the cost of energy required to make it. High hybrid yields are economically justifiable only as long as fertilizers are cheap. *Thus, availability of cheap energy is either directly or indirectly responsible for almost all of the increase in corn yield during the twentieth century.*

Petroleum price increases in 1973–74 caused increases in fertilizer costs. This may spell disaster for the "Green Revolution," which promised to ease chronic food shortages in much of Asia.

While corn yields increased 240 per cent from 1945 to 1970, labor input per acre decreased 60 per cent. Mechanization—made possible by abundant energy—accounted for the reduction in labor input. Mechanization and increased use of fertilizers and pesticides also resulted in a large leap in energy consumption during the same period of time. In 1945, about one million kilocalories of energy were used in United States corn production. By 1970, this figure had reached about three million kilocalories. The output in terms of number of food-calories-out for number of energy-calories-in decreased from 3.7 to 2.8. This means that although total output increased dramatically, energy required to grow the corn increased at an even greater rate.

Furthermore, because energy (obtained primarily from coal, petroleum and natural gas) is such a major factor in corn production, recent increases in energy costs will almost certainly cause increases in cost of corn production. Energy used in corn production in the United States is equivalent to roughly 110 billion gallons of gasoline. To feed the 1973 world population a diet equal to the average American diet would require the equivalent of almost 500 billion gallons of gasoline. Known recoverable reserves of petroleum amount to the equivalent of about 12,000 billion gallons of gasoline, an amount sufficient to provide the required food production energy for only 24 years.

However, petroleum is not the only fuel available, many factors other than energy are involved in world food production and the preceding calculation is little more than an academic exercise. But from it we can draw several important conclusions:

1) Energy plays a major role in food production.

Energy availability affects the amount of food available and the cost to the consumer.

2) Energy availability imposes important restrictions on the amount of food that can be produced worldwide.

Some of the required energy can be supplied at a reasonble cost from fuel other than fossil fuels, but fuel to power tractors and other essential machinery and to make pesticides is closely related to fossil fuel availability.

Energy and Technology

In the preceding section, we observed that energy use is ubiquitous in contemporary society. The role of energy in corn production exemplifies the subtle, often unsuspected ways in which availability of energy affects our lives. In this section we will briefly consider the opposite end of the spectrum and review the correlation between energy, technology and history.

From prehistory until the middle 1800's, man depended upon renewable resources for his energy, which he obtained from burning wood, from animals and slaves and from wind and water. The demand for energy played a large part in the wars of the Roman Empire, because new conquests brought new slaves, new goods and, possibly most importantly, new forests. As the Roman Empire expanded, Mediterranean lands were deforested to supply wood for Roman iron furnaces. Near the end of the Empire, lack of local forests required relocation of Roman smelters to Germany's abundant forests. Romans were forced to import iron from their former provinces. We are often told that Roman civilization collapsed because of moral decay; actually exhaustion of local energy resources was probably a more important factor.

In the early 1800's, the English repeated the Roman mistake and deforested the British Isles to fire their iron smelters. However, the English were luckier than the Romans. They were able to substitute coal for wood, and coal resources were abundant in Great Britain—a fact which prevented the collapse of the infant English iron industry and ultimately provided bone and sinew for the expansion of the British Empire.

In the twentieth century we have seen the United

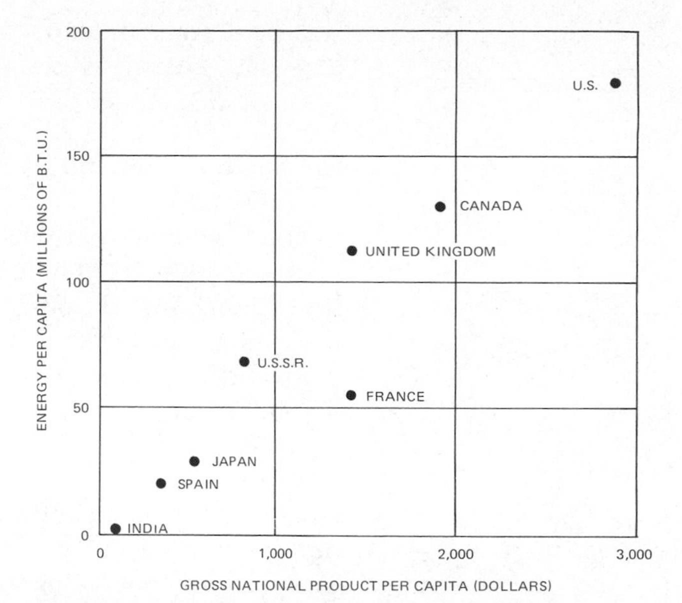

Figure 8–2. Relationship between national consumption of energy and gross national product for 1961. (Adapted from Singer, S. F.: Human energy production as a process in the biosphere. Sci. Am. *223*(3):175–190, 1970.)

States go from energy self-sufficiency to energy dependency. A small percentage decrease in petroleum imports in 1973–74 sent shivers through the economy. Gasoline sales were restricted, heating oil was scarce and iron and other manufactured goods became scarce or temporarily unavailable, all because of an almost insignificant reduction in available energy. Clearly, material, technologic and probably cultural progress are closely related to energy availability.

The correlation between energy and technology is best illustrated by a graph showing per capita *gross national product* as a function of *per capita energy consumption* (Fig. 8–2). Developed countries such as the United States, United Kingdom and U.S.S.R. have a large per capita energy consumption and a large per capita gross national product. (Japan now has both a much larger gross national product and a much larger energy consumption today than in 1961.) Spain and India have small per capita gross national products and small energy consumption. Developing countries, if plotted, would lie in the lower left corner of Figure 8–2.

Correlation between cultural and social levels and energy consumption is more difficult because cultural and social progress cannot easily be defined. Few people would question the fact, however, that undernourished or starving people in less developed countries lack the cultural and social opportunities of their well-fed, well-dressed counterparts in countries with abundant energy.

Energy and the United States

The emergence of the United States as a world power is closely related to energy resources. Energy consumption in the United States increased very slowly during the first few decades following independence (Fig. 8–3). During this time, the United States exerted negligible force in world affairs. In the early 1880's coal replaced wood as the primary fuel in the United States (Fig. 8–4), and the United States began to make its weight felt. The period 1850–1880 might be considered the infant stage of the industrialization of America. The "adolescent" stage extends from 1880 to about 1940. During this period World War I gave a big boost to industrial development in America, but the Great Depression stagnated growth during much of the time between the end of World War I and 1940. The latest and most powerful surge in industrial growth began at the start of World War II and continued almost unabated until the late 1960's.

Barring unexpected breakthroughs in areas of nu-

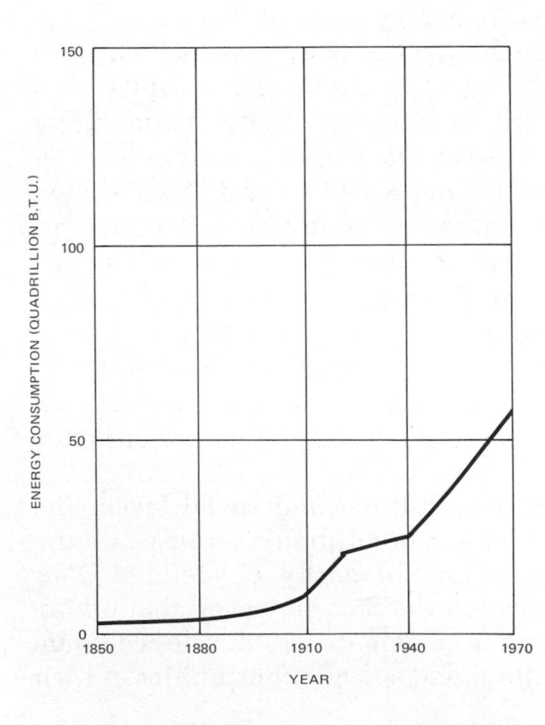

Figure 8–3. Energy consumption in the U.S., 1850–1970. (Adapted from Singer, S. F.: Human energy production as a process in the biosphere. Sci. Am. *223*(3):175–190, 1970.)

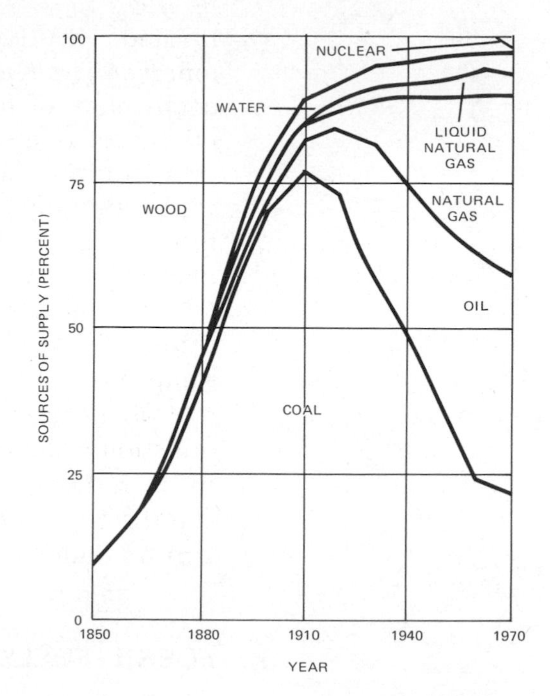

Figure 8–4. Changing patterns in sources of energy in the U.S. between 1850 and 1970. (Adapted from Singer, S. F.: Human energy production as a process in the biosphere. Sci. Am. *223* (3):175–190, 1970.)

clear power technology concerned with *reactor design* and *environmental safety,* the immediate future will be years of either limited growth, no growth or, perhaps as in 1974, years of retrenchment. A strong possibility exists that we are at a turning point. The period of increasing material goods, increasing energy and increasing ease may be ending. If so, new levels of availability of these factors will require economic, political and psychologic adjustments.

The United States now relies on oil and natural gas for about three fourths of its energy supplies (Fig. 8–4). This heavy reliance sharply contrasts with the fact that about 90 per cent of United States fossil fuel reserves are in the form of coal. We must rearrange our priorities of energy use.

Supplies of nuclear fuel are virtually unlimited, but serious technologic and environmental problems remain unsolved. Other sources of power such as running water and geothermal heat can be of value locally but cannot provide enough energy worldwide to replace fossil fuels. The potential of solar energy is great, but so is the capital investment required to convert solar power into stored or useful energy on a large scale.

If coal reserves are far greater than natural gas and petroleum reserves, and nuclear fuel reserves are unlimited, why is use of oil and natural gas still so extensive? The reasons are complex. Some are related to occurrence of fuel reserves and extraction problems, others are related to pollution and still others are geopolitical in nature.

In following sections of this chapter we will discuss the origin, occurrence, extraction, processing and marketing of energy from coal, gas, petroleum, radioactive minerals, the sun, water, geothermal heat and tides. We will briefly discuss a few solutions and scenarios proposed for future consumption.

It is probably not possible to completely answer questions concerning energy usage, but information in the following sections will provide a basis for understanding problems and proposed solutions to the United States energy situation.

2 FOSSIL FUELS

Fossil fuels appear relatively late in the earth's history. For the first 4 billion years in the earth's 4.5 billion year history, plants and animals were so small, primitive and widely distributed that no significant amounts of dead plant and animal debris—the source material of fossil fuels—accumulated in sedimentary rocks. The name, *fossil fuel,* derives from the fossil nature of this plant and animal debris.

About 500 million years ago, plant and animal evolution "exploded," and many new and widespread species appeared (Fig. 8–5). The biologic explosion coincided with and was intimately related to the appearance of large quantities of free oxygen in the atmosphere. Increasing quantities of free oxygen spurred the evolution of oxygen-breathing animals and thereby increased the volume of biologic debris available for entrapment and conversion to fossil fuels. The cause of the sudden increase in free oxygen is not clearly established, but it probably resulted from the evolution of plants and animals that used CO_2 (carbon dioxide) in cell or shell construction. This removed large quantities of CO_2 from the atmosphere and caused a sharp increase in the relative amount of free oxygen in the atmosphere. Much of the CO_2 used in shell formation remains trapped in sedimentary rocks in the form of calcium carbonate ($CaCO_3$) or limestone (Fig. 8–6).

YEARS BEFORE PRESENT	LITHOSPHERE	BIOSPHERE	HYDROSPHERE	ATMOSPHERE
20 MILLION	GLACIATION	MAMMALS DIVERSIFY GRASSES APPEAR		OXYGEN APPROACHES PRESENT LEVEL
50 MILLION				
	COAL FORMATION VOLCANISM			
100 MILLION		SOCIAL INSECTS, FLOWERING PLANTS APPEAR		ATMOSPHERIC OXYGEN INCREASES AT FLUCTUATING RATE
		MAMMALS APPEAR		
200 MILLION				
	GREAT VOLCANISM			
	COAL FORMATION		OCEANS INCREASE IN VOLUME	
		INSECTS APPEAR LAND PLANTS APPEAR		
500 MILLION				
	GLACIATION SEDIMENTARY CALCIUM SULFATE	METAZOA APPEAR RAPID INCREASE IN PHYTOPLANKTON		OXYGEN AT 3-10 PRESENT OF PRESENT ATMOSPHERIC LEVEL
1 BILLION	VOLCANISM		SURFACE WATERS OPENED TO PHYTOPLANKTON	OXYGEN AT 1 PERCENT OF PRESENT ATMOSPHERIC LEVEL, OZONE SCREEN EFFECTIVE
		EUCARYOTES APPEAR		OXYGEN INCREASING, CARBON DIOXIDE DECREASING
2 BILLION	RED BEDS	ADVANCED OXYGEN-MEDIATING ENZYMES	DIFFUSION OF OXYGEN INTO ATMOSPHERE	
	GLACIATION			
	BANDED IRON FORMATIONS OLDEST SEDIMENTS OLDEST EARTH ROCKS	FIRST OXYGEN-GENERATING PHOTOSYNTHETIC CELLS PROCARYOTES ABIOGENIC EVOLUTION	START OF OXYGEN GENERATION WITH FERROUS IRON AS OXYGEN SINK	
5 BILLION	(ORIGIN OF SOLAR SYSTEM)			NO FREE OXYGEN

Figure 8–5. Chronology of the earth. Fossil fuels could not form in abundance until plants and animals were abundant. Plant and animal evolution is closely related to the abundance of free oxygen in the atmosphere. (Adapted from Cloud, P., and Gibor, A.: The oxygen cycle. Sci. Am. *223* (3):111–123, 1970.)

About 400 million years ago, land plants appeared. They evolved rapidly and began to contribute to the amount of free oxygen in the atmosphere and to the inventory of biologic debris available for burial. Although some coal and petroleum occur in older rocks, *all major deposits of oil and coal were formed after the appearance of land plants.*

Formation of fossil fuels is a complex process in

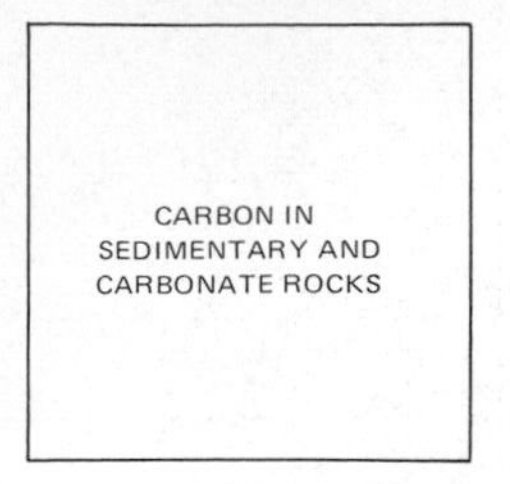

Figure 8–6. Relative amounts of carbon in rocks and in the atmosphere, hydrosphere and biosphere. (Adapted from Cloud, P., and Gibor, A.: The oxygen cycle. Sci. Am. *223* (3):111–123, 1970.)

which the sun is the ultimate source of energy. A small fraction of solar energy is trapped in *photosynthesis*, the process whereby plants fix carbon in their cell structure.

The chemical reaction in photosynthesis, whether it takes place on land or in the sea, is simple and consistent:

$$\underset{\text{(input)}}{CO_2 + 2H_2A + \text{light}} \rightarrow \underset{\text{(output)}}{CH_2O + H_2O + 2A + \text{energy}}$$

Usually $2A$ is free oxygen, O_2, and H_2A is common water, H_2O, but some bacteria use sulfur, an organic radical, or nothing at all in the place of oxygen. CH_2O is formaldehyde, the simplest organic compound and the first step in the ladder of organic chemistry. (Proteins occur near the top of the ladder.)

The term "energy" refers to energy stored by the organism. When the organism oxidizes, decays, or "burns," the energy is released as heat. This heat is the source of all fossil fuel energy.

The process of converting CO_2 to formaldehyde or other organic compounds is called *carbon fixation*, and the rate of fixation is called *productivity*.

Productivity varies widely. Rapidly growing tropical rain forests annually fix 1 to 2 kg of carbon per square meter (kg/m^2) of land surface. Arctic tundra may fix only 0.01 to 0.02 kg/m^2 annually. Land plants fix an estimated 25 billion metric tons of carbon annually and marine vegetation an estimated 40 billion tons annually.

On land and in the sea, animals eat vegetation, transform organic molecules into more complex molecules and release a portion of the CO_2 back into the atmosphere. A fraction of one per cent of decaying animal debris is trapped in sedimentary rocks. Oxygen in waters circulating through near-surface rocks oxidizes much of this trapped organic material. *The small amount that is not oxidized evolves chemically into fossil fuels.*

Coal: Fuel of the Past, Present and Future

The discovery that coal could be used in the place of charcoal to smelt iron ore triggered the industrial revolution. Coal-fired steam engine whistles serenaded the passing of the era of the horse as man's primary source of transportation, which lasted for more than a millennium. Today, coal is to the steel industry what bread is to man. In addition, coal is a major contributor to man's energy pool. If nuclear energy problems remain unsolved, coal may again become man's primary energy source as natural gas and petroleum resources are depleted.

Although coal is currently man's most abundant energy resource, it also presents many problems. Transportation of coal is expensive in comparison with transmission of oil and gas through pipelines. Consequently, coal's thermal energy is usually converted to electric energy. But such conversion is intrinsically inefficient. Fifty-eight per cent of coal's energy is lost as waste heat (other fuels leave behind even greater percentages), and only 42 per cent is transmitted as electricity.

Burning coal emits *particulate material* (fine ash), *nitrogen oxides* and *sulfur oxides*. Sulfur oxides are the most noxious because they combine with water to form sulfuric acid. In the air, sulfuric acid attacks almost everything exposed to it including building surfaces, vegetation and lung tissue. A large fraction of the world's coal reserves contain excessive quantities of sulfur in the form of the mineral pyrite or fool's gold, FeS_2. The cost of removing the sulfur from these coals raises overall production costs above the cost of oil and natural gas, and hence they are little used.

Much coal lies buried deep in the ground. The danger of explosions in mines and respiratory diseases resulting from breathing coal dust make underground mining one of the world's most hazardous occupations.

Coal is a dirty, unpleasant source of energy. The costs and difficulties of obtaining clean energy from coal are the main reasons for its replacement by oil and gas as the world's largest source of energy. As oil and gas become less plentiful and more expensive, coal is the best bet for short-term energy until problems of nuclear and solar energy are solved. But the price of using coal will be high.

Geology of Coal

Coal consists of *compressed remains of land plants*. The earliest preserved land plants occur in rocks of the

Silurian Period (440 to 400 m.y. ago). In the Carboniferous Period (350 to 270 m.y. ago), advancing plant evolution, tectonics and favorable geography combined to form widespread swamps in tropical and subtropical regions. The resulting coal fields are so extensive and ubiquitous in rocks of this period that the name of the period, Carboniferous, derives from the abundant coal.

Approximately 400 million years ago, two or possibly three large continental plates collided, forming the Appalachian Mountains of North America and the Caledonides of Europe, much as the more recent collision of the Indian subcontinent with Asia resulted in the formation of the Himalayas. Tectonic activity associated with the collision continued for 200 million years and culminated in a final spasm of deformation, metamorphism and intrusion called the Appalachian orogeny in North America and the Hercynian orogeny in Europe.

The mountainous *geosuture* of the plate collision trended more or less east-west at the time of the collision. Furthermore, the geosuture lay largely within the tropical, subtropical and warmer portions of the temperate zone. Luxuriant vegetation remains suggest that rainfall was abundant. Slow undulations of the earth's surface in the vicinity of the geosuture caused formation of widespread shallow basins, which became the loci of the great Carboniferous coal accumulations (Fig. 8–7).

Conditions were right again for coal formation during the Cretaceous Period and much of the Tertiary Era (135 m.y. to the present). For reasons which will be

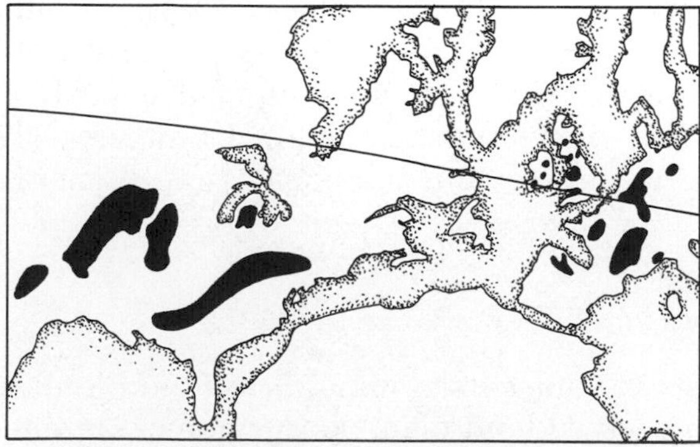

Figure 8–7. North America and Northern Europe in Carboniferous time (350–270 m.y.). The Appalachian-Hercynian seam or *front* crosses the equator at a small angle. Coal fields of northern Europe formed in broad shallow basins along the front. These deposits are the source of much of Euro-American industrial coal.

discussed later, the average quality of Cretaceous and Tertiary coals is lower than that of Carboniferous coals.

Coalification vs. Rotting. In the normal course of events, vegetation dies and rots. In the rotting process, the carbohydrates oxidize, releasing their heat of formation into the surroundings and leaving a chemical residue of water and carbon dioxide.

In waterlogged environments such as swamps, the oxidation process does not go to completion. The limited supply of oxygen in these waters impedes bacterial and fungal activity which promote rotting. Instead of decay, the softer plant material is transformed into a dark brown jellylike mass of humus. The soft brown humic material invades the cells of wood fragments as well as cracks and crevasses between fragments with the result that the whole mass of material is effectively protected from oxygen-bearing circulating waters. An accumulation of humified products along with resin, pollen and other more resistant materials is called *peat* and is the first stage in the coalification process. Peat contains significant amounts of water and will not burn unless it is air-dried for quite some time.

Formation of peat involves chemical reactions in the vegetable matter including the formation of *methane*. Methane is a highly flammable gas that burns with a pale blue flame sometimes seen above the surface of a marsh or bog. Methane also accumulates in coal mines, sometimes in substantial quantities. Many mine explosions derive from inadvertent ignition of methane.

Peat is forming today in the swamps of the Atlantic seaboard of the southeastern United States—Dismal Swamp, Okefenokee, the Everglades and others. Peat averages 7 feet thick in the Dismal Swamp of North Carolina and Virginia. A swamp in northeastern Sumatra has a known peat accumulation of 30 feet. Borings in the Ganges delta of India revealed numerous peat horizons separated by layers of sand and clay. These swamps, all located in tropical or subtropical climates, probably resemble the coal-forming swamps of the Carboniferous and Cretaceous periods except for differences in plant species.

A peat bog becomes a coal seam only if three events occur. First, peat must accumulate for a long period of time. Approximately one foot of peat is required to make one inch of coal. Most seams in United States coal fields range from 2 to 10 feet and thus represent 24 to 120 feet of peat. The Adaville seam in Wyoming is 84 feet thick; approximately 1000 feet of peat went into the formation of this seam.

Second, the area must subside during and after peat formation. Uplift of the peat results in drying, oxidation and eventual destruction; continued subsidence results in burial and preservation of the peat.

Third, the peat must be buried beneath a substantial thickness of sediments. The weight of overlying sediments contributes pressure to compact the peat. Deep burial results in heating of the peat. These two factors ultimately determine the quality or *rank* of the resultant coal.

T-T-P. Time, temperature and pressure. These convert peat into high-rank hot-burning coal.

There are three ranks of coal: lignite, bituminous and anthracite. Lignite is the first stage after peat, and anthracite is the last stage before graphite or pure carbon. In general, as per cent moisture goes down, the C/H ratio (or per cent fixed carbon) goes up, and heat per pound increases from lignite to anthracite.

Lignite and subbituminous coals, which account for 56 per cent of United States coal reserves, are the lowest quality coals. They are soft, tend to be brownish in color and have not yet been widely mined in the United States. Lignite contains less than 40 per cent fixed carbon and up to 40 per cent moisture. It produces only 5500 to 8000 British Thermal Units (BTU) of heat per pound. American lignites are relatively free of sulfur but produce large amounts of ash upon burning.

Subbituminous coals are intermediate between lignite and bituminous coals. They contain 40 to 60 per cent fixed carbon, 20 to 30 per cent moisture, up to 15 per cent ash, and they produce 8000 to 12,000 BTU per pound.

Bituminous coal is the most widely used rank of coal in the world. It is harder and blacker than lower rank coals but lacks the hard glossy sheen of anthracite. In contrast to lower grades, which tend to be very smoky, bituminous coal burns with moderate smoke and produces 11,000 to 15,000 BTU per pound. Many United States bituminous coals contain relatively large percentages of sulfur (Fig. 8–8). Sulfur is the

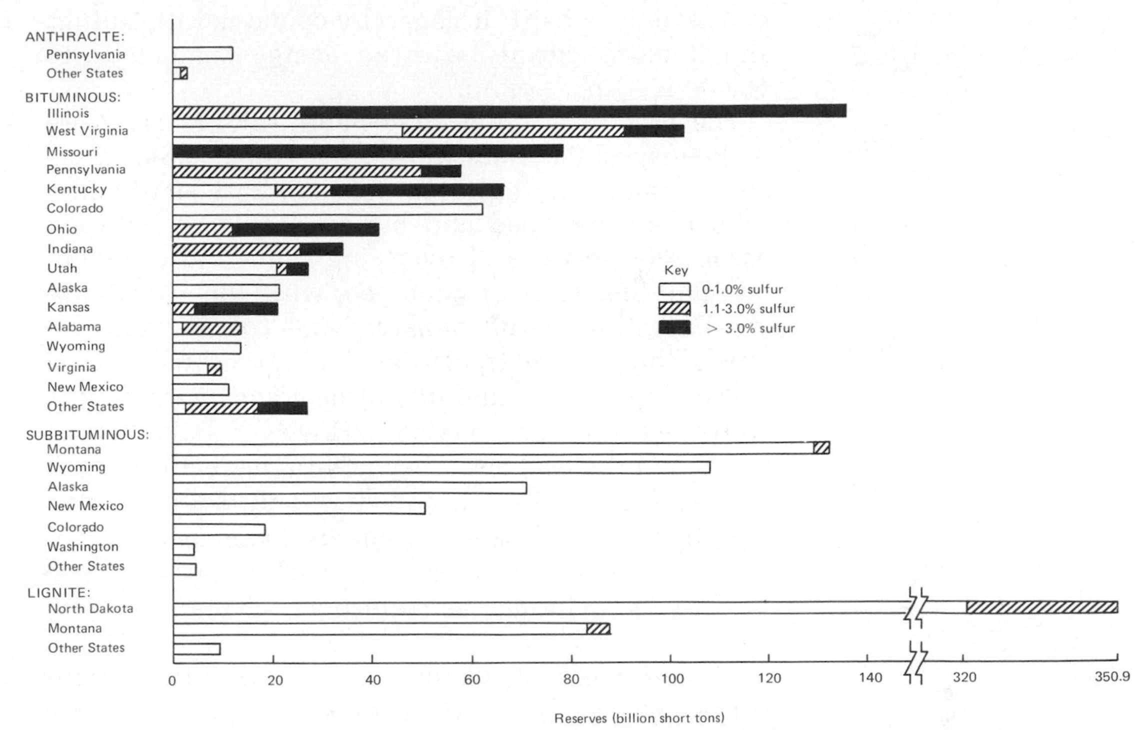

Figure 8–8. Sulfur content of U.S. coal reserves broken down according to states. Eastern and midwestern bituminous coals are disagreeably sulfurous. Eastern anthracite and western coals are relatively free of sulfur. (Adapted from Hammond, A. L., et al.: Energy and the Future. Washington D.C., American Association for the Advancement of Science, 1973, p. 184.)

major source of pollution resulting from use of coal. More than 70 per cent of the United States coal reserves contain more than 1 per cent sulfur, and 43 per cent contain more than 3 per cent. High sulfur content makes burning of these coals either unpleasant or, in some cases, illegal. Nevertheless, sulfurous Appalachian coal fields account for 70 per cent of United States production.

Relatively pure bituminous coal cooked in an oxygen-free oven to drive off moisture and volatiles produces *coke,* which is used in smelting iron. Discovery in nineteenth-century England that bituminous coal could be "coked" permitted rapid expansion of the British iron industry. England had relied first on charcoal, then on anthracite for smelting. Both were relatively scarce, but bituminous coal was abundant. Bituminous coal suitable for use in iron production is called "coking coal."

Anthracite is the highest rank of coal. It contains 86 to 98 per cent fixed carbon and less than 2 per cent moisture. It is hard, it generally contains little sulfur, and it burns cleanly with an energy content up to 16,000 BTU per pound.

The principal United States anthracite reserves are in Pennsylvania. Here during the Appalachian *orogeny*, or mountain-building episode, coal strata were folded, downwarped and buried deeply beneath the earth. The pressure of overlying strata and of mountain-building forces, combined with the earth's internal heat, drove off moisture and volatile constituents of the coal and transformed it into anthracite. Coal subjected to lesser amounts of heat and pressure was converted to bituminous coal. However, only a small fraction of the coal strata lay within the region of high heat and pressure. Consequently, anthracite reserves are limited. Coal of this rank is too expensive for ordinary heating and most industrial purposes. Its use is restricted to industries requiring a hot clean-burning coal.

Western coals are younger than coals of the eastern United States. None have been subjected to heat and pressure sufficient to convert them to anthracite rank, and only a few are of bituminous rank. They are widespread, abundant and constitute much of the United States reserve.

Geography of Coal

As pointed out previously, coal formation requires a favorable combination of geologic and climatic factors. Furthermore, coal is a relatively recent addition to the world fuels. Thus it should be no surprise that coal is irregularly distributed.

During the period of coal formation (economically significant deposits date back only about 350 m.y.), most of Africa, Australia, India and South America lay further to the south than at present. In addition, these continents and subcontinents were involved in relatively little tectonic activity during coal-forming time. North America, Europe and Asia fared much better.

In the United States, Appalachian mountain building resulted in formation of the eastern coal fields. Broad downwarps occurring at about the same time—probably distal effects of mountain building—gave rise to the midwestern coal deposits. Formation of the Oua-

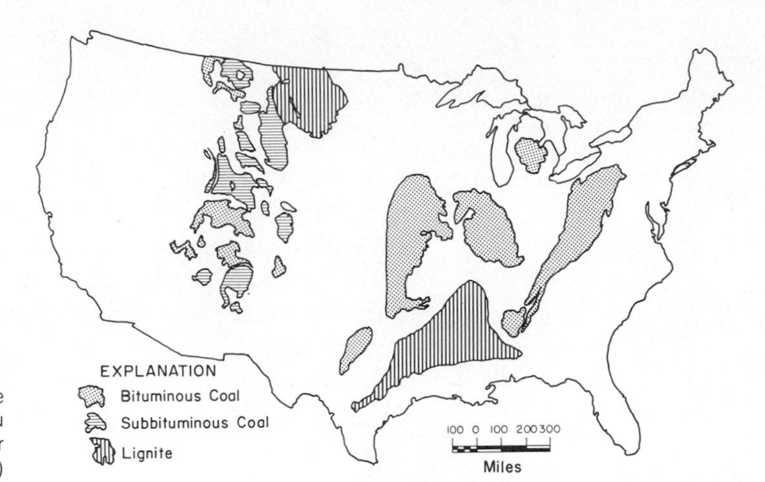

Figure 8–9. Coal fields of the United States. (From U.S. Bureau of Mines Information Circular 8535. U. S. Dept. of Interior, 1972.)

chita Mountains, a southwestward extension of Hercynian-Appalachian mountain building, resulted in formation of coal basins in Oklahoma, Arkansas and Texas. Locations of these coal fields are shown in Figure 8–9.

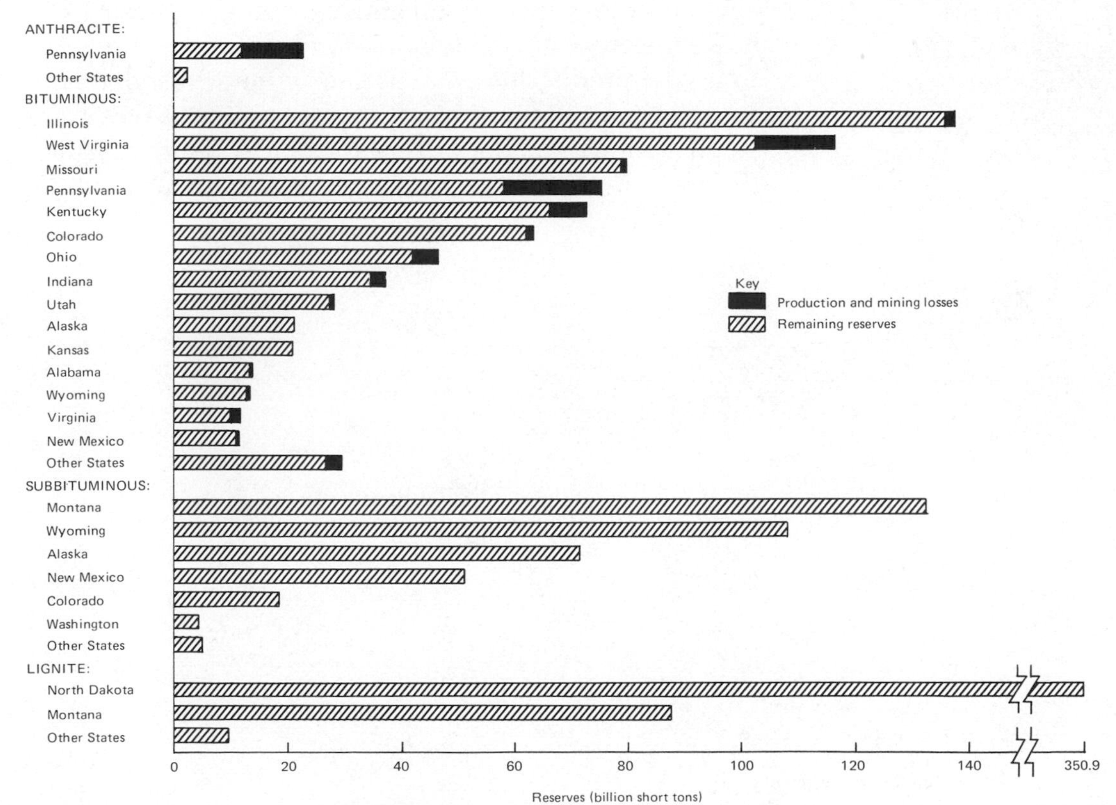

Figure 8–10. United States coal production and coal reserves. (Adapted from Hammond, A. L., et al.: Energy and the Future. Washington, D.C., American Association for the Advancement of Science, 1973.)

Uplift of the Rocky Mountains formed coal basins from New Mexico to Canada. Coals in these basins are mostly low rank, but reserves are large (Fig. 8–10) and sulfur content is low. Arid surroundings, however, pose special environmental problems in their extraction (see later in this chapter).

Small coal basins occur in the far western United States. Alaska also contains significant amounts of coal, some of which is high-rank bituminous, but transportation costs have precluded their economic exploitation.

Vast deposits of high-rank coals in China and Manchuria were in part responsible for Japan's aggressions in World War II, because Japan lacked significant domestic coal resources. Similarly, the German drive into the Caucasus was motived by a desire for iron, petroleum and coal deposits in that region.

Coal Mining

Coal is obtained from deep underground mines and from surface mining, mainly strip mining. Figure 8–11 shows the areas of underground and surface mining. In 1971, coal production from surface mines exceeded for the first time production from underground mines.

Strip mining has the advantage that it obtains virtually all the coal in the area, whereas underground

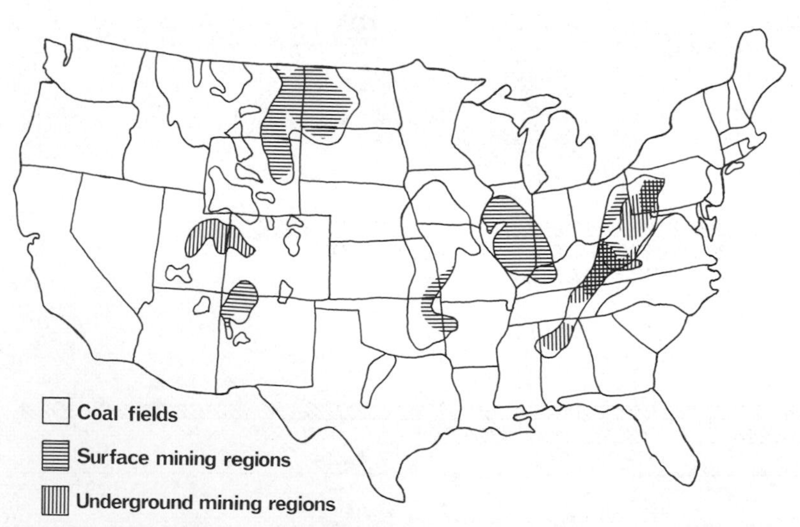

Figure 8–11. Regions of surface and underground mining in the United States. (Adapted from Walsh, J.: Problems of expanding coal production. Science *184*, 336–339, 1974.)

mining leaves a large fraction of the coal underground. Strip mining is also generally cheaper than underground mining. In the early 1960's mammoth drag lines with bucket capacities of up to 220 cubic yards were introduced. The use of these and other large machines, some weighing 8000 tons, has nearly doubled the output per man-day in coal stripping operations.

Underground mining technology has advanced, too. In the 1940's mechanization of cutting, drilling and blasting from the seam face began to replace manual operations. Loading machines gathered up the loosened coal and transferred it to shuttle cars, which carried it to the surface.

The 1950's saw the introduction of a continuous mining machine. A rotating cutter head rips coal directly from the face. Then a conveyor belt transfers the coal to shuttle cars, eliminating the need for a separate loading operation. More than half the coal mined underground today is extracted and loaded by this technique. Coal production has risen from 5 tons per man-day in the 1940's to 20 tons per man-day in the 1970's.

Underground mining is wasteful, however. The most common technique, room-and-pillar mining, leaves behind an average of 43 per cent of the coal, mainly in the form of pillars supporting the roof. Strip mining, on the other hand, recovers almost 100 per cent of the coal.

A promising new technique, *longwall mining*, may lead to substantially increased underground recovery. In longwall mining, a *shearer* travels back and forth 300 to 800 feet across the face of a seam, biting off several feet of coal as it travels. The broken coal falls onto a conveyor belt, which carries it away. A row of mobile roof supports stands parallel to the face just behind the shearer. When the shearer completes a traverse along the face, all roof supports are moved forward and the shearer is adjusted for its next traverse. The wall behind the mobile roof supports is left unsupported and eventually collapses.

Longwall mining has the advantages of greater productivity and safety. Ventilation is simpler, and danger of explosions is reduced. However, in its present stage of development, it cannot be used everywhere. Nevertheless, its introduction should significantly increase overall productivity of underground mining.

Both underground and surface mining result in

significant environmental damage unless active countermeasures are employed. In underground mining, miners suffer four times more fatalities per ton of coal mined than surface miners. Respiratory diseases are much more common. Pyrite (FeS_2) is common in sulfur-bearing coals, and water is pervasive in almost all mines. When combined with oxygen they form iron sulfate and sulfuric acid (H_2SO_4). Pumping water from the mines to prevent flooding dumps acid waters in nearby streams, where they can completely destroy aquatic life. Deforestation to provide roof supports for mines frequently leads to excessive erosion and landslides. Land subsidence, fires in mines and waste heaps add to the problem.

Except for health problems and land subsidence, strip mining suffers the same environmental ills as underground mining. In addition, stripping of overlying strata to reach the coal produces prodigious quantities of crushed rock. In flat, humid areas, reclamation is not overly expensive. Waste rock can be returned to the pit, surface smoothed and grass and trees planted to stabilize the soil and control erosion (Fig. 8–12).

Reclamation of a hillside strip mine presents a more difficult problem. The waste material must be stabilized immediately in order to prevent erosion and landslides. Vertical faces where coal is mined (Fig. 8–13) may require terracing. Steps must be taken to prevent leaching of undesirable chemicals, mainly sulfur, from the waste heap. It may be necessary to haul the waste from the hillside and deposit it in flat areas free from landslide risk.

Reclamation in arid areas poses special problems.

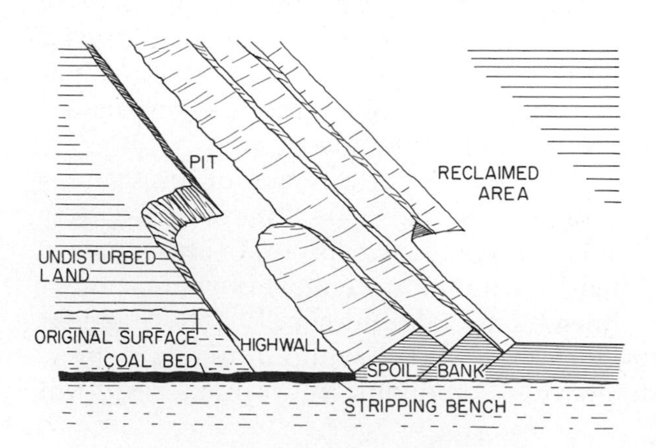

Figure 8–12. Strip mining and concurrent reclamation. (Adapted from McKenzie, G. D., and Utgard, R. O.: Man and His Physical Environment, Minneapolis, Burgess Publishing Co., 1972, p. 338.)

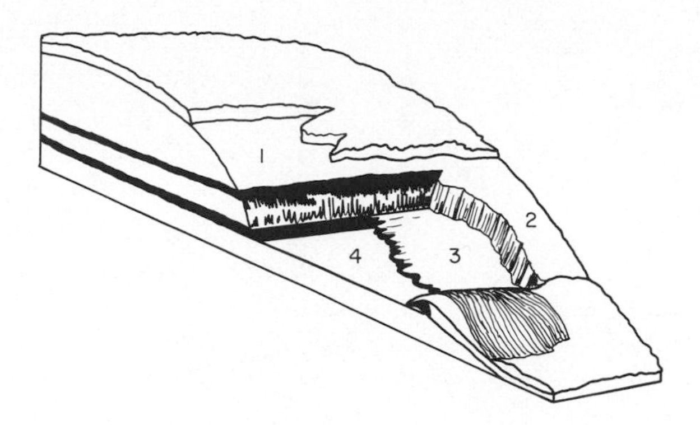

Figure 8–13. Strip mining on a hillside. 1, site preparation; 2, drilling and blasting overburden; 3, removal of overburden; 4, coal excavation. (Adapted from McKenzie, G. D., and Utgard, R. O.: Man and His Physical Environment, Minneapolis, Burgess Publishing Co., 1972, p. 338.)

Destruction of tenuously attached grass cover during surface mining can result in serious erosion. Without adequate water to supply soil moisture, reclaimed strip mines accept new grass cover very slowly. Lack of water may preclude reclamation of western strip mines. Even where water is available, the absence of a commercial source of sage seed, prairie grass seed and seeds of the wild grasses impedes reclamation.

Alternatives to Pollution

The numerous environmental problems associated with coal mining and burning have spurred investigations of alternative, innovative methods of obtaining energy while at the same time minimizing environmental hazards. Many solutions involve new methods of treating coal or its exhaust gases. These include *solvent refining, pyrolysis, magnetohydrodynamic burning* (MHD) and *purification of stack gases.* None of these methods is ready for widespread applicability at the present time.

In solvent refining, the organic components of coal are dissolved, leaving a residue of ash and sulfur. The organic components separated from the solvent form a high-heat clean fuel. Pilot plants may be in operation by the time this book is published.

MHD uses heat from combustion gases to drive a generator. The process is efficient and relatively nonpolluting, but high operating temperatures and the corrosive nature of the gases pose serious metallurgic problems for generator rotors.

Pyrolysis produces char, oil and low-BTU by heating coal in an oxygen-free environment. The feasibility of the process has been demonstrated, but high costs preclude wide applicability.

Underground gasification has also been proposed. Basically, this method entails firing of coal in a controlled manner. Heat drives methane from surroundings into conduits that carry the gas to the surface, where it is trapped and distributed in much the same manner as natural gas. The U.S. Bureau of Mines has pilot operations underway in Wyoming.

The Future

There is very little doubt that use of coal in the United States will increase during the next few decades. Reserves are abundant and relatively cheap to extract. The technologic base (that is, methods of building coal furnaces and coal-fired electric generators) required for use of coal is well developed. No major new technologic developments are necessary, but production of new coal-burning generators will require more than economic expenditures.

The price we must pay for use of coal consists largely of environmental deterioration. In contrast with the well-known technology of burning coal, the technology of suppression of adverse environmental impact is poorly understood and expensive to implement. In addition to the obvious problems of air pollution in the vicinity of coal-burning plants and land deterioration in the vicinity of mining activities, other more subtle hazards exist. For example, the June, 1974, population of the Northern Cheyenne Indian reservation in Montana, which lies in the Fort Union coal basin, was 3190. Coal exploitation will bring miners, storekeepers, engineers and others into the area. Cheyenne culture in the basin will almost certainly be submerged by the flood of new inhabitants, with the result of a net cultural loss to the United States as a whole.

The National Academy of Science recently recognized another hazard associated with coal mining. In a report on potential utilization of low-sulfur western coals, the N.A.S. pointed out that inadequate water supplies in parts of the western United States may inhibit use of coal in certain areas. In some western river basins, water resources are 25 per cent overcommitted already. A single commercial gasification plant producing 250 million cubic feet of burnable gas per day requires 20,000 or more acre-feet of water per year. This amount is not everywhere available in the arid

West. Diversion of water for coal gasification would hurt farmers, ranchers and local industry. A serious conflict of priorities and ethics will develop if gasification is pursued in these regions.

As a general rule, reclamation of strip-mined land is possible only in areas receiving 10 inches or more rain per year because of the difficulty of reestablishing grass and trees in arid regions. Lands located in the Four Corners area of Arizona, New Mexico, Utah and Coloradio receive less than the required amount. Construction of a coal-fired electric generating plant in this area has already been criticized because of probable air pollution. The alternatives appear to be massive infusions of water from distant sources or determination that environmental quality of selected areas must be sacrificed in the national interest. The belated realization of the importance of water in coal gasification and reclamation underscores the need to consider the total geologic environment in resource utilization (see Chap. 15).

The long-term future of coal utilization is difficult to predict. The problem is complex, and many issues excite strong emotional responses. A national coal-utilization policy may be difficult to agree upon and will certainly be controversial, regardless of the final decision.

Petroleum – Our Vanishing Resource

In 1870, the United States had one horse or mule for each four human beings. This ratio had remained constant for many years. If a federal bureau at that time had estimated the horse and mule population of the United States for 1970, it might have predicted a figure of more than 50 million horses and mules – provided that the 1970 population could have been correctly calculated at all. Actually, United States horse and mule population in 1974 was about 8 million, and few of these produced any "horse power" for a plow, mowing machine, wagon or other farm or transportation equipment. Most, of course, are horses maintained for pleasure.

This little story illustrates three points which are of vital importance in any assessment of petroleum and natural gas resources.

The first point is that petroleum, mainly in the form of gasoline and diesel fuel, is to modern transportation what food is to an animal. Modern society lives on energy. It pervades our every activity, from driving automobiles to watching television.

The second point regards the difficulty of projecting future energy or resource consumption patterns. Many complex factors affect energy consumption. In 1870, few serious scientists or engineers would have predicted the automobile, the airplane or widespread use of electricity in 1970, all of which resulted from technologic breakthroughs that were unknown at that time. Today, technologic feasibilities are better understood, but our future still depends heavily on not-yet-achieved technologies such as breeder reactors and controlled fusion.

American resource reserves are poorly studied. In spite of technologic expertise, the United States does not know how large its energy reserves are, particularly its petroleum reserves. In this respect, the United States is like a family madly writing checks on a bank account whose balance it can only crudely estimate. The 1973–74 energy crisis initiated a serious assessment of reserves, but until the assessment has been completed, industry and government are planning in a room dimly lit by information.

The third point is that we cannot "go back in time" without suffering and misery on a grand scale. It is pleasant to recall the peaceful, pastoral life of earlier centuries, but the fact is that our society depends on energy not only for recreation and leisure but also for food and health. Any attempt to return to a horse-drawn society would result in malnutrition and possibly starvation in the United States. It would be impossible to produce and transport food on anywhere near the present scale. Even if adequate pastures, hay and grain existed, the excrement of 50 million horses and mules would convert large areas of the United States into stockyards. Not only would the odor assault the senses but the stockyards would serve as breeding grounds for a variety of biota which would severely menace the health of nearby inhabitants.

We believe that management of petroleum and natural gas reserves is the major technologic problem of the United States today. If we are to efficiently manage these reserves, it is essential that we know the basic facts. Therefore, in the remainder of this section we will discuss the origin, distribution, extraction, use and future of petroleum in the United States and, to a lesser extent, the world. As with many problems in environmental geology, no final solutions are provided, because all necessary data are not available. But if we understand the basic geologic facts concerning petroleum and natural gas, we can participate intelligently in the decision-making process concerning the future of this vital natural resource.

Geology of Petroleum

Petroleum is a naturally occurring, complex mixture of mainly hydrocarbon substances—liquid, gas and solid—which make up crude oil, natural gas and asphalt. Chemically, petroleum consists of complex hydrocarbons (hydrogen and carbon compounds). It typically includes minor amounts of nitrogen, sulfur and oxygen as impurities. In its original state, liquid petroleum, or *crude oil*, consists of various amounts of *dissolved gases, bitumins* and *impurities*. It looks like ordinary motor oil sold at service stations. It floats on water, an important factor in its accumulation in underground reservoirs. Petroleum gas, or *natural gas*, consists of several lighter hydrocarbons, of which *methane* (CH_4) is the most abundant. Solid and semisolid forms of petroleum consist of heavy paraffinic hydrocarbons and bitumins. They are called *asphalt, tar, pitch* and other less common names. The terms "hydrocarbon" and "petroleum" are often used interchangeably.

Important Types of Petroleum Deposits. From the point of view of energy resources, there are three important types of petroleum deposit—*liquid petroleum, oil shales* and *tar sands*. Virtually all present-day petroleum products derive from *liquid petroleum* deposits, which are found by drilling and then are pumped from the ground. Abundant reserves of oil shales occur in the western United States. Until recently, the cost of extracting and refining petroleum from oil shales precluded their development as a major source of petroleum products. Increases in the price of crude oil in 1973–74 have made exploitation of oil shales a more attractive prospect, but many problems remain to be solved. Tar sands, while less abundant than oil shales, are an important reserve, but their development has been impeded by high costs and other problems. For example, Athabascan tar sands (in Canada) are climate dependent, because winter snows cut off access to the deposits.

Origin of Petroleum. In spite of intense research efforts, the origin of petroleum is not completely understood for a number of reasons. The most important of these derives from the fact that petroleum usually occurs as a liquid and hence is capable of migration. It is possible to demonstrate in many instances that petroleum has migrated from the site of its formation. Because these original rocks cannot be identified, other aspects of petroleum genesis also remain unclear.

The following known facts constrain any hypotheses regarding the origin of petroleum:

1). Virtually all petroleum occurs in sedimentary rocks. The few places where it is present in nonsedimentary rocks are close to sedimentary rocks that could easily have been the source of the petroleum.

2). Most sedimentary rocks containing petroleum are of marine origin. This probably results from the fact that non-marine, or continental, sediments containing organic debris are more likely to be oxidized and the petroleum destroyed before they are adequately protected by burial.

3). No two petroleums are exactly alike in organic molecular make-up. This probably means that primary source material, environment of formation, migratory history and environment of reservoir accumulation differ from deposit to deposit. These variations complicate the problem of understanding the origin of petroleum.

4). Petroleum occurs in rocks ranging in age from Precambrian (older than 600 m.y.) to Pleistocene (0 to 2 m.y.), but economic deposits have been found only in post-Precambrian rocks (younger than 600 m.y.). Fifty-eight per cent of all known deposits occur in rocks of Tertiary age (0 to 70 m.y.). Deposits decrease with increasing age. Mesozoic rocks (70 to 225 m.y.) contain about 27 per cent of known reserves, and Paleozoic rocks (225 to 600 m.y.) contain the remaining 15 per cent. There is some question as to whether Pleistocene rocks contain mature crude oil. For many years, geologists thought that Pleistocene rocks were too young to permit maturation of petroleum-forming substances, but offshore drilling has shown that some Pleistocene sediments contain petroleum. It is still not clear whether the petroleum formed in older rocks and migrated to its present position or formed in place.

Correlation of petroleum occurrence and *age of reservoir rock* suggests that occurrence is related to the evolutionary explosion of plants and animals during the past 600 m.y. Although a few geologists in early years believed that petroleum may have formed inorganically, all now agree that animal and plant remains are the original source of hydrocarbon compounds comprising petroleum. Some uncertainty regarding the relative importance of plants versus animals still exists, however.

5). Soluble liquid petroleum hydrocarbons occur in three of the major sedimentary rock types: shale, sandstone and limestone. Thus it appears that petroleum source material is widely distributed.

6). Temperatures of petroleum reservoirs seldom exceed 100°C., although a few deeper reservoirs reach 140°C. The presence of *porphyrin* pigments (hydrocarbon compounds formed from red coloring matter of blood or green coloring matter of plants) indicates that the petroleum has not been hotter than 200°C., because porphyrins are destroyed at a slightly lower temperature. This means that most petroleum forms at low temperatures.

7). Petroleum, like coal, forms in an oxygen-deficient (*anaerobic*) environment. Otherwise porphyrins and other easily oxidized hydrocarbons found in petroleum would have been destroyed.

8). The time required for petroleum to mature, migrate and accumulate in pools is probably less than 1.0 m.y. Recent sediments in the Gulf of Mexico recovered in shallow cores contain measurable amounts of hydrocarbons, which were dated by carbon-14 isotopic techniques as being 11,800 to 14,600 years old. In eastern Venezuela a sand deposited less than 10,000 years ago contains immature petroleum. At present rates of migration and accumulation, this sand will constitute a rich reservoir in less than one million years.

9). Abundant evidence supports an organic origin for petroleum. We have mentioned the correlation of occurrence of petroleum and evolution of plant and animal life. The occurrence of porphyrins and nitrogen compounds unique to organic matter adds further support. Most petroleum is optically active; that is, it rotates the plane of polarization of light passing through it. This is an exclusive property of organic fluids or solutions. In organic oils it is thought to result from the presence of *cholesterol* or other proteins which occur both in animal and vegetable matter.

Organic debris from both terrestrial and marine organisms probably contribute to the formation of petroleum. Rivers transport terrestrial debris to a water-filled basin (usually the ocean), where currents carry it to its burial site. Marine organic debris, which probably provides most of the source material, is buried along with the terrestrial debris, marine fungi, bacteria, algae (all plants), foraminifers, radiolarians, sponges, corals and other animals, all of which contribute to the flux of debris.

Conditions favoring accumulation and preservation of petroleum-forming material are as follows:

a) abundant supply of organic matter;
b) rapid accumulation of inorganic matter—sands, silts, clays—to insure quick burial; and
c) limited supply of oxygen in waters in contact with the organic debris. This prevents oxidation prior to burial.

Organic debris can be transformed into petroleum by one of several processes: bacterial action, heat and pressure, catalytic reactions or radioactive bombardment. Bacterial action causes the everyday decomposition of organic matter in addition to emitting methane, a principal constituent of natural gas. Bacteria are present in large quantities on and immediately below the sea floor. In highly anaerobic environments, bacteria convert plant and animal debris into a petroleum-like substance. For these and other reasons, many investigators suspect that bacteria are the prime agents in petroleum formation.

Heat and pressure may also convert organic matter into petroleum. Most organic matter found in non-reservoir rocks consists of *kerogen*, a solid pyrobitumin that breaks down into petroleum and other products at high temperatures (350 to 400°C.). The presence of porphyrins in naturally occurring petroleum indicates that most petroleum never experienced temperatures high enough to break down kerogens. Some investigators, however, have suggested that given enough time and a few kilobars (approximately 1000 atmospheres) of burial overburden pressure, kerogens will break down at temperatures low enough to preserve porphyrins.

Heat and pressure probably affect the composition of petroleum. Petroleums from deeper reservoirs in Gulf Coast rocks contain more naphtha and gasoline and fewer heavy residues than petroleums of comparable age found in shallower Gulf Coast reservoirs.

Catalysts (substances which aid or accelerate chemical reactions) are well known and widely applied in refinery operations. Clay minerals and certain metallic elements known to be present in small amounts in most rocks are the most likely natural catalysts in petroleum formation.

Radioactive elements occur widely in sedimentary rocks, especially in black organic shales. Laboratory experiments have shown that petroleumlike compounds can be formed by radioactive bombardment of organic compounds. Many geologists dispute this mechanism, however. Laboratory experiments indicate that the hydrogen-carbon ratio of petroleum should decrease with time; however, the opposite appears to be true. In addition, older radioactive black shales include little free petroleum. If radioactive bombardment were an effective agent in the formation of petroleum, significant amounts of free petroleum should be present in these shales.

In summary, petroleum is derived from alteration of organic debris trapped in sediments. The liquid fractions migrate and accumulate in traps. Marine rocks, especially those found along continental margins, contain the overwhelming majority of petroleum deposits.

Kerogen, a solid hydrocarbon, is probably not ancestral to significant quantities of liquid hydrocarbons. However, large deposits of kerogen enclosed in shales (usually called "oil shales") are important as future sources of petroleum products.

Asphalt consists of a mixture of heavy non-hydrocarbon solids and semisolids. It occurs along with hydrocarbons and is left behind as the lighter hydrocarbon components seep from the ground and evaporate. Asphalts in rocks are an important fossil fuel reserve, although they are far less abundant than liquid petroleum and kerogen.

Geography of Petroleum

Petroleum is not uniformly distributed on earth. Two regions, the Middle East and the Gulf of Mexico–Caribbean region, contain two thirds of all known reserves. The recent discovery of petroleum in the North Sea and North Slope of Alaska is not likely to significantly alter this figure.

Proven reserves (reserves which have been proven by drilling) are largely concentrated in the Middle East (Fig. 8–14), which has proven reserves of 356 billion barrels. Saudi Arabia (137 billion bbl.), Kuwait (74 billion bbl.) and Iran (62 billion bbl.) each have greater proven reserves than the entire North American continent (47 billion bbl.). The United States (36 billion bbl.) has limited reserves in comparison with the U.S.S.R. (42 billion bbl.) and North Africa (53 billion bbl.). Only South America (26 billion bbl.) and western Europe (10 billion bbl.) have smaller reserves than the United States.

The Middle East. The existence of large deposits of petroleum implies large volumes of sedimentary rock. As a general rule the largest volumes of sedimentary rock occur along the margins of continents. This is true on the Gulf Coast of the United States, the North Slope of Alaska, the North Sea, Nigeria, northern Venezuela and western Peru, for example.

During early Mesozoic time, open ocean separated much of present-day Arabia (Africa) and Iran (Asia). Large sedimentary basins fringed each continental block. In the Cretaceous period (70 to 135 m.y.), the two continental plates converged, crushing rocks along the margin and raising the Zagros Mountains (Fig. 8–15) along the western border of Iran. The depression of the downgoing slab (Arabia) formed the Persian Gulf. The vast basin formed between the two continental plates was subsequently infilled by sediments eroded from surrounding highlands. It contains over half the world's known petroleum reserves.

Folding and faulting of rocks in the region and upward migration of salt deposited before convergence helped form *oil traps,* or areas where migration is blocked. It has been suggested that migration of petroleum up the downgoing lithospheric slab is responsible for the staggering quantities of oil found in individual traps.

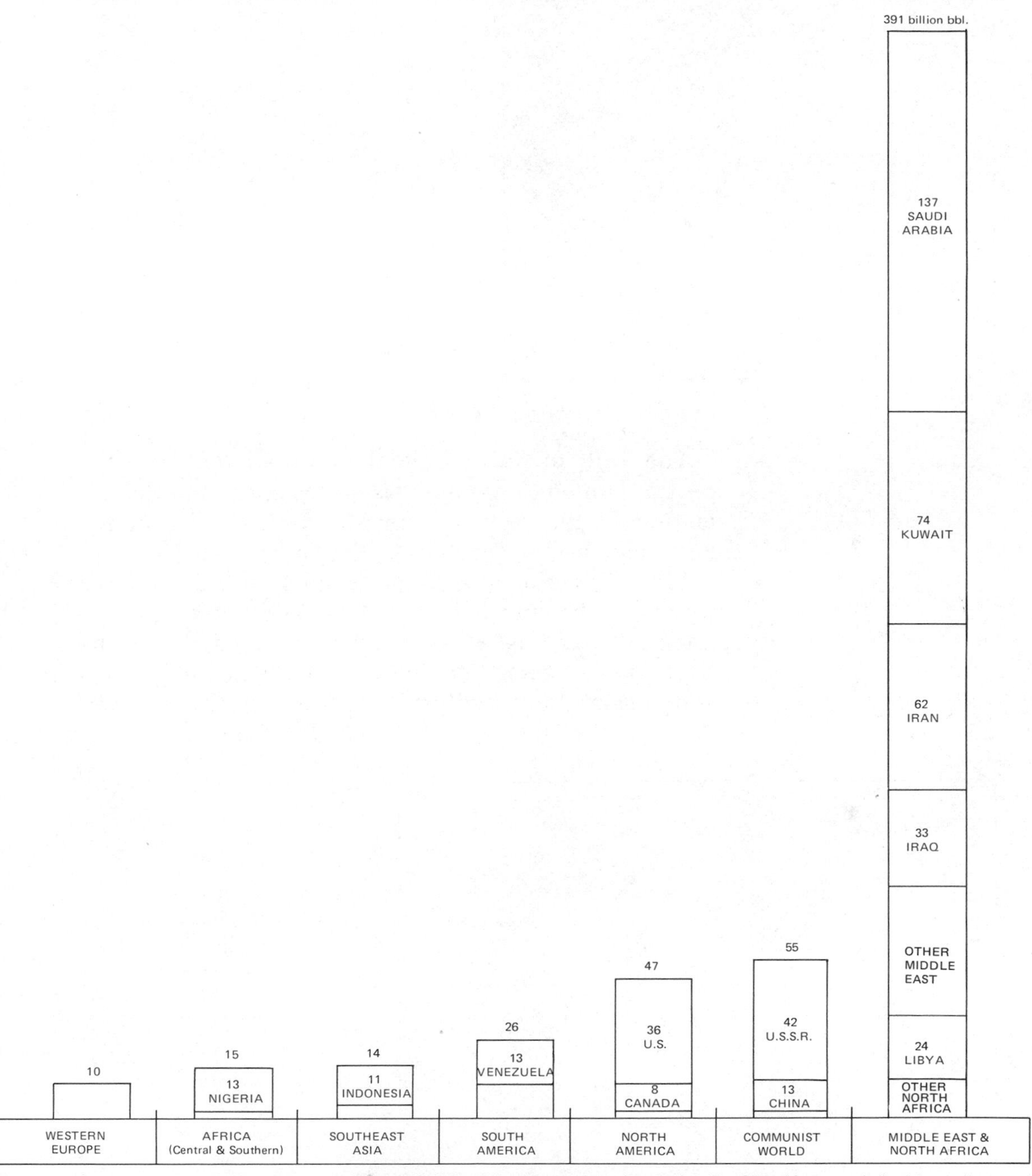

Figure 8–14. Proven reserves in crude oil, 1972. (A potentially large discovery made in southern Mexico in 1974 may add significantly to North American reserves.) (Adapted from Time, November 19, 1973.)

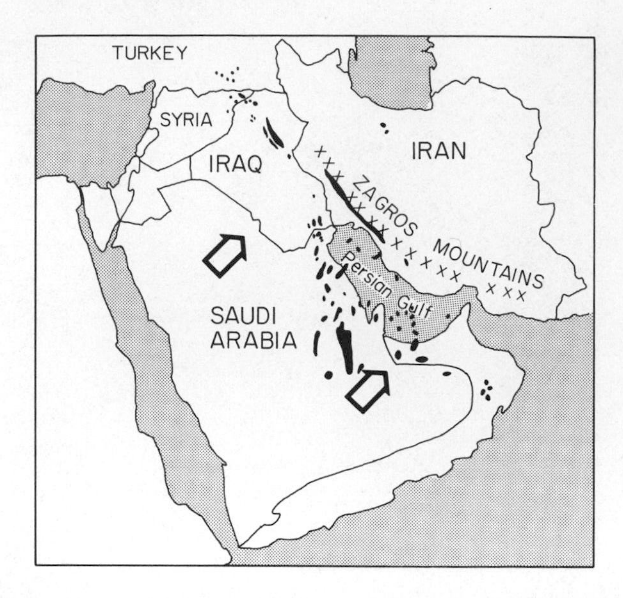

Figure 8–15. The Middle East. Major oil fields are shown in black. Arrows indicate motion of down-going (Arabian) crust. (Adapted from International Petroleum Encyclopedia. Petroleum Publishing Co., 1974.)

The Gulf of Mexico. Sediments eroded from the central United States and eastern Mexico after the formation of the Appalachian Mountains have for the most part been deposited in the Gulf of Mexico. The volume of sediment is huge. Deposition in near-shore waters of the Gulf has caused the shoreline and continental shelf (Fig. 8–16) to advance steadily into the Gulf for more than 200 million years. Beneath the shelf and coastal plain, sediment thickness is known to be

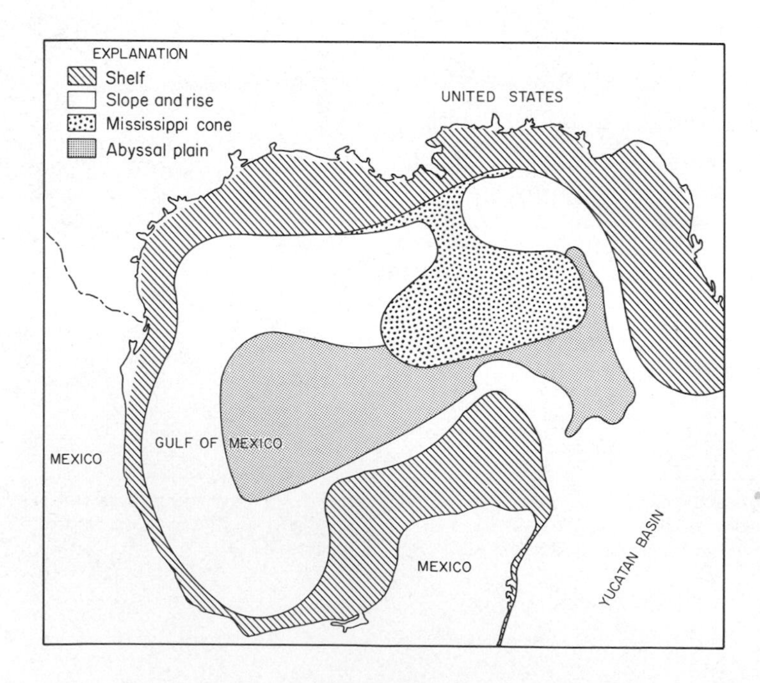

Figure 8–16. Geologic provinces of the Gulf of Mexico.

Figure 8–17. Semisubmersible drilling platform in northern Gulf of Mexico. Crews live in dormitories on the platform. The boat alongside the platform is delivering supplies and drilling steel.

as much as 17 km (55,000 ft). Further offshore, up to 7 km of sedimentary rock overlies the basaltic oceanic crusts of the abyssal plain, in contrast with 1 to 2 km overlying normal oceanic crust.

Faults due to slumping and subsidence within this wedge of sediments in the northern Gulf and rising masses of low-density salt deposited during the Jurassic period (135 to 180 m.y.) form many traps.

Rocks of the western Gulf have been compressed into folds and intensely faulted. These features are responsible for many of the traps in Mexico and beneath the Mexican shelf.

Rocks of the coastal plain and the continental shelf have produced petroleum in abundance. The shelf is the site of many offshore platforms (Fig. 8–17), from which a number of wells may be drilled. Drill pipe is slanted outward from the platform in order to extract oil from deposits at various depths within the surrounding area.

The Gulf of Mexico probably contains reserves exceeding the proven reserves of the remainder of North America. Much of the shelf (600 feet or less water depth) has not yet been drilled, and there is evidence of oil in sediments in areas of deeper water. A wide slope extending outward from the shelf (see Fig. 8–16) is known to contain large bodies of salt, which may trap oil. Holes drilled into abyssal plain sediments by the

Glomar Challenger oceanographic vessel yielded cores containing methane and immature crude oil. At the present time it is not economically feasible to drill and complete wells in water deeper than 600 feet, but it is technically feasible to do so at depths of 1500 feet or more. If the price of oil continues to rise, the deep Gulf could prove to be an important oil province.

Oil Extraction

As indicated earlier, petroleum can be obtained from three major sources: petroleum liquids, oil shales (mainly kerogen) and tar sands (mainly asphalt). At the present time, the world gets almost all of its petroleum from deposits of liquid petroleum found in underground reservoirs. Therefore, we will devote most of the discussion in this section to extraction of petroleum liquids.

Extraction of petroleum is relevant to environmental geology for two reasons. First, geologic structures control the accumulation of petroleum and therefore play a major role in extraction technology. With shallow accessible oil in short supply and in great demand, it is important to know where the remaining reserves lie and to understand the geologic and technical problems of extraction.

Secondly, oil spills occur from time to time during extraction. When most production was on land, drilling firms simply bulldozed a dam around the runaway well to contain the escaping oil. Now many fields and most of the world's reserves are located offshore, where it is more difficult to contain spills. For this reason we will also discuss the physical basis of "gushers" and describe measures which can be taken to minimize oil spills in the marine environment.

Traps. Petroleum is lighter than water. It is also immiscible with water. As a result of its immiscibility, petroleum separates from rock pore fluids and accumulates in small globules, which rise because they are lighter than water, just as a piece of ice rises in a glass of water. As they rise, they encounter other globules and coalesce into larger bodies of petroleum. These larger bodies continue to seep upward.

Not all rocks are *permeable;* that is, holes in the rocks are either nonexistent or too small to permit passage of fluids. Sooner or later, rising oil globules either encounter *impermeable* strata or else seep out at ground level. If the globules encounter an impermeable layer of rock, the petroleum seeks out the highest place within the permeable layer and accumulates there. This high spot is called an *oil trap*. The permeable rock containing the trapped petroleum is called the *reservoir rock*. The rock in which the oil originated is called the *source rock*.

During seepage and entrapment, gas separates from the liquid. Gas, being lighter than both liquid petroleum and water, accumulates at the very top of the trap.

Traps fall into two broad groups: *structural* traps and *stratigraphic* traps. Structural traps usually result from folding or faulting. Two common structural traps are *anticlinal* traps and *salt dome* traps. Anticlinal traps form when alternating impermeable and permeable strata are folded. The concave-down folds are called *anticlines*, and the concave-up folds are called *synclines*. Petroleum migrates through permeable rock toward the crests of the anticlines and accumulates, as shown in Figure 8–18.

Salt is a very mobile rock. It deforms easily and flows under minor differences in pressure. Often salt finds a weak zone in overlying rock and pushes upward in fingerlike masses called *salt domes*. As the salt rises to the surface, it drags surrounding strata upward. Salt is largely insoluble in oil and is also impermeable, so oil migrating upward in adjacent rocks can go no further when it encounters the salt. Petroleum accumulations often "collar" salt domes, as shown in Figure 8–19.

Fault traps (Fig. 8–20) are also common. In a fault

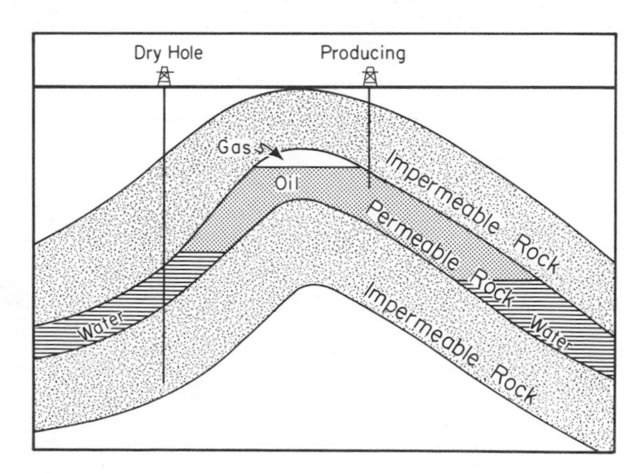

Figure 8–18. Anticlinal trap. Oil has migrated through a permeable horizon (reservoir rock) and accumulated in the crest of an anticline. An overlying impermeable horizon prevents further upward migration.

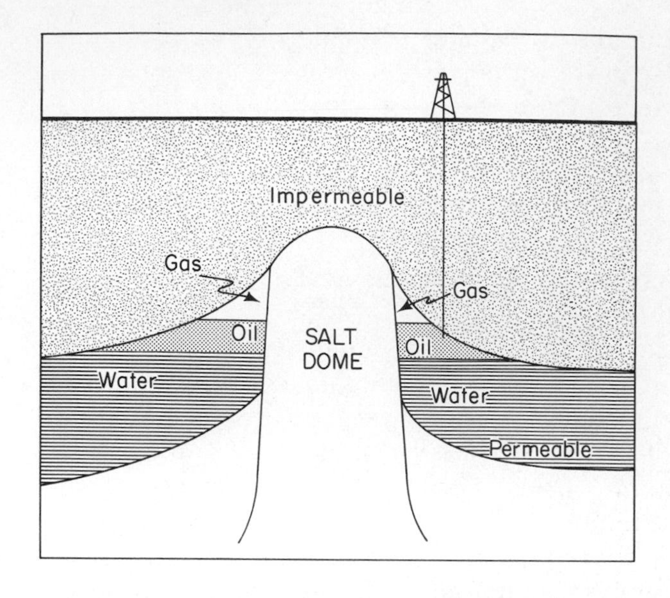

Figure 8–19. Cross-section of a salt dome, showing strata bent up against the side of the dome. Petroleum is trapped in the collar of upwarped sedimentary rocks.

trap, permeable strata displaced by faulting come to rest against impermeable strata.

Structural traps are easier to locate than stratigraphic traps because structural traps are associated with some dislocation of strata. Stratigraphic traps, on the other hand, result from changes in permeability within strata and are not related to tectonic dislocation.

Strata thicken and thin, depending on their depositional and erosional history. Sometimes a permeable horizon between two impermeable strata will simply *pinch out*, or thin and disappear. This can result in a trap such as the one shown in Figure 8–21.

The permeability of strata varies from place to place. A permeable sandstone horizon may grade into an impermeable shale horizon. If strata above and below are

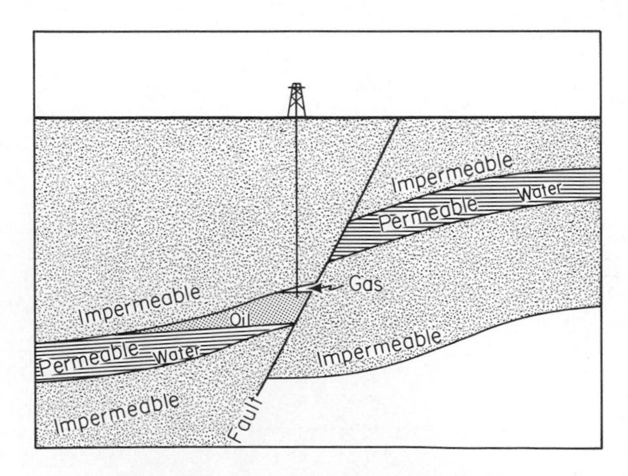

Figure 8–20. A fault trap.

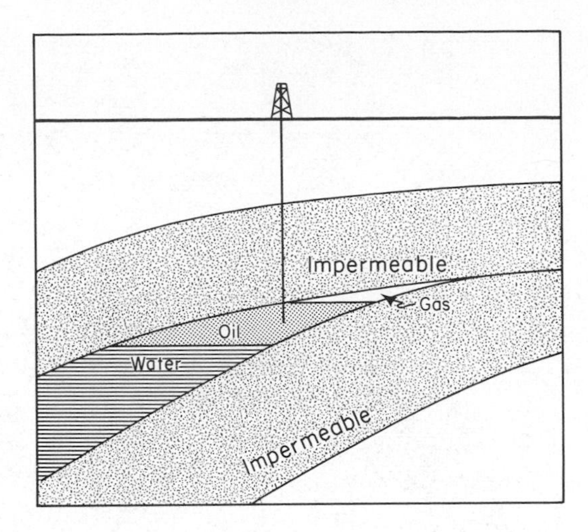

Figure 8–21. A pinch-out. A permeable stratum pinches out between impermeable strata. Petroleum is trapped in the pinch-out.

impermeable, oil may be trapped at the point where permeability becomes so small that the petroleum can no longer flow through the rock.

Finding the Oil. The geologist uses a variety of techniques to find oil. In the early days of petroleum exploration, it was often sufficient to map the surface distribution of rocks until an anticline or salt dome was found (Fig. 8–22). As it became necessary to probe more deeply into the earth for petroleum, geologists adopted a number of indirect methods of investigating the earth's interior. The most widely used and effective method in current use is *seismic reflection*. In seismic reflection, seismic waves bouncing off layers within the earth are recorded. The source and receivers are moved along a few feet at a time until a complete *line* has been recorded. The data from individual "shots" are combined into a single cross-section (Fig. 8–23). The seismic reflection technique as now practiced

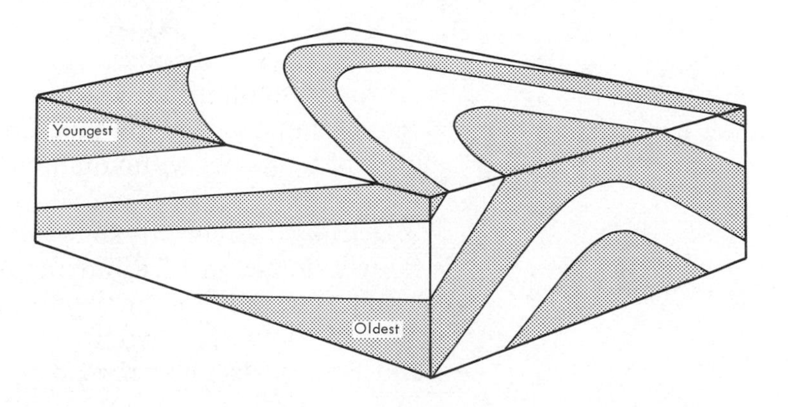

Figure 8–22. Block diagram of an anticline. Anticlines are easily identified because the oldest rocks outcrop in the center.

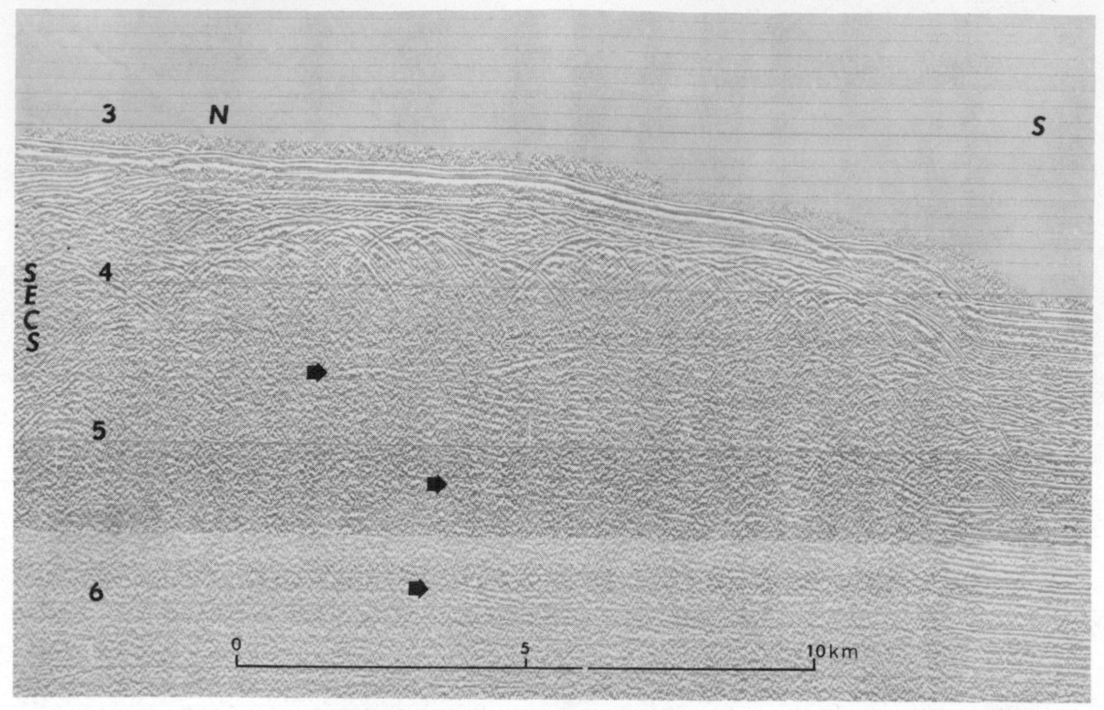

Figure 8–23. Seismic cross-section of the Sigsbee Scarp in the Gulf of Mexico. In this section, a landslide containing salt has flowed laterally over sedimentary layers of the abyssal plain. Scattering of energy at the fractured top of the landslide causes the deterioration of reflection quality beneath the salt. Reflections from 11 km sub-bottom were obtained in the abyssal plain (extreme far right). Arrows point to reflecting horizons beneath the landslide.

costs about $250 to $400/mile (1974) at sea and several times that amount on land. Its ability to penetrate deeply into the earth and precisely record details of structure justifies the expenditure of millions of dollars each year.

Recent data-processing improvements now make it possible to "see" oil and gas deposits on seismic cross-sections (Fig. 8–24). This discovery, called *"bright spot"* analysis, results from differences in reflectivity of rocks containing gas accumulations. Guided by bright spots, oil companies now hit oil in more than nine of ten holes in favorable areas.

Drilling. Figure 8–25 schematically shows a drilling rig. Rotation of the drill pipe causes the bit (Fig. 8–26) to break the rock at the bottom of the hole. "Mud"—actually a mixture of heavy minerals, water and chemicals—is forced down into the hole via the

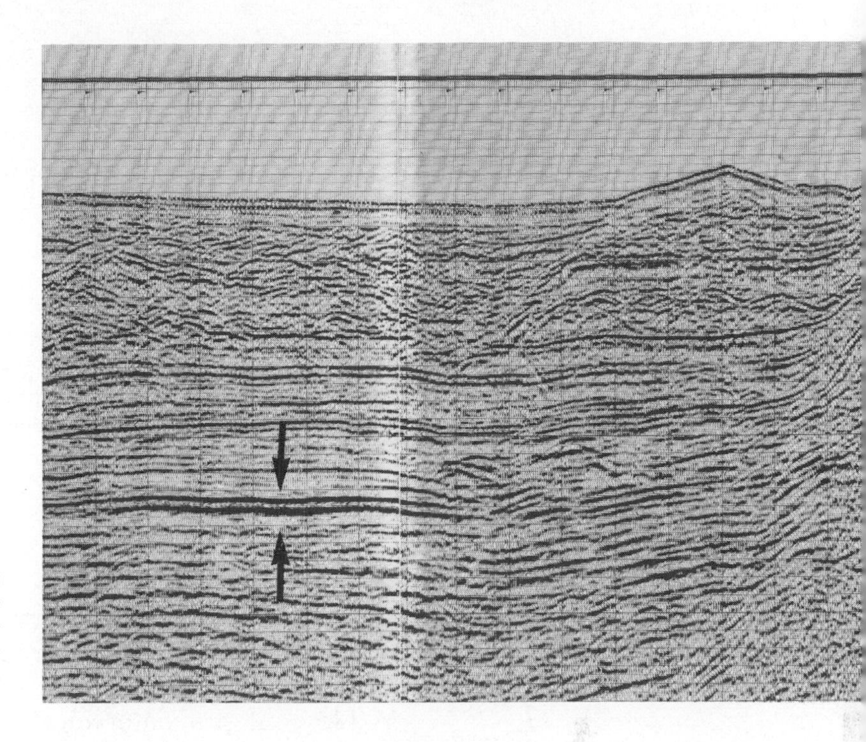

Figure 8–24. "Bright spots" on seismic cross-sections indicating probable oil and gas accumulations. (From Western Geophysical Company, Houston, Texas.)

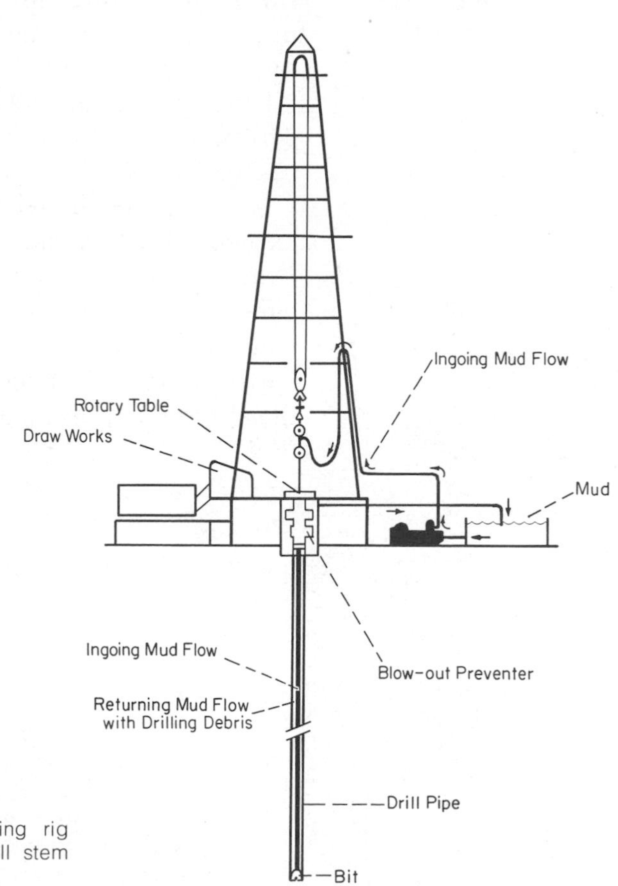

Figure 8–25. Schematic drawing of drilling rig showing mud tanks, blowout preventer, drill stem and bit.

Figure 8–26. A tricone bit. This type of bit and its variations are used worldwide. Rotation of the drill stem causes the teeth to revolve and "chew" the rock. The tricone bit was invented by Howard Hughes's father and is the basis of the Hughes fortune. (From Hughes Tool Company, Houston, Texas.)

hollow drill stem. The mud extrudes through a hole in the center of the bit and flows back up the hole, carrying the cuttings.

A blowout preventer located at the top of the hole can shut off flow in the event of an emergency. The upper portion of the hole is cased; that is, a large-diameter pipe is set in the hole, and the drill pipe and bit are thereafter lowered through the casing during drilling. The lower part being actively drilled is not cased.

Blowouts occur when gas pressure in the formation exceeds mud pressure. The effect of gas pressure is countered by varying the weight of the mud. An exact balance is sometimes difficult to achieve. If the mud weighs too much, it escapes into the rock and the cuttings are not raised to the surface. If it weighs too little, the gas pressure pushes the mud out of the hole (Fig. 8–27).

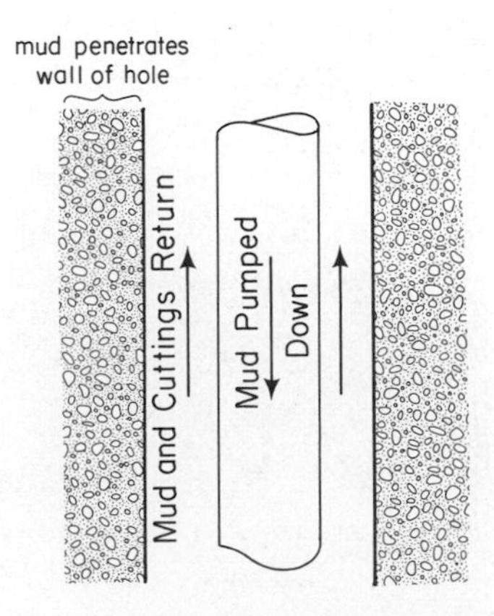

Figure 8–27. Detail showing mud circulation within a drill hole. (See Fig. 8–25.)

Pumping. If gas pressure within the reservoir rock is high enough, it will push out the oil. Eventually, however, the pressure drops, and pumps are installed to push the petroleum up the hole.

The percentage of oil and gas in the reservoir recovered by pumping seldom exceeds 50 per cent. Additional recovery can be effected in a number of ways. Sometimes fluid is pumped into the well under high pressure to crack rocks around the well, thereby increasing the permeability of the rocks. The additional cracks make it easier for the petroleum to flow into the well. Inadvertent *hydraulic fracturing,* as this procedure is called, was partially responsible for the Denver earthquakes (see Chap. 2).

Hydraulic fracturing has been attempted on a large scale in a effort to release gas from low-permeability standstones of the Fort Union and Mesa Verde formations in Colorado. These formations, lying at depths between 5000 and 7000 feet in Rio Blanco County, Colorado, contain an estimated 100 trillion cubic feet of recoverable natural gas. Unfortunately, the permeability of the rocks is so low that the gas will not flow freely into wells. Of 50 wells drilled by Equity Oil Company, only one has been successful.

The AEC (Atomic Energy Commission), as part of its program to utilize nuclear technology for peaceful purposes, has demonstrated that gas flow can be stimulated by detonation of nuclear devices in low-permeability formations. In 1967, the AEC detonated a 29-kiloton device in the Gas Buggy experiment in New Mexico, and in 1969 a 43-kiloton shot was detonated in the Rulison area. Gas production increased ten times at the Rulison well.

The Rio Blanco experiment is on a much greater scale. Eventually as many as 1000 nuclear devices may be detonated in a number of wells distributed throughout the field. The objective is widespread fracturing of rocks throughout the producing zone If successful at Rio Blanco, the technique could be used in similar low-permeability formations in the western United States (Fig. 8–28).

The efficacy of the experiment, however, is as yet unproven, and many important questions regarding radioactive pollution and other undesirable side effects remain unanswered.

Production can also be improved by pumping water

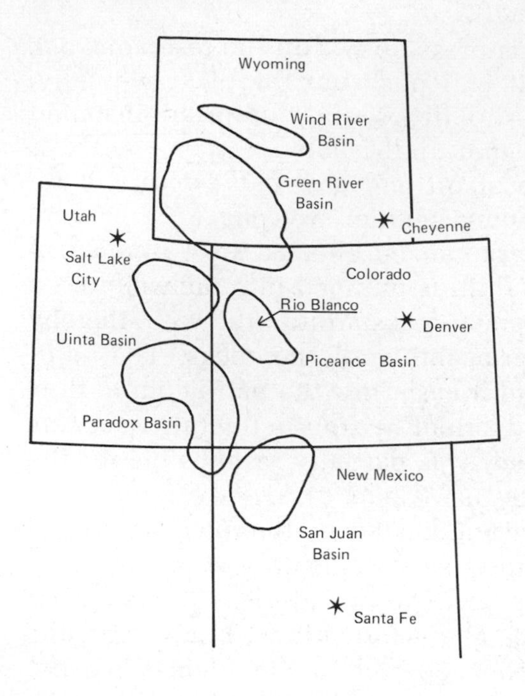

Figure 8–28. Basins of Wyoming, Colorado, New Mexico, Arizona and Utah containing deep, low-permeability gas-bearing formations. (Adapted from Carter, L. J.: Rio Blanco: Stimulating gas and conflicts in Colorado. Science *180,* 844–848, 1973.)

into the reservoir rock to raise the pressure and push petroleum to the well and by pumping chemicals into the reservoir rock which "wash" the rock and remove residual oil. The chemicals and residual oil are then pumped out. Such techniques, called *secondary recovery* methods, if widely applied, might result in a doubling of United States proven reserves. As with other aspects of petroleum production, cost is a determining factor. It has often been cheaper to drill new wells in new fields than to attempt recovery of additional crude from old fields. Doubling and tripling of crude oil prices in 1973–74 will make secondary recovery much more attractive in the future.

Pollution

On January 29, 1969, Union Oil Company drillers were completing the fifth well to be drilled from Platform A on Federal Tract OCS P-0241 in the Santa Barbara channel (Fig. 8–29), when drilling mud began to flow up the drill stem. This was followed by a 13-minute blowout of a heavy gaseous mist accompanied by a deafening roar. The drilling crew was then able to drop the drill stem and close the blowout preventer.

Within minutes, a myriad of small gas bubbles began rising through the water around the platform. From 1½

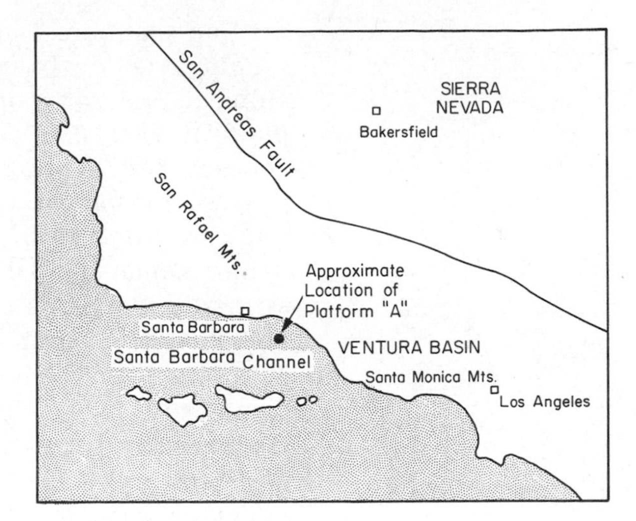

Figure 8–29. The Santa Barbara channel and vicinity. (Adapted from Vedder, J. G., et al.: Geology, petroleum development, and seismicity of the Santa Barbara Channel region, California. U.S. Geological Survey Prof. Paper 679, 1969.)

to 2 hours after the blowout preventer was shut, the crew observed turbulent boiling of the water about 800 feet east of the platform (Fig. 8–30).

The area of boiling expanded and intensified, and another boil developed near the northeast corner of the

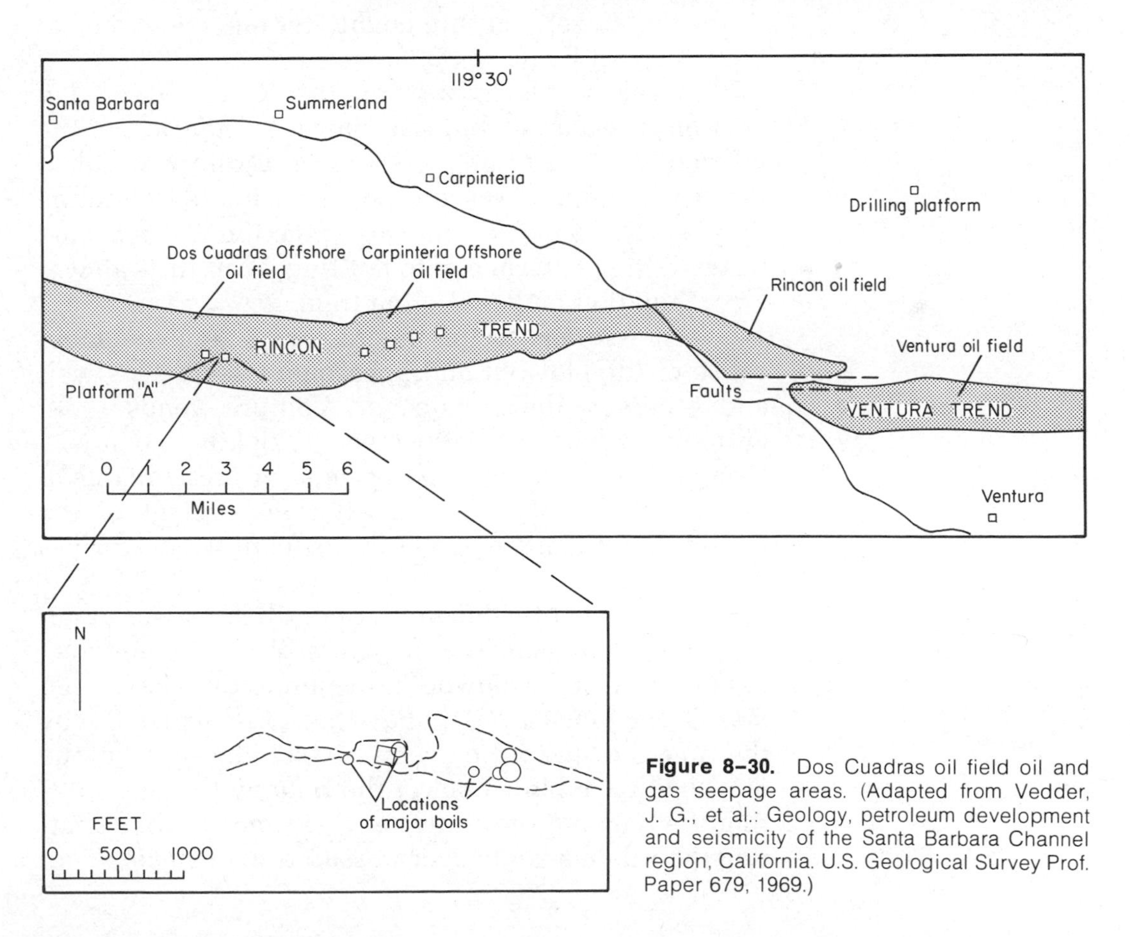

Figure 8–30. Dos Cuadras oil field oil and gas seepage areas. (Adapted from Vedder, J. G., et al.: Geology, petroleum development and seismicity of the Santa Barbara Channel region, California. U.S. Geological Survey Prof. Paper 679, 1969.)

drilling platform. Within 24 hours, oil and gas were seeping vigorously from numerous seafloor fractures along an east-west zone, extending 250 feet west of the platform and 1050 feet to the east. The oil flowed unchecked until February 7, 1969, when the well was plugged by cement. Oil continued to seep at reducing rates until late 1969.

As estimated 10,000 barrels of crude oil washed ashore on the Santa Barbara beaches, triggering one of the greatest environmental debates of all time. The Santa Barbara Channel was closed to new drilling, commissions were formed to investigate and volunteers swarmed the beaches, cleaning filthy goop from birds and trying to cleanse the beach of oil.

Three years later, a little-noticed article reported that the beaches, flora and fauna had regained their prespill balance. About the same time, the channel was quietly reopened for drilling.

In the emotional climate following the oil spill, recriminations flew from side to side like ricocheting bullets, usually without benefit of supporting data. With the benefit of hindsight, let us review what happened on Platform A.

Platform A was located on the Rincon *trend* (an area of oil accumulation), a seaward extension of the oil-rich Ventura trend in Southern California. Oil of the Rincon trend was trapped in a faulted anticline (Fig. 8–31). Part of the oilbearing section was repeated beneath the platform due to faulting. Thus the trap was a combination fault–anticline trap.

Overpressured gas in the reservoir was the main cause of the blowout and subsequent seep. This well and others of the Rincon and Ventura trends—both anticlinal zones—had a history of "kicking"; that is, a tendency of the mud to rise because of pressure due to gas in the formation. Normally, the weight of the drilling mud is increased, and this counteracts the gas pressure.

At the time of the blowout, the drill stem was out of the hole, and there was no quick way to counteract gas pressure. If the blowout preventers could have been closed, the blowout might have been prevented. But by the time the blowout preventers were closed, all of the mud had been blown out of the hole by the gas. Shutting the blowout preventers at this time was like shutting the lid on a kettle overpressured with steam. In an

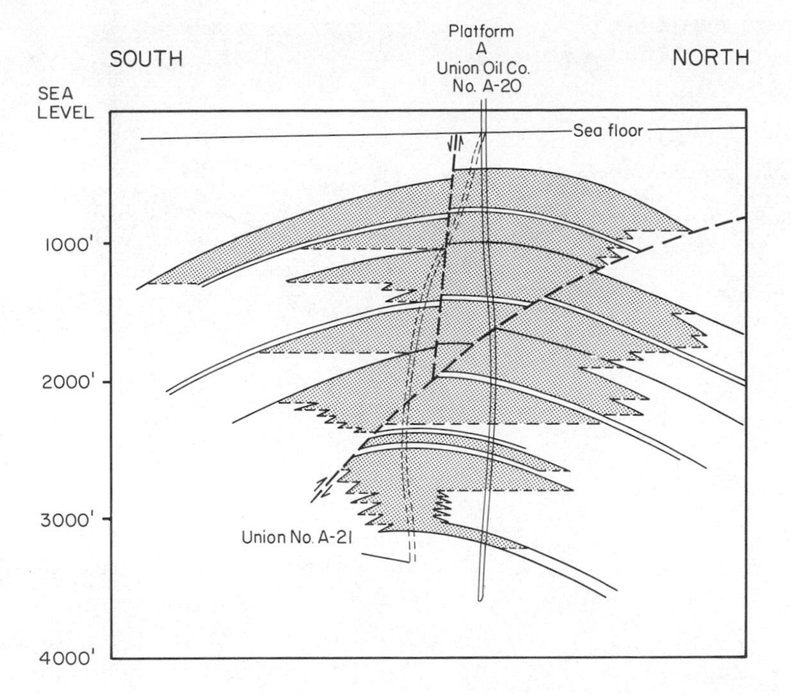

Figure 8–31. Cross-section showing oil bearing strata in the vicinity of Platform A in the Santa Barbara channel. Faulting of the anticline caused a portion of the oil-bearing rock to be repeated beneath the platform. (Adapted from Vedder, J. G., et al.: Geology, petroleum development, and seismicity of the Santa Barbara Channel region, California. U.S. Geological Survey Prof. Paper 679, 1969.)

overpressured kettle, the seams will burst, and this is what happened beneath Platform A.

The gas pressure exceeded the strength of the overlying rock, which simply cracked. The fault going through the axis probably localized the cracking, but pressure was nevertheless sufficient to crack the rock. It was a case of massive hydraulic fracturing.

In retrospect, the crucial error was probably the depth of casing. Casing extended only 238 feet below the sea floor. In an area where surface rocks are solid and well cemented, this might have been sufficient to prevent fracturing. Beneath Platform A, however, near-surface rock consisted of weak, poorly cemented sandstone and claystone. This rock, which was so friable that cores could be crumbled by hand, was easily fractured by the pressure of the gas. An additional 500 feet of casing would have provided much greater protection, since the gas pressure would then have had to fracture 738 feet of rock rather than 238 feet.

The Torrey Canyon supertanker wreck of 1967 spilled 100,000 barrels of oil on English beaches. The Santa Barbara channel contains at least 60 known natural seeps. An area of 900 square miles in the northwestern Gulf of Mexico contains an estimated 12,000 gas seeps, which are thought to be responsible for

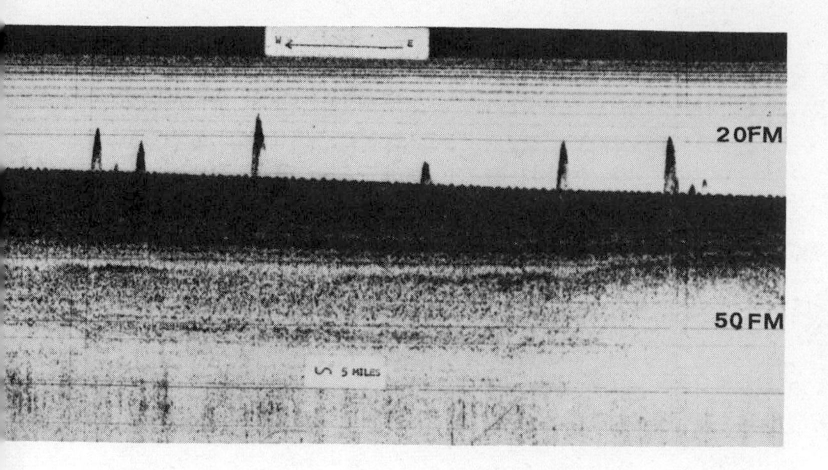

Figure 8–32. Gas seeps in the northwestern Gulf of Mexico. Each spike represents gas seeping from the sea floor into the overlying water. An area of 900 square miles contains an estimated 12,000 seeps.

many of the asphalt blobs washed up on Texas beaches (Fig. 8–32). These examples are more serious from a pollution standpoint than the Santa Barbara spill was. The Torrey Canyon spill caused widespread protest; the seeps have been virtually ignored. Both the Santa Barbara channel and English beaches have largely recovered from the effects of the spills. The tendency to "cry wolf" at every environmental mishap without carefully examining the facts makes it more difficult to direct public and governmental attention toward genuine major crises and their solutions.

The publicity surrounding oil pollution in the ocean is an example of "crying wolf." During a cruise in 1973, the author (JSW) and the late Maurice Ewing observed a few small globules of oil floating in the western North Atlantic. Dr. Ewing, one of the deans of marine geology, observed that in his many cruises, spanning a period of four decades, he had observed oil pollution only rarely, and furthermore, in his opinion no perceptible increase in oil pollution had occurred during his years of marine investigation.

Many oils contain sulfur, which produces undesirable gases during combustion. Pollution problems arising from use of high-sulfur oils resemble those discussed previously in the section on coal.

The Future of Petroleum

The world's supply of oil and natural gas is finite. If properly managed, the supplies can last for a few more centuries. The United States' reserves are small in comparison with those of the Middle East. There is little doubt that availability of Middle East petroleum will continue to be uncertain at best. Natural gas, which is more expensive to transport by sea than liquid petroleum, is in even more limited supply.

The reserve potential of the marine sedimentary rocks of the world remains unknown. Certainly, there is reason to hope for substantial reserves if only because marine sedimentary basins cover an immense area (Fig. 8–33) and comprise an immense volume of rock. Gulf of Mexico rocks of the abyssal plain, rise and slope—excluding the shelves—probably contain more recoverable petroleum than is known to exist in the remainder of North America. Three large sedimentary basins, Georges Bank, Baltimore Canyon and Blake Plateau, fringe the eastern seaboard of the United States (Fig. 8–34). These basins are presently areas of active seismic exploration. A large number of producing wells have been drilled in the Gulf of Mexico.

The north slope of Alaska contains an estimated 10 billion barrels of recoverable petroleum. While this is an important supply, equal to roughly 20 per cent of present United States reserves, it is not as important as the Gulf or Atlantic offshores areas, which may con-

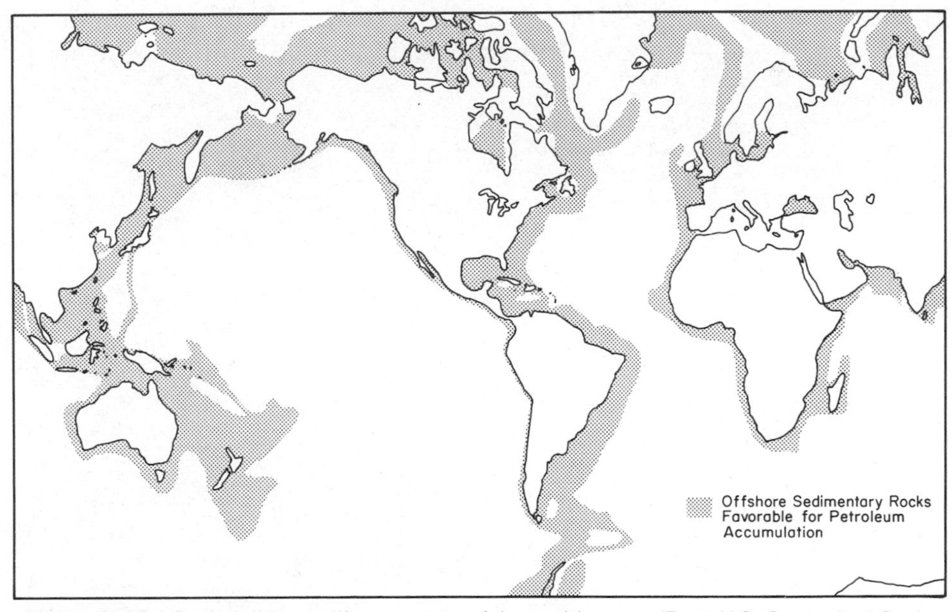

Figure 8–33. Potentially petroliferous areas of the world ocean. (From U.S. Geological Survey Investigations Map I-632, 1970.)

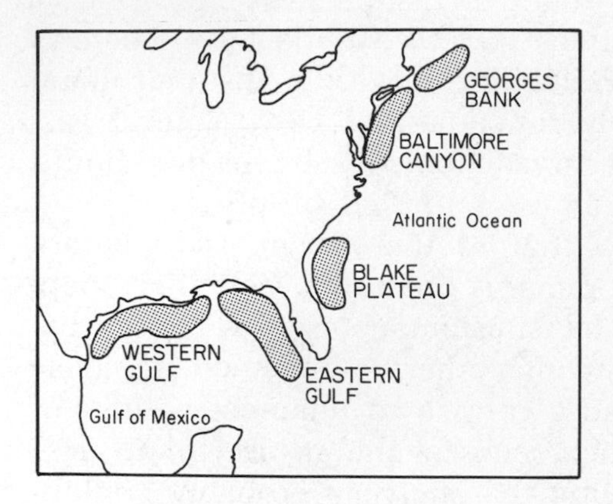

Figure 8–34. Major sedimentary basins of the shelves and slopes adjacent to the eastern and southern United States. These basins are the most promising targets for oil and gas exploration. (Adapted from U.S. News and World Report, p. 57, December 24, 1973.)

tain well over 10 times this amount of recoverable crude oil.

The reserve potential of continental shelf rocks equals or exceeds the proven reserves of North America in 1973 (about 47 bbl.). In addition, reserves of the slope, rise and abyssal plain may be two or three times as great as North American proven reserves.

As drilling proceeds to deeper waters, costs rise and blowout prevention may become more difficult. Equipment necessary to complete wells at great depths is still in the design stage (Fig. 8–35). The ultimate design will be complex and expensive.

Oil shale reserves are abundant, amounting to 1600 billion barrels in the United States, compared with 47 billion barrels of known liquid petroleum reserves (in the United States) and 352 billion barrels proven re-

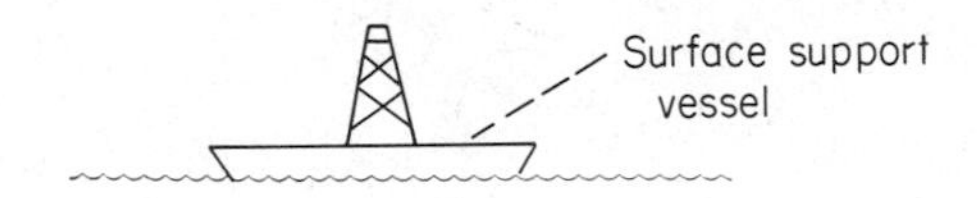

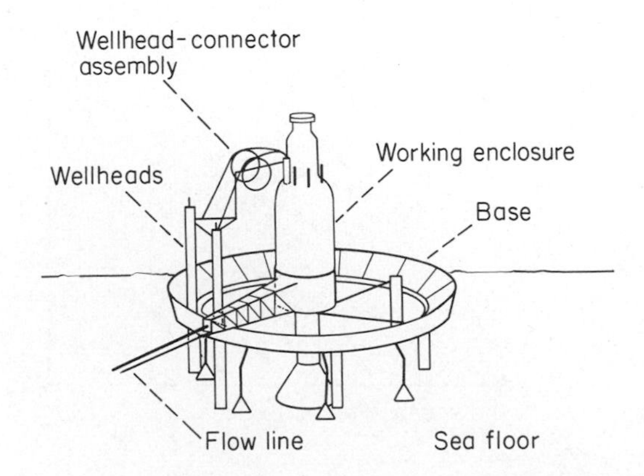

Figure 8–35. Subsea wellhead assembly. (Courtesy of Lockheed Company.)

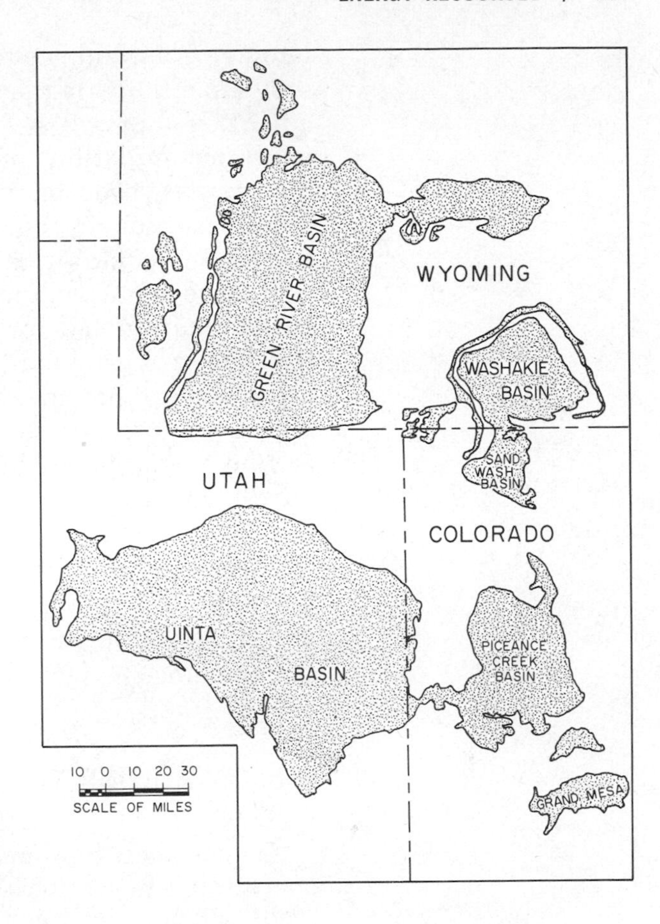

Figure 8–36. Location of major oil shale deposits in the United States. (Adapted from Cook, G. L.: Oil shale—an impending energy source. Journal of Petroleum Technology, 1325–1330, November 24, 1972.)

serves in the Middle East. Most American oil shales are located in Wyoming, Utah and Colorado (Fig. 8–36). Extraction of the oil from shales may pose serious problems. Large quantities of water are required to extract petroleum from oil shales, and water is in short supply in the region where the oil shale deposits occur. A recent report estimates that by using all available water in Colorado, Utah and Wyoming, only 3 to 5 million barrels of oil per day could be produced. More realistically, a maximum production rate of 1 million barrels per day might be maintained leaving adequate water supplies for other purposes. This rate of production would supply 4 to 5 per cent of United States petroleum needs and could be fully operational by 1983. In addition, disposal of solid debris, mainly crushed rock, is a serious obstacle to oil shale development. Careful planning of the development of oil shales is essential.

The United States has adequate reserves of petroleum and natural gas to last for many years, provided they are efficiently managed. But management is difficult. Development of the most likely sites for petroleum accumulation, off the east coast of the United States, has been blocked by environmentalists on grounds which, in the opinion of the author (JSW), are geologically, biologically and technically unsound. The issue is an emotional one, however, which may remain unresolved for some years. In the meantime, the danger of geographic polarization appears to be increasing. Bumper stickers in the Gulf Coast states proclaim: "Let the Bastards Freeze in the Dark." The feeling on the Gulf Coast is that northeasterners want the oil but also want the pollution hazards to remain on the Gulf Coast.

Thus many problems remain unsolved. At the present time, the future course of petroleum production and distribution in the United States looks anything but smooth.

3 NUCLEAR ENERGY

"No one knew how long tank 106-T had been leaking, or how much of its caustic, boiling contents had seeped into the sandy soil near the center of the reservation. As a matter of fact, no one was certain how much liquid had been in the tank in the first place."

From a report by Robert Gillette describing a radioactive leak near Hanford, Washington, in *Science*, the official publication of the American Association for the Advancement of Science, August 24, 1973.

"*. . . not only regrettable, but disgraceful.*"

Dixie Lee Ray, Chairman, Atomic Energy Commission.

"Enough work has been done to permit a reasonable assessment of the major issues of nuclear power. Most of the recent fluctuations in energy patterns tend to reinforce what seemed evident even several years ago: *a massive switch to nuclear power for energy generation*" (italics added by the author).

From a paper by David Rose, Science, April 19, 1974.

The preceding quotations illustrate the risks and the realities of nuclear power. Nuclear power is a chained dragon. Released, its hot radioactive breath devastates all in its path. Controlled, it promises to solve man's energy problems for many years. Can we control it without sinking out of sight in a quagmire of radioactive waste? Or will waste heat upset the earth's climatic balance, bring on glaciation or turn the earth into a hothouse?

Answers to our questions depend on two pieces of hardware and on two related environmental problems. The hardware items are the *breeder reactor* and the *fusion reactor*. The environmental problems concern adequate *nuclear fuel* and *pollution*. We will only briefly discuss hardware problems because they are not directly relevant to geology. The issues of nuclear fuel and pollution will be examined at greater length because of their geologic aspects.

Fission and Fusion

There are two types of nuclear processes, fission and fusion. In *fission*, the nucleus of the radioactive isotope of an element, commonly uranium-235, *splits* when bombarded with neutrons (Fig. 8–37). Splitting produces new elements (barium and krypton in the case of uranium-235), additional neutrons and *energy*. If enough uranium (or other isotope), called a *critical mass*, is present, neutrons released during splitting will hit other nuclei and split them. The reaction will then proceed spontaneously without additional neutrons from an external source. This is called *chain reaction*.

In *fusion*, nuclei of light atoms, typically a hydrogen isotope, combine to form a heavier element, typically a helium isotope. This reaction also releases neutrons and energy (Fig. 8–38).

The quantities of energy released in these processes are tremendous. The heat potential of one pound of uranium-235 equals that of more than 6000 barrels of oil. Deuterium (heavy hydrogen) in a *cubic yard* of sea

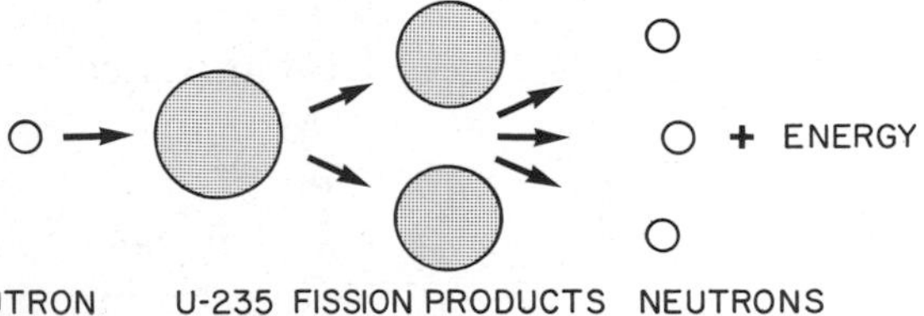

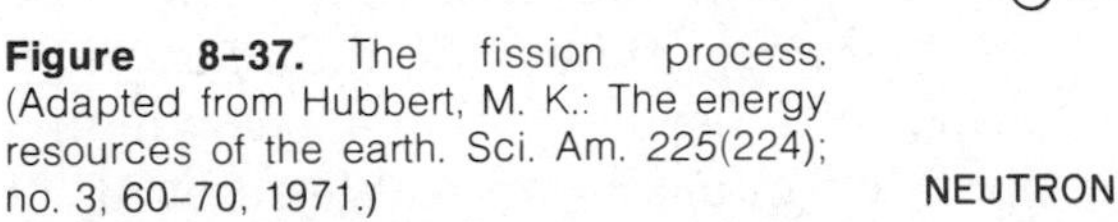

Figure 8–37. The fission process. (Adapted from Hubbert, M. K.: The energy resources of the earth. Sci. Am. *225*(224); no. 3, 60–70, 1971.)

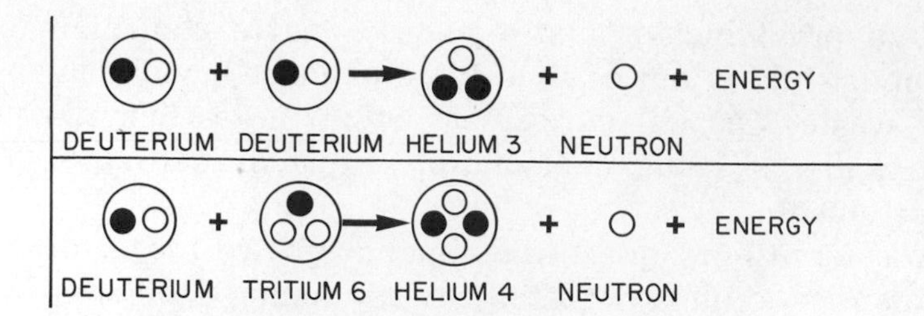

Figure 8–38. The fusion process. (Adapted from Hubbert, M. K.: The energy resources of the earth. Sci. Am. *225*(224), no. 3, 60–70, 1971.)

water has a heat potential equal to that of 1000 barrels of oil—and only 1/5000 of the hydrogen atoms in sea water are deuterium atoms. The potential energy in a cubic mile of sea water is about the same as the energy available from all of the world's petroleum. Clearly, the stakes in the nuclear power issue are high.

Problems in Nuclear Energy

Fuel resources present a major problem in development of nuclear energy. Their availability is more or less inversely proportional to the level of hardware development. For example, the simplest type of nuclear reactor is a *burner*. Burner reactors are well developed and widely used. The fissionable material used in a burner reactor, usually uranium-235 or plutonium-239, is totally consumed. Uranium found in nature is principally in the form of uranium-238; uranium-235 comprises only 0.7 per cent of naturally occurring uranium. Easily extractable natural supplies of uranium-235 are sufficient for a few decades at most. After these are consumed, costs of uranium-235 will rise greatly.

The *converter* is a more efficient reactor. In a converter, excess neutrons convert relatively abundant uranium-238 and thorium-232 into fissionable isotopes, which can be used as fuel. If a converter makes more fissionable material than it consumes, it is called a *breeder* reactor. Breeder reactors, when perfected, will be able to use uranium-238. Domestic reserves of uranium-238 are sufficiently abundant to supply United States needs for hundreds of years (Fig. 8–39).

Although breeder reactor development has been identified as a national goal of the United States, progress has been slow, and problems have been many. Without a breeder reactor, easily mineable reserves of

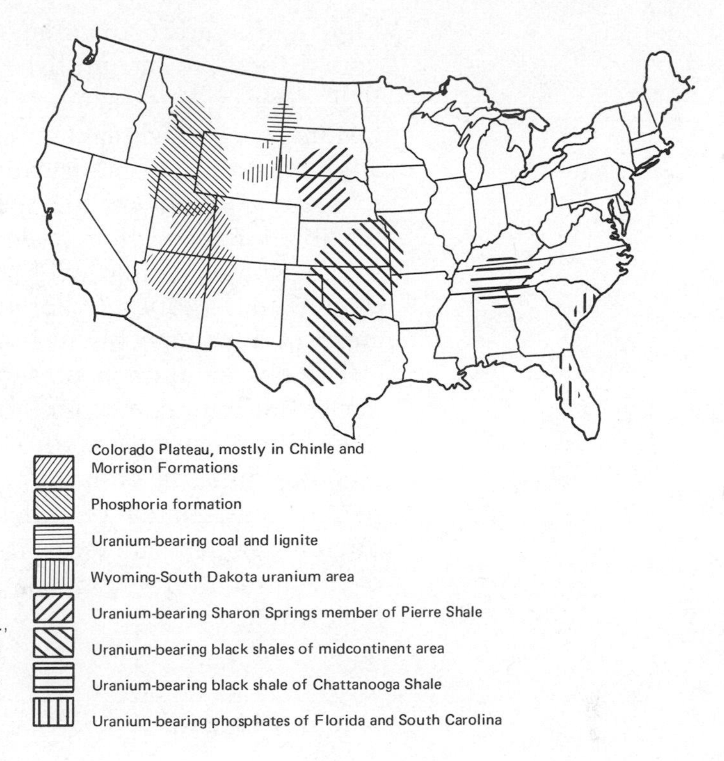

Figure 8–39. Uranium deposits of the U.S. (Adapted from Cargo, D. N., and Malloy, B. F.: Man and His Geologic Environment. Reading, Mass., Addison-Wesley Publishing Co., 1974.)

uranium-235 will be consumed by about 1990, and we will be forced to use ores costing 1.5 to 3 times as much as those now in use.

Measured against quadrupling costs of crude oil, increases in uranium ore costs seem less staggering, but it should be remembered that increased petroleum costs entail increased mining costs, which may affect the ultimate price of uranium ore. Although the cost of future uranium-fueled reactors may remain competitive with costs of fossil fuel–burning generators, the price will be high by pre-1973 standards.

Two primary types of fusion reaction, tritium-deuterium and deuterium-deuterium, appear to be feasible. The tritium-deuterium reaction seems most promising for early use. However, the main source of tritium, an isotope of hydrogen, is through nuclear bombardment of lithium, which is a relatively rare element. Unless another tritium source can be found, the tritium-deuterium reaction will not be a great help in solving the world's long-term energy needs.

As previously mentioned, there is plenty of deuterium. But the deuterium-deuterium reaction, in which deuterium nuclei must be squeezed together at high temperature in order to sustain the reaction, is very difficult to accomplish in controlled surroundings. It is done in nuclear bombs by surrounding the deuterium with fissionable material and setting it off. High heat and pressures caused by fission are adequate to start a fusion reaction. But this technique is not feasible for generation of electric power. A method has to be found which will squeeze small volumes of deuterium isotopes. Otherwise the result is a bomb instead of a manageable heat generator.

The best method at present utilizes an intense magnetic field to compress hot ionized gas containing deuterium. Another method receiving much attention recently uses laser beams concentrated on a small pellet. The energy of the laser beams squeezes the nuclei together. However, it seems unlikely that either of these techniques will be in commercial use in less than 20 years.

Nuclear Hazards

As opera and literary buffs know, Faust sold his soul to the Devil in return for money, power and Marguerite's charms. Are we in the twentieth century selling the lives and happiness of our descendants to the Devil in return for a few years of unlimited nuclear power? Are radioactive wastes generated from our reactors a serious threat to the future of mankind and much of terrestrial life? Some scientists fear that this may be true. The long *half-lives* (the time required for half of a given sample to decay) of some waste products, the Hanford leaks in Washington and the increased levels of strontium in the atmosphere and in the food chain all support this view. However, closer examination of the facts indicates that this prognosis, while possible, is not necessarily inevitable if proper precautions are taken to dispose of radioactive waste.

There are two major nuclear reactor hazards: *reactor safety* and *waste disposal*. Reactor safety is a short-term hazard resulting from possible explosions at reactor sites or theft of nuclear material for subversive purposes. Antitheft security mechanisms are presumably in force, and explosion is a technical problem which has been discussed at length in popular and scientific journals.

Radioactive waste disposal is a geologic problem. Reactor wastes fall into two categories: fission products of *intermediate atomic weight* (strontium-90, cesium-137 and krypton-85) and the *heavy-element residues* (isotopes of plutonium, neptunium, curium, americium, and others). The principal intermediate-weight isotopes all have half-lives of 30 years or less. Within 700 years, radioactivity due to these isotopes will decrease to $1/10,000,000$ of its original intensity. Heavy-element isotopes typically have much longer half-lives. Plutonium-239, for example, has a half-life of 26,000 years. These isotopes are all very toxic because of their tendency to collect in bone and other body sites. If merely stored, dangerously high levels of radioactivity will persist for one million years or more.

At the present time, plutonium and uranium are extracted from the waste for use as reactor fuel. The level of extraction, 99.5 per cent, is determined by profitability. Extraction of 99.9 per cent of uranium, neptunium and plutonium and 99 per cent of americium and curium would reduce radioactivity by a factor of 100 and cost an estimated 0.02¢ per kilowatt-hour. Such additional extraction would also greatly decrease the storage problem. A million-year hazard becomes a more manageable 700-year hazard. Clearly, the social benefits of additional extraction justify the extra cost.

The *form* of the waste is an important factor in its disposal. At Hanford, the waste is a rusty-brown fluid, stored in tanks, which boils continuously owing to heat released by radioactive decay. In the past, some waste fluid has been pumped into deep disposal wells. This is considered to be a less desirable method because of possibilities of groundwater contamination and other hazards (see the discussion of the Denver earthquakes, Chap. 2).

Most presently envisaged disposal techniques are based on the premise that the radioactive waste will be reduced to solid cakes. A number of disposal sites have been suggested including outer space, Antarctica, oceanic abyssal plains and natural or man-made underground caverns. In addition, it may be possible to convert long-lived isotopes into short-lived isotopes by nuclear chemistry.

The AEC seems to favor burial in bedded salt. Bedded salt deposits in the continental interior—not the mobile salt domes of the coastal regions—are one of the most stable geologic structures known. Bedded salt strata are free of water, are good radiation shields, and their plasticity under stress tends to anneal any cracks developed as a result of earthquakes or other tectonic stress.

An early plan to store radioactive wastes in bedded salt near Lyons, Kansas, was abandoned when it was discovered that owners of a mine less than one half mile from the proposed disposal mine had experimented with hydraulic fracturing (a means of increasing recovery from oil and gas wells, discussed earlier in this chapter). The owners of the adjacent mine had "lost" 175,000 gallons of water and possibly compromised the geologic integrity of the proposed Lyons disposal site. Other sites in Kansas and New Mexico are being investigated.

Required storage volumes for nuclear waste are small. If all electricity in the United States were generated by nuclear power for the next 350 years, the waste could be stored in a cavern 200 × 200 × 200 feet. This size is negligible in comparison with the amount of ground made unusuable for long periods of time by strip mining.

Health hazards are also minor in comparison with those due to other fuels. For example, recent data indicate that SO_2 (sulfur dioxide) and other air pollutants released by burning fossil fuels reduce the life span of people in urban areas by roughly three years. This is in addition to material and social losses such as deterioration of buildings and works of art.

These are some of the hazards involved in the use of nuclear fuels, but it must be remembered that hazards exist in the use of fossil fuels, too. Pollution of air and water are well known. From time to time disastrous fires have swept refineries. Present-day monitoring techniques for nuclear equipment are more effective than those of earlier decades. On the whole, problems arising from nuclear fuel use have probably been fewer than those arising from obtaining comparable amounts of energy from fossil fuels.

Nuclear fuels are very probably the next link in the chain of man's major energy sources. Man used wood, then coal, then petroleum; soon radioactive isotopes will be his main source of fuel. The issue of nuclear energy has excited strong emotions because of limited public knowledge due to the newness of atomic technology. As public awareness and understanding increase, it may become clear that the real and serious problems of nuclear technology are outweighed by its benefits.

4 OTHER ENERGY SOURCES

Basically, all sources of energy available for human consumption derive either from the sun or the earth-moon system. We use solar energy in the form of direct solar energy, fossil fuels, wind power and water power. We use earth-moon energy in the form of nuclear power, tidal power and geothermal power.

The energy potential of both solar and terrestrial resources far exceeds our present rate of use. The earth intercepts over 250,000 trillion kilowatt-hours of solar energy each day. Humans consume only a tiny fraction of this amount. Similarly, worldwide radioactive heat released into the earth's interior exceeds by many times the amount man is likely to consume during his stay on earth.

Estimates of potential power are academic, however, because the key to the problem is not total energy but *recoverable energy*. Recovery is a complex, expensive and often technologically difficult process. Historically, man has used the most easily recoverable fuels and then progressed to fuels which were harder to recover. Wood, man's first fuel, is the easiest to recover. Wind power, which is also easily converted to useful energy, was used early in history to push ships. Coal can be mined using simple technology, petroleum extraction requires a moderately complex technology and nuclear power requires high technology. Direct solar energy is a widely available, clean source of energy, but the technology of converting solar energy into useful electricity is a primitive stage of development, so it is little used.

The cost of recovery is an important factor. One reason man relies on fossil fuels is that they are relatively cheap to recover and utilize. The future of less used energy sources depends largely on development of economically viable recovery techniques. Plenty of coal remains deep within the earth, and plenty of petroleum lies buried beneath the sea floor, but both are too expensive to recover at the present time.

Transportation is an important aspect of energy consumption. Coal and petroleum can be transported inexpensively by ship. Petroleum and natural gas can be piped from one region to another. Most other sources of energy are converted to electricity, and the energy is "transported" by power lines. But transmission losses due to resistance of transmission lines makes this one of the most expensive means of energy transport. It is slightly more expensive than transporting coal by train, and 70 per cent more expensive than transporting natural gas by pipeline. Due to inefficiencies in generating electricity and power-line lossage, it may cost 20 times as much to heat a house with electricity made from natural gas and transmitted by overhead power lines as it would to heat with the same natural gas transported by pipeline.

Given the costs of conversion to electricity and transmission of electric energy, it is easy to see the effect of *geographic location* on energy resources. Tidal power is restricted to coastlines, geothermal power is restricted to volcanic regions and water power (dams) is restricted to favorable topographic regions. Transportation cost of radioactive reactor fuel is low because of its high energy/weight ratio, a factor which has contributed to the growth of the nuclear power industry.

With the preceding discussion in mind, let us examine some of the alternatives to nuclear and fossil fuels.

Solar Power

Solar power is one of the most intriguing alternatives to fossil fuels and nuclear energy. Solar power has been used in small quantities for many years as, for example, salt produced by evaporation. Fresh water can be obtained cheaply by evaporation of salt water beneath a transparent dome. In the United States, Japan, Australia and Israel, sun-heated water circulating through pipes warms houses. Archimedes is supposed to have set fire to a Roman fleet attacking Syracuse in 212 B.C. by focusing the sun's rays on the ships with concave mirrors. Today solar furnaces employing the same technique focus solar rays into an oven to obtain the high, clean heat required to produce ultrapure crystals and high-temperature refractory materials.

Use of solar energy requires a method of conversion into electricity. Solar cells used in the space program are an example of such a method. Sodium solar cells, the most widely used variety, have an efficiency of only 10 to 12 per cent, which means that a large collecting area is needed to generate a useful amount of electricity. Solar cells are also expensive. An optimistic estimate of the capital cost of cadium sulfide cells for generation of electricity on a large scale is $250 per kilowatt-hour. This is about 5 times as expensive as electricity from fossil fuel–fired generators, and more than 3 times as expensive as electricity from nuclear reactors.

Transmission of solar energy is also a problem. The best sites are in the arid West, but the greatest need is in the cloudy East. Solar generators in the eastern United States would operate at substantially lower efficiencies and higher costs owing to smog and clouds. Electricity from western solar energy generators would be too expensive to transmit to the East Coast.

Figure 8–40. A geothermal power plant on the island of Guadeloupe, French Antilles. This island is near Martinique, the site of the disastrous 1902 eruption of Mount Pelée discussed in Chapter 3.

Geothermal Energy

Like solar energy, the potential of geothermal energy is vast. The U.S. Geological Survey estimates that United States potential for geothermal energy is 350,000 times greater than energy from all sources used in the United States in 1970. (Sources of geothermal heat are discussed in Chapter 3.)

The average increase in temperature with depth in the earth is about 30°C. per kilometer. In most areas, geothermal wells must penetrate many kilometers to find rock hot enough to generate significant amounts of geothermal power. Because hot rocks occur at shallow depths in volcanic regions, all planned or operating geothermal power plants are in these areas (Fig. 8–40), including the Geyser field in northern California, the Larderello field in Italy, the Valles Caldera in northern New Mexico (see Chap. 3 for a discussion of calderas) and fields in Mexico, Japan, New Zealand and Iceland. It seems likely that future development of geothermal energy will be restricted to volcanic regions of the world.

Geothermal energy reservoirs can be divided into two major groups: *wet* and *dry*. In wet reservoirs, circulating ground water is converted to either *dry-steam* or *superheated water;* that is, water heated to temperatures above 100°C. but remaining liquid because of high pressures. In superheated water fields, the water "flashes" to steam as it rises to the surface, the exact point depending on temperature and pressure.

In dry-steam fields, the water is generally under lower pressure than in superheated water or wet-stream fields. When tapped, steam from the reservoir is fed into steam turbines, which turn electric generators. The Geyser and Larderello fields are of this type. The energy potential of superheated water appears to be much greater than that of dry-steam fields. Fields of the superheated type are producing in New Zealand, for example (Fig. 8–41).

Another type of geothermal power resource is dry (not dry-steam) geothermal power. This type of installation is still in the experimental stage. A test well was recently drilled 5 km into hot dry granite in the Valles caldera of New Mexico. Rocks at the base of the well were cracked by hydraulic fracturing. Cold water was pumped into the base of the fractured zone, where temperatures are about 300°C. (Fig. 8–42). Resulting superheated water escaped into a second well drilled into the top of the fractured zone. This experiment may herald a new trend in geothermal energy because zones of dry hot rock are probably more widespread than zones of naturally heated ground water.

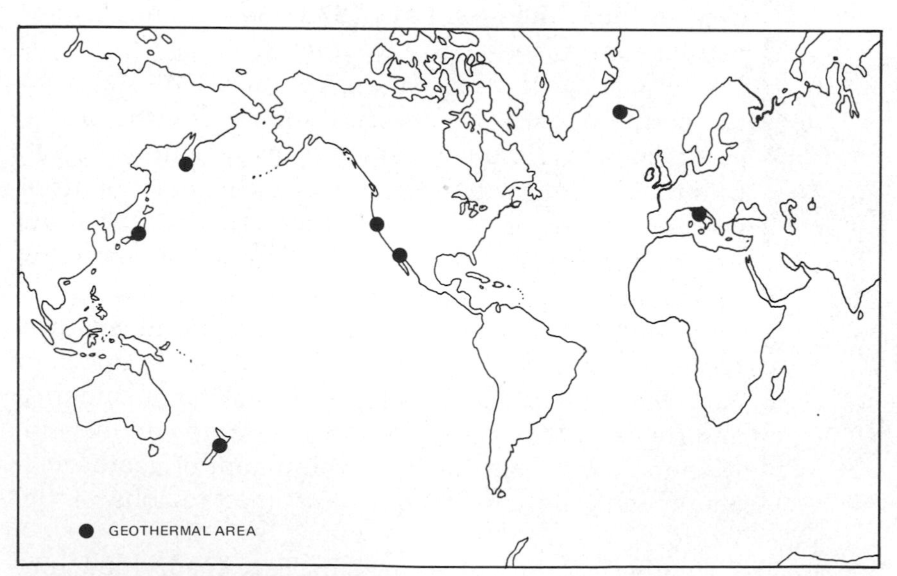

Figure 8–41. Geothermal power sites. (Adapted from Cargo, D. N., and Malloy, B. F.: Man and His Geologic Environment. Reading, Mass., Addison-Wesley Publishing Co., 1974.)

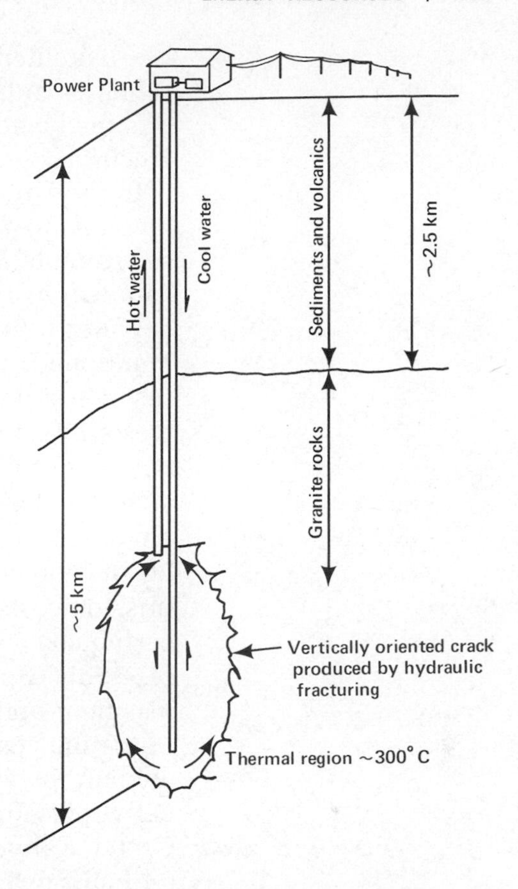

Figure 8–42. The Valles caldera dry-field geothermal experiment. (Adapted from Hammond, A. L.: Dry geothermal wells: promising experimental results. Science *182*, 43–44, 1973.)

Geothermal power is not without its limitations. Geographic limitations have been previously discussed. In addition, costs of geothermal power are not well established. Existing generators are at locations where steam seeps from the ground, and development costs may be lower at these sites. Hot water in the Imperial Valley of Southern California contains a high percentage of dissolved salts. These hot brines severely corrode metal and make power generation difficult. Removal of salt is costly, and disposal of salts and chemicals removed from the brines creates further difficulties.

The potential of geothermal power is as yet poorly known. It has been considered as a major source of energy only in the last decade. Much remains to be discovered about the geology of hot rocks and the technology and economics of exploiting geothermal power.

Water Power

Water power is a form of solar energy. The sun evaporates water from the oceans, the water vapor rises, condenses and falls as rain. Where topographic relief is sufficient, water running downhill can be dammed and made to turn turbines, which generate electricity.

In the United States, installed water-power capacity amounts to 45,000 megawatts; the estimated maximum recoverable is 161,000 megawatts, or about four times as much as is currently being utilized.

Water power has several limitations. As previously mentioned, *transmission costs are high*. Most large reservoirs (and favorable dam sites) are located in the western United States. Transmission of western electric energy to the heavily populated East would be expensive. The best reservoir sites in the United States have already been developed. New reservoirs and dams will be in more remote areas, which means higher transmission costs; in geologically questionable sites, where earthquake or other hazards exist; or in areas where reservoir capacity, height of the dam or water flow is less than optimal. All of these factors tend to increase operating costs.

Many potential dam and reservoir sites are aesthetically pleasing. Damming of a section of the upper Grand Canyon inundated beautiful rock formations and cliff dwellings. Proponents of the dam argued that the recreational value of the reservoir was of greater importance than the aesthetic value and scientific importance of the original site.

Water power is *not a renewable resource*. By trapping sediment behind the dam, reservoirs reduce the energy-producing life of a dam to only a century or two. Sometimes sediment entrapment has other destructive side effects. For example, sediment now being trapped by the Aswan Dam in Egypt formerly replenished the fertility of the Lower Nile each year. The blockage of the silt by the dam will seriously affect Egypt's food production in years to come.

The potential of water power in the United States is not great enough to rival fossil fuel or nuclear energy. Water power will never be more than a valuable but minor part of the United States energy complex. It is important to note, however, that the largest unused water potential in the world is in South America and Africa—areas deficient in fossil fuels. The water power potential of Africa is 780,000 megawatts and that of South America is 577,000 megawatts. Development of water power resources could help relieve chronic energy deficiencies of these continents.

Minor Sources of Energy

Of the existing minor sources of energy, tidal and wind power are the most widely discussed.

Tidal power derives from the rotational energy of the earth-moon system. As tidal energy is dissipated, the speed of rotation of the earth decreases. Astronomic data indicate that the present rate of decrease is equal to an increase in the length of the day by 0.001 second/century.

The *tidal power potential of the world* is an estimated 13,000 megawatts, which is about 8 per cent of United States water power potential and less than 1 per cent of the world potential water power. Consequently, tidal power stations do not now nor will they ever contribute very much to the world's energy resources.

The two best known tidal power stations are the Passamaquoddy Bay Station on the United States–Canadian border and La Rance estuary in northern France. Both have relatively high tidal ranges (5.52 m at Passamaquoddy and 8.4 m at La Rance). The La Rance station, which opened in 1966, was the world's first major tidal-electric plant. The dam encloses an area of 22 km^2, and its ultimate power production capability is 0.32 megawatts. The Passamaquoddy capacity is slightly less.

Although the world potential for *wind power* is much greater than for tidal power, exploitation of this resource presents a serious problem. Other types of power generation can, to some extent, be turned on and off, and the output of power generator can be matched to power demand. Wind power is intermittent and uncontrollable. This means that large amounts of power are available during part of the year (or month, or week or day) and unavailable at other times. So far the technical and economic difficulties of storing electric power have precluded widespread utilization of wind power. If a cheap and efficient method of storing electric power could be found, use of wind power would be more feasible.

5 ENERGY IN 1984

What will the energy situation be in 1984? Will Big Brother peer at us from countless TV monitors, telling us when we're using too much energy? Almost certainly not—because of the cost of the electricity to run the TV monitors if nothing else!

On the other hand, the freewheeling days of limitless gasoline and an excess of energy have almost certainly joined the Model-T Ford (which incidentally was not a bad automobile from an energy point of view) on Memory Lane. Gasoline wars and TV ads urging the use of bigger, more energy-consuming appliances will be replaced by ads urging conservation. There is even reason to believe that such a slow-down in our frantically paced society may prolong and improve the quality of life.

As we have so often observed, specific answers to questions regarding the future are difficult or impossible to obtain. However, several important facts appear well established. Let us review these.

It is clear that any short-term (0 to 10+ years) solution to the energy problem must involve production of additional quantities of fossil fuels, mainly oil and natural gas. All alternative sources of energy require long lead times for development of extraction and production technologies and development of an adequate energy distribution system. It required about 50 years to convert the United States economy from a wood-burning to a coal-burning energy base; approximately the same amount of time was required to convert from coal to petroleum and natural gas. Twenty years have elapsed since the development of a feasible nuclear reactor, and in spite of present intensive efforts another twenty or thirty years may elapse before technologic, environmental and logistic problems are solved and nuclear energy is a full participant in the national energy network. This goal will be reached only if a successful breeder reactor or fusion reactor can be developed.

To a lesser extent, these delays apply also to reactivation of coal as a primary fuel. Coal-burning locomotives are not built in a day, nor are new coal-burning electric power plants. Undesirable environmental effects of high-sulfur coals can be overcome, but time is required to improve the technology and to install cleansing units. Mines must be opened or reactivated and the distribution system expanded before abundance of coal in the ground can be translated into abundance of energy at the consumer level.

Analogous problems face development of oil shale deposits. In addition, government antipollution regulations will probably further delay utilization of this resource.

Tidal, water and geothermal energy do not have the potential necessary to replace fossil fuels and nuclear energy except in specific regions where these resources can be tapped easily. If our 50-year rule of thumb can be applied, solar energy may be a realistic alternate energy source by 2025. It certainly will not be feasible much before 2000 even if a crash program of research and development were initiated today.

The problems of meeting immediate needs are serious. A National Academy of Science panel concluded recently that it was unlikely that the United States could achieve energy self-sufficiency by 1985. They estimated that the following program would be required to maintain a modest 2 per cent per year increase in energy supplies between 1974 and 1985 (recently, energy use in the United States has been increasing at a rate of 4 per cent per year):

1) capital costs—$600 billion (much more than the total United States budget for 1974)
2) manpower—almost 600,000 new engineers, construction workers, equipment operators and coal miners
3) water—shortages are foreseen due to energy development requirements.
4) transportation—8000 new locomotives, 150,000 new hopper cars and greatly increased pipeline capacity.

The actual timetable of energy development will also depend heavily on planning. The United States in recent times has moved from crisis to crisis—civil rights, urban decay, hunger, drugs, campus unrest, insufficient medical care facilities, environmental deterioration and now energy. The crisis atmosphere must be conquered by a systematic approach to problems.

Most contemporary planning concentrates on only a few aspects of a given problem. "Total environment" study and planning is of vital importance. A new factory must not be judged solely on the basis of goods produced; economic benefits must be weighed against environmental costs. On the other hand, a particular project cannot be summarily discarded because of environmental hazards.

It is too late to "return to the farms" where most of our ancestors originated. Three positive alternatives exist: modest growth, zero growth or negative growth. Historically, civilizations have risen, overused their resources, abused and polluted their environments and fallen. We may have a chance to deviate from this pattern. Only time will tell whether or not our civilization has evolved sufficiently to do so.

9 MINERAL RESOURCES

High-pressure water spewing from this water "gun" washed down entire hills of gold-bearing sand and gravel during the heyday of gold mining in California.

1 INTRODUCTION

In July, 1853, Commodore Matthew Perry, USN, led a small squadron of ships to anchor into Tokyo Bay. His mission was to convince the Japanese that trade with the United States would benefit both nations. An observer accompanying Perry would have given Japan little chance of achieving eminence among the world's nations. True, the people were vigorous, and soon the astute, progressive Emperor Meiji would ascend the throne. But the islands of Japan lacked extensive areas of arable land and were almost totally deficient in iron ore and coking coal, the twin mineral foundation piers of the Industrial Revolution.

In 1894, Japan attacked Korea, then a province of nearly defunct Manchu China. The Japanese overran Korea, Formosa and southern Manchuria, but pressure from Russia and European nations forced them to relinquish territory occupied in Manchuria. Good quality iron ore in northwestern Korea assured Japanese industry of a domestically controlled supply of iron.

On February 8, 1904, a Japanese squadron attacked the Russian Pacific fleet at Port Arthur (now Lushun), which fell to Japanese landing forces on December 31, 1904. Mukden, the capital of Manchuria, fell on March 16, 1905. The treaty of Portsmouth, which ended the war, contained concessions that made Manchuria virtually a Japanese province. The Japanese gained control of low quality but usable iron ores in southern Manchuria and, far more importantly, of the coking coal they desperately needed in order to smelt iron ores.

By 1941, Japan, a country virtually barren of iron ore in 1853, was a major world power, producing about 2.5 million metric tons of iron and steel. Korean mines contributed 675,000 metric tons, Manchurian mines 1,157,000 metric tons and the recently occupied Yangtze Valley of China fed 1,529,000 metric tons of iron ore into Japanese blast furnances.

In 1941, Japan attacked the largest iron and steel producing country in the world—a country with a production rate about 20 times that of Japan. From the point of view of mineral economics, the outcome of the war was predictable and inevitable.

2 MINERALS IN ANCIENT SOCIETIES

Many factors contributed to the rise of civilization. The development of agriculture and the scientific method, the invention of writing and numbers—these are some of the steps ascended by the human race during recent millennia. Underlying these innovations, however, are four or possibly five great technologic advances that were fundamental to the rise of civilization. Without them, many other discoveries would have been delayed or not made at all.

First, humans learned to use stone implements. This gave them an advantage over other animals. No longer mere hairless apes, they could now kill and feed on animals larger, stronger and faster than themselves (although fossil bones showing healed wounds indicate that even equipped with stone weapons, human beings lost a number of battles with other beasts).

Second, *Homo sapiens* discovered how to refine metals that have simple chemistries and low reduction temperatures. This discovery allowed man to make more sophisticated implements—for agriculture, hunting, art and, of course, war.

Third, man learned to refine several less tractable metals, notably iron, which occurs much more abundantly than metals used previously. Farmers, who had not greatly benefited from previous metal discoveries, could now use improved metal implements, raise more crops and feed more people. Iron's greater hardness makes iron implements superior to those made of copper and copper alloys. With abundant hard iron axes, the forests of Europe could be cleared and colonization begun.

Fourth, humans learned to harness quantities of energy far greater than could be provided by animal power. Roman engineers built truly fantastic structures, but even they could never rise above the limits imposed by the lack of large energy sources. Until fossil fuel energies were effectively harnessed, energy limitations and the need to produce food tied most of the earth's population to the land.

Although it is too soon to say with certainty, the fifth great discovery seems to be that of nuclear power. If pollution and technical problems can be overcome, atomic power will provide an almost unlimited supply of energy.

In this chapter, we will examine aspects of the first, second, third and fourth discoveries, which formed the foundations of the Stone Age, Bronze Age, Iron Age and Industrial Revolution, respectively. Energy discoveries and sources of energy are discussed in Chapter 8.

Prehistoric Man

Primitive man's first use of minerals consisted of picking up stones and using them to break nuts, throw at lions and perhaps threaten recalcitrant companions. About one million years ago, humans discovered that stones could be shaped. This discovery and the use of fire were the first technologic inventions of the human race (Fig. 9–1). Some rocks, notably *obsidian*, a volcanic glass found in some lavas, and *flint*, a fine-grained variety of quartz, are more easily shaped than others. Both of these flake readily and form sharp edges suitable for knives, axes and arrowheads. The irregular

Figure 9–1. Stones used by primitive man at Olduvai Gorge, Tanzania. Bones in the picture are those of animals killed by the primitive man. He cracked the bones in order to eat the marrow.

occurrence of flint and obsidian led to one of humanity's earliest commercial ventures. Tribes with supplies of flint or obsidian traded them to neighboring tribes, who in turn traded them with more distant tribes. Flint and obsidian artifacts found far from any known source demonstrate the wide range of these early trading ventures. However, trade was not restricted to flint and obsidian. By 25,000 BC, gold, copper, salt, amethyst, quartz, jade, iron oxide and many other mineral materials were being bartered.

At the time of consolidation of the Upper and Lower Kingdoms of Egypt (about 3400 BC), Egypt's domestic resources consisted of soil, water, clay, stone and gold. The gold-mining region lay between the Nile and Red Sea. Egyptians mined domestic gold for many years, but later in their history, Nubia (Sudan) and Ethiopia supplied great amounts of gold to Egypt.

To supplement their meager domestic copper and turquoise resources, Bronze Age Egyptians attacked and conquered the more primitive peoples of the Sinai Peninsula. A later invasion of Cyprus brought rich copper deposits under Egyptian control. Readily available copper provided Egyptians with abundant bronze weapons. (At this time bronze was too expensive to be made into agricultural implements, and farmers continued to use wooden tools.)

Egypt and Mesopotamia

The earliest "civilized" communities of the Middle East and Mediterranean evolved in the fertile alluvial valleys (see Chaps. 5 and 10) of great rivers such as the Tigris, Euphrates and Nile. Availability of mineral resources determined the direction of technical evolution of these societies. Easily quarried limestone and sandstone outcroppings along the banks of the Nile encouraged development of the famous Egyptian stone architecture. The Pyramids, the Sphinx and many other examples of Egyptian stonework survived the ages and can be seen today. Mesopotamian (Tigris-Euphrates) cultures, lacking readily available stone, developed a magnificent ceramic technology. Their brick structures, while not as massive as the pyramids, were nevertheless marvels of Bronze Age engineering.

At the peak of their development, Mesopotamians irrigated and farmed vast areas. Eventually, salt accumulated in the soil and destroyed its fertility, and the Mesopotamian cultures decayed. The Nile flood, which washes away salt and renews Egyptian soils yearly, was a major factor in the long life of Egyptian culture.

The development of Egyptian civilization was limited by Egypt's dependence on foreign mineral sources. The loss of foreign copper mines was particularly damaging, because Bronze Age technology was based on copper alloys. (Bronze is an alloy of copper and tin.) Without domestic iron ores, Egypt became a technologic backwater after the beginning of the Iron Age.

Greece

Among ancient civilizations, Greece perhaps best exemplifies the role of minerals in development of civilization.

Phoenician sailors living in what is now Lebanon, motivated by a desire to locate sources of minerals, explored the Mediterranean and surrounding seas. One of their more important discoveries (circa 1000 BC) was the silver deposits of Laurium in Greece, southeast of Athens. Greeks later assumed control of the Laurium mines.

During the three centuries of the classical period in Greece, Laurium mines produced an estimated two million tons of silver-rich lead ore containing 40 to 120 ounces of silver per ton. Slaves burrowed into the mountains via 2000 shafts, some nearly 4000 feet deep. The demand for trees for use as mine timbers and as fuel for smelting the ores denuded surrounding hills and valleys.

Revenue from the mines was so great that citizens of Athens received periodic dividends. The wealth from the mines simultaneously underwrote Athenian cultural development and the Athenian war machine. It is ironic that slaves toiling in miserable subterranean holes provided the mineral wealth of the fabled Athenian democracy.

Greeks quarried marble, sandstone, limestone and other rock types for use in buildings, statuary and esthetic purposes. Ceramic arts and glass working were highly developed. In reality, Jason and the mythical argonauts in search of the Golden Fleece were sailors who visited Colchis on the east shore of the Black Sea, where miners extracted fine particles of gold from sand and gravel by washing gold, sand and gravel over sheepskins. The fleece trapped the gold but allowed sand and gravel to escape.

Bronze weaponry comprised the major portion of the Greek arsenal until the sixth century BC, when iron weaponry become more abundant. The Greeks do not seem to have been great exploiters of iron, though. Perhaps sparsity of domestic mines or of timber resources needed to reduce iron ore inhibited Greek development of iron technology. Depletion of Greek mineral resources and the rise of the vigorous iron-oriented Romans contributed to the gradual decline of Greek culture.

Rome

The development of early Roman civilization was based on Roman discoveries and on assimilation of culture, technology and resources of nearby peoples. Contributions of the Etruscans, who preceded the Romans as the dominant culture of west-central Italy, were particularly important.

The mists of time obscure the origins of the Etruscans.

Their language is not of Indo-European derivation, and their inscriptions remain incompletely deciphered. Herodotus suggested an Oriental origin for these people and hypothesized that they were drawn by mineral resources to the west coast of Italy between the Arno and Tiber rivers sometime before 800 BC. The Colline Metallifere, the area between Siena and Livorno, the peninsula around Populonia and the island of Elba offered copper, lead, tin and, most importantly, iron. Great slag heaps around Populonia show that the Etruscans exploited the mineral deposits on a grand scale; it has been estimated that the Populonian mines yielded 10,000 to 12,000 tons of iron annually for more than four centuries. Etruscan agricultural technology based on the availability of iron implements and fertile volcanic soils yielded some of the largest crops of ancient times.

The earliest Romans arrived in Italy before 1000 BC, settling along the Italian west coast immediately south of the Etruscans. Six hundred years elapsed before they conquered their first strongly fortified Etruscan city, Veii, in 396 BC.

The Romans were the greatest exploiters of mineral wealth in the ancient world. Stone quarries disgorged basalt, limestone, sandstone, granite, marble—never before were so many materials available to craftsmen.

Figure 9–2. The Roman Colosseum, built largely of brick.

Figure 9–3. Bricks of the Colosseum. Roman brick and mortar remain solid and firm after centuries of exposure to weathering.

The Romans mixed Neapolitan volcanic ash and lime to obtain a hydraulic cement that would harden underwater. In addition to stone, they built extensively with brick (Figs. 9–2 and 9–3). Abundant remains of iron furnaces throughout the empire indicate that Roman colonists produced large quantities of iron locally for their own use. Lead also was widely used in pipes, sheathing, weights and household utensils.

After the collapse of the empire, Roman ironmaking techniques were lost in some regions, and the superiority of "fossil" Roman weapons such as the Sword of Roland and King Arthur's Excalibur contributed to medieval mythology.

In the millennium between the fall of Rome and the onset of the Industrial Revolution, minerals continued to play an important role in the economies of various countries. The most spectacular mineral-related event was the introduction of New World gold into the Old World (Fig. 9–4). New World gold financed a glittering age in Spain in much the same way that silver financed the classical age of Greece.

Mining techniques gradually improved. In 1627, explosives first broke rock in European mines. Better hoists, pumps and rock crushers were developed. Perhaps most importantly, knowledge of ore-forming processes advanced, notably through the work of Georgius Agricola. Born in Saxony in 1494, he wrote *De Re Metallica,* one of the most significant contributions in the history of metals and mining.

On the whole, however, mineral technology ad-

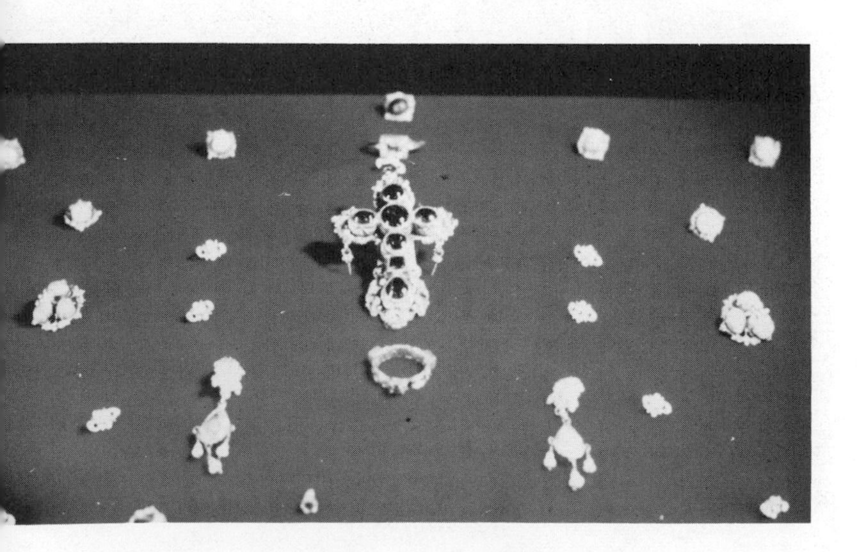

Figure 9–4. Gold jewelry recovered from a Spanish galleon that sank near Bermuda.

vanced very slowly in the years between the Roman Empire and the Industrial Revolution. Civilization marked time until the advent of the next major innovation in the mineral industries. This discovery was to result in cheap iron and triggered the almost unbelievable advances of the past two centuries.

3 MINERALS TODAY

Modern technology stands on the twin foundations of energy and minerals. At least 100 different minerals commonly contribute to the well-being of the human race. Minerals are essential to many processes. They are used in the production of food, foodstuffs, rubber, petroleum; they conduct electricity; they dam water in reservoirs; and they conduct water from the reservoirs to home and industries. With the advent of nuclear power, minerals supply energy.

Strictly speaking, a *mineral* is a naturally occurring chemical compound with a specific atomic arrangement. In this chapter, however, we will conform to industrial usage of the word and include a number of nonminerals such as alloys (e.g., steel), rocks (e.g., limestone) and other materials (e.g., coke) in our definition. The definition of mineral used in this chapter is roughly equivalent to "economic minerals, alloys, rocks and mineral products."

Minerals can be classified into six groups according to their functions in modern technology.

The *iron group* is the most important. It includes iron, iron ores, coke and the iron fluxes, mainly limestone. Iron is the basic metal of the industrial age. (A later section in this chapter describes occurrence, smelting and uses of iron.)

Iron alloys including manganese, tungsten, chromium, nickel, vanadium, titanium and others comprise the second group. Small quantities of these metals alloy with iron making it harder, stronger, tougher and more resistant to heat, corrosion and abrasion. Without the iron alloys, automobiles would not run as fast, machine tools would cut metal more slowly and generators would produce less electricity.

Aluminum, copper, lead, zinc, tin and magnesium comprise the *nonferrous metals*. Each metal in this group exhibits one or more special properties. Copper, for example, conducts most of modern society's electricity. Lead is used in batteries and as metal sheathing. Each of the other metals occupies a similarly distinctive niche in contemporary technology.

Aluminum, magnesium and titanium comprise the *light metals*. (Note that some metals appear in more than one group.) Because of the abundance of these metals in nature, some people have suggested that they may eventually rival iron as the principal industrial metal. At the present time, however, man produces only one light metal, aluminum, in large quantities.

Construction "minerals" include sand, gravel, gypsum, limestone, clay, stone, asbestos and other similar minerals and mineral aggregates. Huge tonnages of these materials are used daily.

The sixth group consists of *chemical and industrial minerals*, including the fertilizers (nitrate, phosphate, potash and sulfur), mica, feldspar, mercury, platinum, silver, antimony and several other less well known minerals.

A thorough discussion of all of the mineral groups far exceeds the scope of this book. In the following sections, we will discuss a few of the more important and representative minerals.

From the Land . . .

Iron and Steel

A small amount of carbon (up to 1.6 per cent) added to iron greatly increases the strength of the iron. We call this iron-carbon alloy *steel*. Steel is the single most important mineral product in contemporary society. Skycrapers, automobiles, airplanes, refrigerators and most other "utensils" of modern man are possible because of abundant low-cost steel. Steel is the wonder metal of modern society.

Occurrence. Most production of iron comes from four minerals: *magnetite*, a magnetic iron oxide containing 72.4 per cent iron by weight; *hematite*, the most common iron ore, a nonmagnetic iron oxide containing 70 per cent iron by weight; *limonite*, essentially hematite combined with water; and *siderite*, an iron carbonate, occurring only rarely in commercial quantities.

Ores of the Lake Superior region account for most domestic United States iron production. Weathering processes slowly convert beds of iron-rich sedimentary sandstone (see Chap. 5) deposited two billion years ago into rich ores. Ground water percolating through the standstone gradually dissolves the sandstone, leaving behind deposits of almost pure hematite. The richest deposits have been depleted, and mining has begun in zones where ground water has not yet removed all of the standstone. The iron-sandstones, know as *taconite*, require concentration prior to smelting.

About 420 million years ago, streams eroded the flanks of a high mountain range rising in the present site of the Appalachian Mountains and Piedmont of the eastern United States. The westward-flowing streams deposited their load in a long narrow sea that paralleled the mountains. Some of the sediments were very rich in iron.

Subsequent earth movements uplifted the sea floor, and erosion has now exposed the sediments deposited in the old sea. Hematite-bearing sediments now outcrop in a narrow zone extending from Alabama to New York and around to Wisconsin.

The low iron content of the sediments (36 per cent) would ordinarily preclude utilization of these ores. Owing to a lucky geologic accident, however, they support a flourishing iron industry in central Alabama. Limestone needed for flux was deposited immediately prior to deposition of the iron-rich sediments, and coal needed for coke production formed in the same area somewhat later. Subsequent folding and faulting deformed the rocks, bringing together coal, limestone and iron ore outcrops in a zone 30 miles wide. Proximity of these mineral materials makes costs of mining and smelting competitive with costs of mining/smelting richer ores in other areas where transportation costs are greater.

A deep road cut in the Red Mountain Expressway near Birmingham exposes the "Big Seam," which has provided much of the Alabama ore (Fig. 9–5).

Most other important ores in the world also consist of hematite. The French mine siliceous hematite in the Lorraine region, the British mine hematite and limonite in eastern England and the Russians mine hematite in the Ukraine. The principal nonhematite ore mined today is a magnetite ore that occurs in

Figure 9-5. The Red Mountain Expressway road cut near Birmingham, Alabama. Iron ore and underlying limestone are exposed in the cut. Coal to fuel Alabama's iron industry outcrops nearby. (From LeMoreaux, P. E., and Simpson, T. A.: Birmingham's Red Mountain Cut. Geotimes, 10-11, 1970.)

Sweden. Large deposits of iron ore also exist in South America, Africa, Asia and Australia. Except where they are accessible to water transport (for example, in Venezuela), costs of transporting the ores to coal-bearing regions or vice versa limit their economic potential.

Pig Iron, Wrought Iron and Steel. The most common iron ores consist of chemical compounds of iron and oxygen. In about 1100 BC it was discovered that iron ore placed in an oven and heated with charcoal produces free iron. This reaction takes place because burning charcoal gives off carbon monoxide (CO). At high temperature CO takes oxygen from the iron ore and forms carbon dioxide (CO_2). The porous sponge-like residuum of free iron is called a "bloom." Iron blooms are heated and hammered (or *wrought*) into the desired shapes. Wrought iron is superior to *pig iron* (unhammered) because of strength imparted to the iron by hammering.

Although numerous improvements in iron-ore smelting techniques were effected during the Iron Age, availability of trees for charcoal production limited iron production. Then in England in 1730, a discovery was made. The importance of this discovery proved to be second only to the development of agriculture.

Industrial Revolution. England in the eighteenth century was richly endowed with mineral deposits. Phoenicians had discovered tin in Cornwall sometime after 600 BC. Romans mined and exported lead and iron after their conquest of the island. Roman soldiers and colonists mined gold, silver, tin, copper and iron for domestic use. English tin miners were granted special courts and special privileges in the twelfth century. A developing iron industry flourished in the seventeenth and early eighteenth centuries. As had been the case in Greece 2000 years earlier, demand for charcoal and mine timbers was rapidly deforesting the land.

A Shropshire Quaker family, the Darbys, owned a substantial iron industry. In the early eighteenth century, depletion of timber reserves was threatening their business and others like it. After years of experimentation, the Darby family discovered that when heated under suitable conditions, volatile impurities were driven out of coal, leaving a charcoal-like residue that could be substituted for charcoal in the smelting of iron. The coal residue is called *coke*. Following the Darbys' discovery, coke and iron ore formed the basis of subsequent steel technology. Today, one ton of iron ore, one ton of coke and one half ton of limestone are used to make a ton of pig iron. (Limestone combines with impurities in the melt, floats to the top and is skimmed off.) Coke contributes heat and carbon monoxide to the process.

England had (and still has) large quantities of coal and limestone. Suddenly it was possible not only to produce more iron than ever before but also to produce it more cheaply. The threatened disaster of deforestation had resulted in a discovery of tremendous importance. Availability of cheap iron led to the invention and widespread use of the steam engine. Steam power gave man access to previously impossible quantities of concentrated energy. The discovery of coke and the invention of its offspring, the steam engine, gave rise to the Industrial Revolution.

Iron Today. In 1870, horses and mules supplied most of the power on American farms. Feeding and transportation of Americans required one horse for every four human beings. The same ratio today would require more than 50 million horses and mules. Actually, the United States has about 8 million horses and mules, few of which are used for agriculture. Steam and gasoline-generated power was the reason for the sudden "negative population explosion" among horses and mules. Without inexpensive iron and steel, widespread use of cheap power would not be possible, and many Americans would still ride horse and mule drawn conveyances.

In 1900, Americans produced about 300 pounds of steel per person; production rose to 660 pounds per person in 1919, then to more than 900 pounds in World War I and peaked at over 1300 pounds during World War II. Since World War II, production has leveled off at about 1200 pounds per person. Americans now utilize about 1400 pounds of steel per person; iron ore is imported, primarily from Canada and South America, to make up the difference between consumption and production. The United States leads the world in steel consumption, but Japanese and Russian consumption levels are increasing rapidly (Fig. 9–6).

Two factors, availability of iron ore and availability of coking coal, determine the locations of the great steel production centers. Both iron ore and coal are heavy and bulky. Long-distance rail transport of either iron ore or coal greatly increases cost of iron and steel.

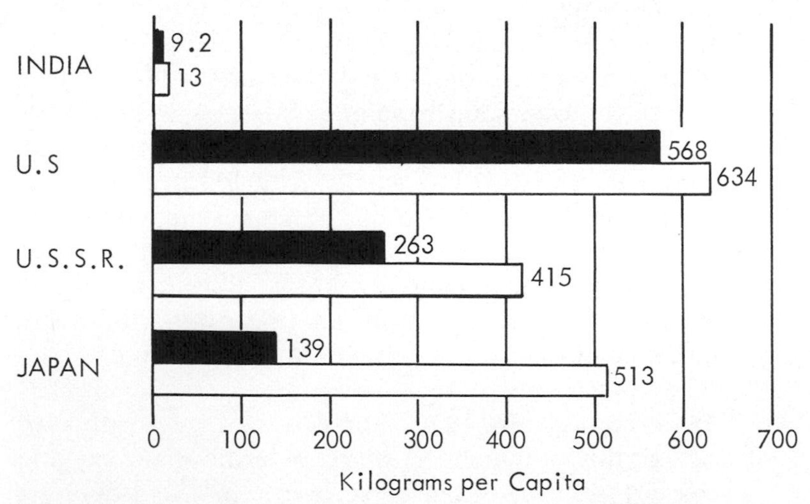

Figure 9–6. Steel consumption of selected countries in 1957 and 1967. Japanese and Russian consumption is rising much faster than U.S. consumption. (Adapted from Brown, H.: Human materials production as a process in the biosphere. Sci. Am. *223*:194–208, 1970.)

This fact alone accounts for the reduced usefulness of large iron ore deposits in the African and South American interiors.

Water transportation of iron ore is much cheaper, an important factor in the growth of the Great Lakes steel industry, where coal from Pennsylvania smelts iron ore from Minnesota. Venezuelan iron ore now finds its way into the blast furnaces of the East Coast and Alabama. Ships from Venezuela traverse the Caribbean and the Gulf of Mexico and then complete their journey to Birmingham via navigable rivers. Steel production in Japan, a country with insignificant domestic iron-making resources, depends on cheap water transportation to provide ore and coking coal.

Although immense reserves of iron ore exist in the interiors of India, Brazil, Canada, Newfoundland, Venezuela, Cuba and Africa, only a few of these are accessible to water transport and thus commercially useful at present.

The five major steel production centers in the world are all located in areas where coal and iron ore occur together in abundance. The major centers are as follows:

1) Great Lakes region of the United States
2) Great Britain
3) Lorraine-Ruhr region of France and Germany
4) The Donets basin of the U.S.S.R.
5) The Kuznetsk basin of the U.S.S.R.

Japan is a lesser center of steel production (Fig. 9–7).

In 1848, opening of iron mines near Marquette, Michigan, marked the beginning of the iron industry in the Great Lakes region. Following the opening of the Pittsburgh coal seam in 1860, the center of United States iron production, then located along the East Coast, moved westward to the Lakes region. Considering the fact that Great Lakes iron and steel formed the backbone of the Union war machine during the Civil War, it is interesting to speculate that if South Carolinians had fired on Fort Sumter ten years earlier, the Civil War might have ended differently.

Great Britain now imports ore from Spain, Sweden and Africa to supplement domestic low-grade ores. Two centuries of intensive coal mining to feed British blast furnaces has seriously reduced Great Britain's supply of low-cost coal, and the British steel industry faces difficult times in forthcoming decades.

Asia

China(mainland)	27,600
Hong Kong	142
India	25,744
Iran	NA
Japan	2,184
Malaysia	5,350
North Korea	6,400
Philippines	1,482
Republic of Korea	687
Thailand	540
Turkey	1,462

Oceania

Australia	18,517
New Caledonia	201

Figure 9–7. World iron ore production in 1967 plotted on a predrift reconstruction of the Atlantic continents. Availability of coking coal determines location of major production centers. The southern hemisphere and India contain little coking coal because climate was too cold during the earth's major coal-forming epoch. Japan ranks as a major production center because of cheap water transportation. (From United Nations Survey of World Iron Ore Resources. New York, United Nations Publications, 1970.)

The largest deposits of iron ore in continental Europe occur in the French province of Lorraine. (A geologic extension of the Lorraine deposit occurs in Luxembourg.) The principal European steel production center, however, lies in the Ruhr Valley of Germany near extensive coal deposits. Following the 1871 war with France, Germany obtained control of the Lorraine deposits, but the Versailles Treaty ending World War I returned the Lorraine to France. This and other concessions cost Germany 80 per cent of its iron ore.

German strategy prior to and during World War II was motivated in part by mineral economics. Germany first gained control of small iron deposits by annexing Austria; then they took over the Bohemian and Slovakian deposits in Czechoslovakia; and they regained control of the Lorraine deposits with the defeat of France. Germany was reaching for Russian-Ukrainian deposits when the Russians stopped and reversed the German advance at Stalingrad.

Russian iron ores include extensive deposits of hematite near Krivoi Rog in the southern Ukraine and the Crimea. Excellent coking coal and limestone lies in the Donets basin 250 miles from the Krivoi Rog deposits.

Kuznetsk basin coal smelts iron ore mined in the Ural Mountains of Russia. The most important Ural deposit is at Magnitogorsk. To reduce transportation costs over the 1250 miles separating the Ural iron ores and Kuznetsk coals, ores are smelted at both locations. Trains running east carry iron ore to Kuznetsk furnaces; trains running west carry coking coal to Magnitogorsk furnaces.

Elevation of the standard of living in underdeveloped countries is one of the most serious tasks facing modern civilization. The standard of living of a given country is closely related to its ability to produce iron and the per capita amount of iron in use. At the present time, 18 nations with a combined population of 680 million people consume steel at rates varying from 650 to 1400 pounds per person per year. By comparison, 1800 million people of underdeveloped countries consume less than 55 pounds per person per year. About 500 million people consume steel at intermediate rates.

It would seem that the first step in raising the

standard of living in an underdeveloped country would be to provide sufficient steel for automobiles, roads, industrial equipment, communication equipment and so forth. Unfortunately the situation is much more complex. From a mineral economics point of view there are two fundamental problems.

First, the amount of steel involved is huge. Supplying underdeveloped nations with enough steel to raise their per capita inventory up to the average level of the 10 richest nations would require the total output from all the world's facilities operating at their present levels for *the next 60 years.* To raise the per capita steel inventory up to the United States level of 10 tons per capita would require the output of the world's iron-making facilities for *the next 200 years.*

Second, uneven distribution of steel-making resources hampers development of poor countries. The United States and the U.S.S.R. control the overwhelming majority of the world's coal and thereby control the world's iron and steel-making resources. Even Europe is relatively deficient in comparison with the United States and U.S.S.R. (Fig. 9–8). South and Central

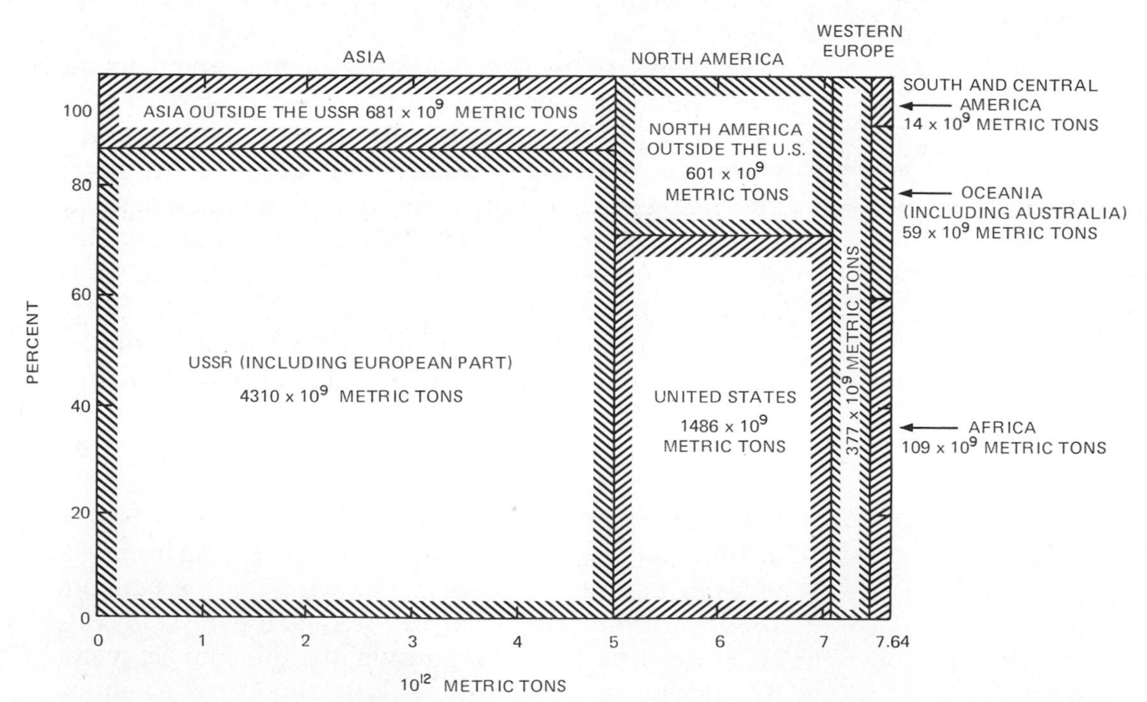

Figure 9–8. Mineable coal in the world. (Adapted from Hubbert, M. K.: Energy resources. *In* Committee on Resources and Man: Resources and Man: A Study and Recommendations. San Francisco, W. H. Freeman Co., 1969, p. 203.)

America, Oceania (including Australia) and Africa have small to negligible steel-making capabilities. It seems unlikely that countries in these areas will ever produce enough steel for their own needs. The main sources of steel in the future are likely to be the same sources we have today.

The problem of supplying adequate steel to underdeveloped countries seems insurmountable. Without a major technologic breakthrough, the lack of adequate steel (and other mineral and fuel resources) required for agriculture, transportation, manufacturing, communications and commerce seems to condemn peoples of most underdeveloped countries to a grim future.

Aluminum

Occurrence. Aluminum is the most abundant commonly used metal in the earth's crust, making up a little over 8 per cent by weight. In comparison, iron makes up 5 per cent, magnesium 2 per cent, and other commonly used metals comprise less than one per cent each.

Weathering processes in hot, humid climates dissolve and carry away silica and other undesirable constituents from aluminum-rich rocks. Eventually, weathering removes virtually everything except aluminum oxide combined with water. This residuum, called *bauxite* (after deposits in the Baux region of France), forms the principal aluminum ore in use today. Recent technologic advances have produced competitively priced aluminum from high-alumina clays. Eventually, it may be possible to produce aluminum from high-alumina rocks also.

Major deposits of bauxite occur on all continents. The largest American deposits occur near Little Rock, Arkansas, where weathering of nepheline syenite, an aluminum-rich rock, produced a thick mantle of bauxite. However, the Arkansas deposit contributes only a small percentage to total United States production. The United States imports bauxite from the Caribbean region, with Jamaica supplying the greatest quantities. The Dominican Republic, Haiti, Costa Rica and Panama have significant reserves. In South America, British Guiana, Surinam, Brazil and Venezuela have large reserves. In Surinam, for example, the coastal plain contains many square miles of bauxite. The deposit is about 15 feet thick, contains almost no impurities and can be "mined" with a bulldozer.

Cryolite and Electricity. We have discussed how a lack of coking coal needed to reduce iron ore limited the development of many iron ore deposits. An analogous situation prevails in the aluminum industry: aluminum ore is abundant, but two necessary ingredients in the smelting process, a relatively rare mineral called *cryolite* and cheap electricity, determine the location of aluminum production centers.

Virtually all aluminum ore is reduced to aluminum metal by the Hall-Heroult process, in use since 1896. Before processing, bauxite is treated to remove chemically combined water and undesirable impurities. The cleaned ore, consisting of pure *alumina*, Al_2O_3, is pounded, mixed with cryolite, Na_3AlF_6, and placed in a steel cell lined with carbon. Electric current passing through the cell melts the alumina-cryolite mixture and electrically separates aluminum and oxygen. The negatively charged carbon lining attracts positively charged aluminum ions while a positively charged electrode in the center of the cell attracts negatively charged oxygen ions. The oxygen escapes into the air. Liquid aluminum collects in the bottom of the cell, where it is drawn off and cast into *pigs*, or ingots. Aluminum obtained in this way is 99.7 per cent pure.

Two tons of alumina (obtained from approximately four tons of bauxite), one half ton of electrodes, 200 pounds of cryolite, and 100 pounds of aluminum go into the production of one ton of aluminum. Reduction of one ton of aluminum consumes about 20 kilowatt-hours of electric energy, the equivalent of about 8 tons of coal. Extensive demands for electricity made by the aluminum industry favor location of aluminum reducing plants near large hydroelectric installations. In the United States, aluminum reducing plants are located in the Niagara Falls region, the TVA region of the southern Appalachians and the Pacific Northwest. The Texas-Louisiana-Arkansas region produces electricity for its aluminum plants from abundant natural gas supplies.

Cryolite is a relatively rare sodium fluoride mineral. The principal known deposit of cryolite is at Ivigtut, Greenland. Cryolite is now being synthesized from fluorite, a calcium fluoride mineral, but fluorite is not very abundant and is already in demand as a source of fluorine gas. Cryolite availability may ultimately impose a ceiling on aluminum production, and scrap aluminum is currently in demand for recycling.

Aluminum Today. The light weight of aluminum makes it useful for many purposes, particularly in the transportation industry. Its strength is much less than the strength of iron or steel. Alloying improves the strength of aluminum but also increases the cost. Iron costs significantly less because of its simpler smelting requirements.

Aluminum conducts electricity but not so well as copper. The far greater abundance of aluminum and the lighter weight of aluminum wire may someday result in aluminum's replacing copper as the world's major electric conductor.

Copper

Copper was the first metal to be widely used by man. Heating readily reduces many copper ores, a fact which accounts for primitive man's early discovery and use of copper. In the Bronze Age, bronze, which is a harder and stronger metal than either copper or tin alone, was obtained by alloying these two elements.

England, which seems to have led the world in production of almost all metals at one time or another, produced the greatest amount of copper annually from 1800 to 1850, at which point Chile assumed the position of the world's foremost copper producer. In the United States, copper was mined in Connecticut as early as 1705. In the eighteenth and early nineteenth centuries, Americans mined numerous deposits of varying size in the eastern and central United States. In 1874, mining began in Butte, Montana, "the richest hill on earth." The Butte deposit has produced more mineral wealth than any other single area except for the Witwatersrand gold deposits in South Africa. The United States has led the world in copper production since 1883.

Occurrence. Copper ores consist of native copper (naturally occurring, relatively pure copper), sulfides, oxides and complex copper compounds. Sulfide

Figure 9–9. Porphyry copper is taken from Twin Buttes mine near Tucson, Arizona. (Courtesy of the Anaconda Company.)

ores are the most important. Native copper was first mined in northern Michigan by Indians who used the copper for ornaments, trade and weapons. Copper from northern Michigan accounted for most domestic copper production from 1845 until the early 1880's, when the Butte and other deposits in the West began to contribute substantially to United States production.

Oxidized copper ores, which usually occur near the surface, are rich and easily smelted. Unfortunately, oxidized deposits tend to be small.

Sulfide ores found beneath the oxidized caps produce most of the world's copper. In the western United States, the so-called porphyry copper deposits have produced large tonnages of copper ore (Fig. 9–9).

The porphyry copper deposits formed during and

after intrusion of large bodies of molten granitoid rock containing copper sulfides in solution. Cooling and solidification of the molten rock released the copper sulfide "juice," which seeped into cracks in the surrounding host rock and deposited copper sulfide minerals. Copper sulfides also precipitated in the outermost portions of the cooling granitoid intrusion.

Grade or "richness" of porphyry deposits tends to be low except where secondary processes have locally enriched the ores. At Butte, weathering processes removed copper from the weathered zone and redeposited it immediately below the water table, resulting in a zone of extremely high-grade ore. The large volume of mineralized rock in the main porphyry mass offsets the low grade of the unenriched portion of the ore zone.

The western United States, Chile, Peru, central Africa, central Canada and the Ural Mountains of the U.S.S.R. contain important copper reserves. Large deposits in Peru and Chile are thought to have been formed by distillation of copper-rich "juice" rising upward from rocks of the Pacific sea floor sliding into the mantle beneath the west coast of South America.

Contemporary Uses. More than half of the world's copper production is used for electric appliances. Copper conducts electricity better than any other common metal. Electrically conducting copper and magnetic iron make possible the generation of electricity.

Other copper is alloyed with tin, lead, zinc and antimony; the construction industry utilizes copper for sheathing and esthetic effects; copper goes into a variety of chemical compounds, electrolytes and manufactured products.

Gold

Gold was probably the first metal to be used by primitive man. Although it is relatively rare, it almost always occurs in the native state, and it is easy to shape and work.

Gold is often associated with granitic intrusions, filling cracks and fissures in and around the intrusion. Weathering frees the gold from the enclosing mineral fabric of the intrusion and allows running water to carry away gold and concentrate it. Streams concen-

Figure 9–10. Discovery site of gold on the American River in California, 1848. James Marshall, examining the tail-race of a water-powered saw-mill he was building for John Sutter, discovered the first gleaming yellow particles here.

trate gold because the transporting power of a stream depends on its velocity and on the size and density of the mineral material being transported. Swift-running streams on mountainsides can transport gold particles, but upon reaching the foothills, the slope of the stream bed flattens. The flow of water slows abruptly, and the slower stream waters can no longer transport gold, which accumulates near the point where stream bed slope diminishes. Gold deposits formed in this manner are called *placer* gold deposits.

One of the most historically famous placer deposits was discovered on January 28, 1848, on the American River in east-central California by James Marshall, who was building a sawmill for John Sutter (Fig. 9–10).

John Sutter was an interesting character. An Austrian, he married at 23, fathered a large family, ran up great debts and fled to America to avoid debtors' prison. He arrived in Monterey, California, in 1839, soon bought land near the confluence of the American and Sacramento rivers in east-central California and began to establish a colony. In 1847, Sutter and a man named James Marshall agreed to build a sawmill on the American River farther upstream in the foothills of the Sierra Nevada. Sutter provided men and money but remained with his colony while Marshall supervised construction of the sawmill. The sawmill was almost finished when Marshall discovered gold at the site.

Ironically, neither Marshall nor Sutter benefited from the gold discovery. Miners drove Marshall off the property rather than pay the commissions he demanded. He attempted to mine elsewhere but failed. For a time, the state of California paid him a pension, but he spent the last eleven years of his life existing on handouts and doing odd jobs.

Sutter's eldest son, Augustus, spurred by the publicity caused by his father's role in the gold discovery, came to America. He took over his father's businesses, straightened out the finances and for a while the elder Sutter lived prosperously. Augustus left for Mexico in 1850, and the old man's poor business sense soon cost him all of his newfound affluence. After squatters burned down his house, he fled to Pennsylvania, where he died in 1880.

"Gold rush" miners initially worked placer deposits in a number of stream valleys descending through the foothills. Prospectors panned for gold (Fig. 9–11) in

Figure 9–11. Modern "prospectors" pan for gold at the site of the original California gold discovery. The man on the right points to a small gold flake.

stream gravels until they found a workable deposit, then they usually set up more elaborate equipment, which extracted gold from the gravels much faster than pans.

The miners fanned out, attacking stream gravels deposited on hilltops and hillsides (Fig. 9–12) by streams during earlier periods of erosion. They dug into the country rock itself in search of gold-bearing quartz veins—the "Mother Lode." Water spewing at high velocity from nozzles up to 9 inches in diameter literally washed away hillsides composed of gold-bearing sands and gravels. Sediment derived from this type of mining, called *hydraulic mining,* choked streams and increased danger of flooding. The miners ravished the hillsides, leaving them bare and ugly. California first banned hydraulic mining in 1884, then permitted it again in 1893 provided that "hydraulickers" prevented sediment from clogging nearby streams. Excessive cost of sediment traps prevented reestablishment of large-scale hydraulic mining.

Hard-rock mining required a substantial workforce, equipment and capital. The day of the single miner striking a rich placer was soon over. Shafts sunk into California rock yielded quartz ore, which was crushed and chemically processed to recover the gold. The element of adventure had been in the placer diggings, but ultimately hydraulic mining, dredging and hard-

Figure 9–12. Stream gravels on a hilltop in the California gold country. Streams eroded the Sierra foothills long ago and deposited these gravels in what was then a stream valley. Later the stream changed its course and cut away the rock and soil around the hill, leaving silent evidence of its former location.

rock mining produced most of the great California gold fortunes.

A visit to the foothills of the Sierra Nevada provides a glimpse of a fascinating period of American history (Fig. 9–13).

The Witwatersrand deposits near Johannesburg, South Africa, have produced more mineral wealth than any other single area in the world. Here sands and gravels washed from nearby highlands and deposited in a primeval sea more than 2 billion years ago contain immense quantities of fine-grained, widely dispersed gold.

The Witwatersrand resembles a vast placer deposit in which gold has accumulated along an ancient beach. Cobbles of quartz and pea-sized particles of pyrite (iron sulfide) mingle with smaller particles of magnetite and gold. This assemblage was in equilibrium with the force of waves and currents washing the ancient beach. The waves and currents could move pebbles and small cobbles of quartz, the lightest mineral present (Fig. 9–14), but were unable to transport pyrite grains larger than peas. The largest transportable grains of the even denser magnetite were even smaller, and gold, the densest mineral of the assemblage, could be transported only as very fine grains.

Witwatersrand mines now reach 10,000 feet into the earth's interior (Fig. 9–15) to extract a very lean ore.

South Africa's economy depends on mining. In addition to gold, South Africa produces a large portion

Figure 9–13. Ruins of "Mayer & Son" store in Mokelumne Hill, California, an important mining town. Miners first mined placers, then went underground into the Mother Lode. The store is built of blocks of rhyolite extruded from ancient volcanoes in the vicinity.

Figure 9–14. Geologist inspecting Witwatersrand gold ore. The light-colored rocks are quartz cobbles and pebbles. Smaller mineral grains are not visible. The black object on the left is a drill, which is cutting a hole in the ore. Dynamite placed in the hole will break up ore to be transported to the surface.

of the world's chromite, gem-quality diamonds, platinum and several other minerals. Without cheap native labor, many mines could not produce profitably, and the South African economy would collapse. Rhodesia's mineral-based economy resembles that of South Africa. The intransigent racial attitudes of South

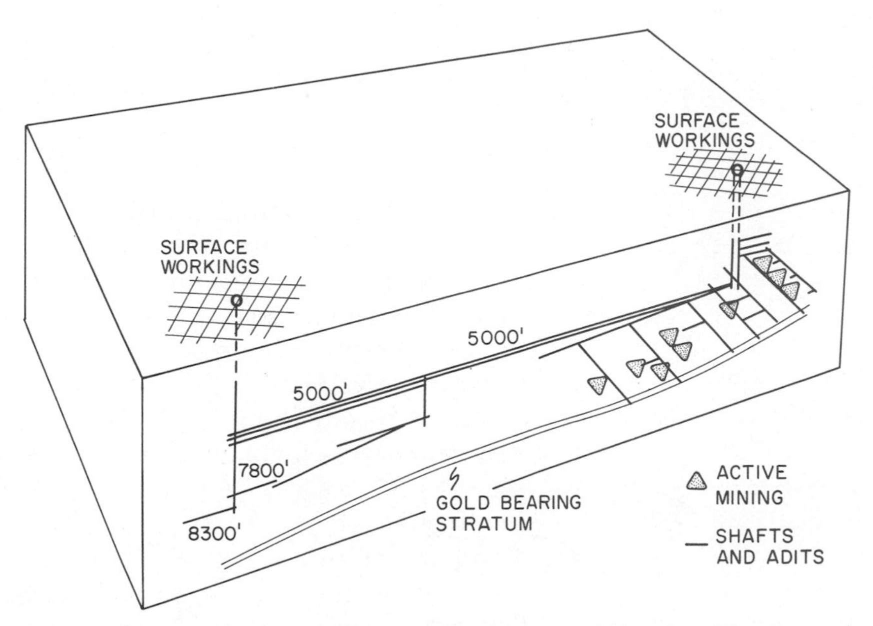

Figure 9–15 A three-dimensional sketch showing gold-bearing Witwatersrand quartzites, mine shafts, adits (horizontal shafts) and surface processing plants.

Figure 9–16 Landscape of tailings pyramids near Johannesburg, South Africa.

African and Rhodesian whites are rooted in the knowledge that without large domestic fuel supplies their continued existence in southern Africa depends on cheap black labor.

All mines produce "tailings," or crushed rock from which the ore minerals have been extracted. Tailings constitute a form of environmental pollution. Near Johannesburg, flat-topped pyramids of tailings cover large areas (Fig. 9–16).

Diamonds

Most people associate diamonds with romance rather than with industry. However, most diamonds are used in industrial processes; only a few of the clearest and most nearly flawless diamonds find their way to a jeweler's shop.

Africa accounts for 90 per cent of world industrial diamond production, and the Congo leads other African nations by a substantial margin. South America, U.S.S.R., India, Borneo and Australia produce small quantities. Arkansas contains the only known diamond deposit in the United States. This deposit was mined briefly but was soon closed owing to high recovery costs. It is now open as a tourist attraction, and gem-quality stones are occasionally found.

Figure 9–17. Diamonds as found in kimberlite. Tons of kimberlite are mined and crushed for each diamond recovered.

Diamond production comes from placer deposits (see section on Gold) and from hard-rock mines. All hard-rock diamond deposits occur within *kimberlite* pipes. In addition to diamonds, kimberlite contains several other minerals that form only at high pressures, indicating that kimberlite originates deep within the mantle. Kimberlite minerals seem to have been solid rather than molten during emplacement. Gas under great pressure probably carried the minerals up into the pipe. Not all kimberlites contain diamonds. In South Africa, only 11 of 150 kimberlite pipes contain economic quantities of diamonds.

Kimberlite weathers readily into yellow colored soil, the "yellow ground" of early diamond prospectors. Less weathered "blue ground" lies beneath the yellow ground, and at greater depth the rock containing the diamonds (Fig. 9–17) is a dark gray-green.

Diamond pipes are mined first as open pits (Fig. 9–18), then by underground tunneling. Crushing the ore frees the harder diamonds from the softer enclosing rock. Crushed rock and diamonds are washed over racks covered with grease. The grease adheres to diamonds but not to the rock. The racks are cleaned periodically, and diamonds (Fig. 9–19) are sent to sorters, who pick out gem-quality stones.

Industrial diamonds are used as tips for tools that

Figure 9–18. The pit of the Premier Mine in South Africa. The man points to the place on the far side of the pit where miners discovered the famous Cullinan diamond.

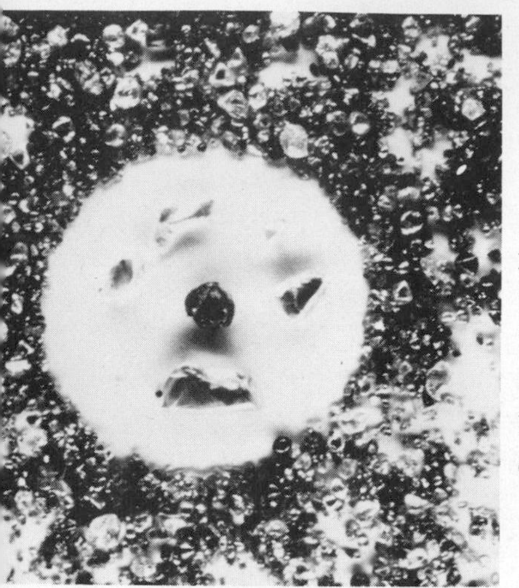

Figure 9–19. Diamonds taken from a grease rack at the Premier Mine, South Africa. Most of the diamonds in this run are suitable for industrial use only. Clearer, lighter diamonds are used for jewelry.

cut stone, glass, quartz and the hardest metals. Diamond-tipped drilling bits core rocks in the search for new mineral deposits, cut test plugs from concrete and bore holes in hard rubber. Diamonds were synthesized for the first time in 1955, and synthetic diamonds now supply almost half of the United States demand of 15 million *carats* (a unit of weight equal to 1/5 of a gram; there are about 2300 carats in a pound) per year. Synthetic diamond production so far consists entirely of small stones. Reclaimed and natural diamonds supply the demand for larger stones. Recycled diamonds amount to almost 20 per cent of current demand.

Conclusion

Calculation of the amount of iron, copper, cement, gravel, building stone and other minerals and mineral aggregates used each year in the United States shows that Americans take a staggering 25 tons of raw resources from the earth annually for each man, woman and child in the United States. These materials are fed into the maws of the United States industrial machine. Extraction and transportation of this large quantity of mineral material earn man a place alongside rivers, glaciers, the wind and the ocean as a geologic force.

How long can the human race continue to exploit the planet's mineral wealth at this rate? Will human evolution peak and then degenerate as did the great dinosaurs? Will future space explorers landing on earth find a worldwide layer of iron-rich (and perhaps radioactive) sediments marking man's meteoric rise and brief burst of brilliance? Nothing so dramatic is likely. Present trends suggest, however, that mismanagement of resources may carry Americans along the same route traveled by the Greeks who deforested the Attic peninsula to mine the Laurium silver and the English, who have now exhausted the best of their iron ore and coal reserves. The time for planning and action is now. It may soon be too late.

From the Sea . . .

Marine Mineral Cycle

Within the past few years, the public spotlight has focused on the enormous mineral resource potential of the sea. The sea provides man directly and indirectly with many resources. Man takes salt from sea water, dredges sand, gravel and heavy minerals from the floor of the sea and mines petroleum, salt and other substances entombed and preserved by sea floor sediment.

At present, sea water supplies only a small percentage of our mineral needs, but increasing world population and new technologic achievements will cause the search for minerals from the sea to be intensified. The immense quantities of mineral wealth in the sea can support our needs far into the future if the problems of extraction chemistry and adequate energy for extraction can be solved.

Ownership of marine resources is also a difficult problem. Some countries claim 12-mile limits; others claim 200-mile limits. Some enforce their claims with gunboats, as United States fishermen off the coasts of Ecuador and Peru have found to their grief. Countries without access to the sea, especially the developing countries, demand a voice in the control of exploitation of sea floor minerals.

Who is to say who owns the sea floor? In 1974, representatives of many nations met in Caracas, Venezuela, to discuss the Law of the Sea. In the meantime Howard Hughes continued to develop a ship capable of harvesting manganese nodules from a rich deposit near Hawaii. Kennecott Copper, Bethlehem Steel and Kaiser have postponed further marine mining developments pending results of the Caracas conference. As this is being written, the question of ownership is unresolved. Like the law of the Old West, which could be applied only within range of the marshall's guns, the law of the sea extends to the range of a country's gunboats.

The history of the earth as recorded in rocks goes back only 3.5 billion years. Theoretical considerations and rocks returned from the moon (the moon is a very primitive planet, whose evolutionary level is comparable to that of the earth more than 4 billion years ago) suggest that the earth's crust, mantle and core differentiated very early in the earth's history, between 4 and 5 billion years ago. The tremendous upheaval of molten rock during the differentiation process released most of the constituents composing the earth's atmosphere and sea. Since this "great catastrophe," exhalations of *juvenile* water (water released from within the earth's interior) accompanying formation of new crust in the mid-ocean ridges (Chap. 2) have augmented the waters of the seas. Exhalations accompanying volcanic eruptions contain small amounts of juvenile water but consist primarily of heated ground water.

All elements probably occur in sea water, but many occur in such small concentrations that even the most precise modern techniques cannot detect them. The presence of certain elements is known because of the organic and inorganic processes that concentrate them. For example, manganese nodules concentrate thallium and platinum to detectable levels, sponges concentrate the isotope silicon-32, and iodine was detected in marine algae 14 years before it was found in sea water.

Juvenile waters originally introduced minerals into the sea, but physical and biologic processes subsequently modified initial abundances. Current levels of abundance seem to have been relatively constant for hundreds of millions of years.

Sodium, chlorine, magnesium, sulfur, calcium, potassium, carbon and bromine are the most common substances in sea water. Each of these substances participates in a cycle of removal and replenishment. Small quantities of salt (sodium chloride) evaporate with water and form raindrop nuclei. Eventually this salt returns to the sea via rain and stream run-off. Other salt evaporates in closed basins and is buried by subsequent sedimentation, thus leaving the sea perhaps forever. Weathering of rocks slowly releases salt, however, which increases the ocean's supply.

Potassium reacts with clays washed into the sea by rivers. The clays sink to the bottom, where the spreading sea floor will eventually carry the clays into a subduction zone beneath a continental margin. The potassium will either be carried down into the mantle or will rise upward as a constituent of newly formed igneous rocks distilled from old crust and mantle. The potassium concentration in sea water was originally about the same as the concentration of sodium, but we now find only $1/28$ as much potassium as sodium because of the potassium-clay reaction.

Organisms remove calcium, magnesium and carbon from sea water, mainly to build their shells. More animals make shells of calcium carbonate than of magnesium carbonate. This preference accounts for greater depletion in sea water of calcium than magnesium over geologic time. Originally, concentrations of these elements were about equal, but now magnesium is three times more abundant than calcium. Bacteria are thought to be responsible for removal of most sulfur from sea water. Evaporation also helps remove sulfur. Most sulfur deposits occur with minerals formed by evaporation of sea water.

Concentration of elements by organic, inorganic and physical processes is important because, without concentration, extraction of these elements would be far more difficult and costly. In the following paragraphs we will discuss extraction of minerals from sea water, dredging of minerals concentrated by wave and current action on the sea floor and extraction of "fossilized" marine resources, that is, minerals concentrated by the sea and buried beneath the sea floor.

Minerals From Sea Water

Fresh Water. Fresh water is an important potential resource of the sea, but it is difficult to extract economically. About 700 desalination plants are in operation or under construction today. Cost of desalinated water has decreased to about 85 cents per thousand gallons, but this cost is still prohibitive in most areas of the world. By comparison, fresh water in the United States costs on the average about 35 cents per thousand gallons. Key West, Florida, is the only city in the United States which derives the major part of its fresh water from desalination. Combination of large-scale nuclear power plants and widespread desalination processing promises significantly lower water costs in the near future. American use of desalinated water will rise in proportion to cost reduction. It is unlikely, however, that cost of desalinated water will ever decrease to a level feasible for use in agriculture. Cost will limit desalinated water to industrial and home use.

Salt, Magnesium and Bromine. In addition to water, man extracts four elements from sea water in significant quantities and at economically competitive prices. These elements are sodium, chlorine, bromine and magnesium. Low concentrations of other elements prohibit profitable extraction at the present time.

Most sodium and chlorine are extracted in the form of common salt. Since the early Stone Age mankind has extracted salt from the sea by evaporation in shallow ponds or *pans*. Although salt extracted from the sea constitutes only a small part of the world's salt production (the rest comes from mines), it nevertheless is a thriving business in some parts of the world. Elemental sodium and chlorine are also taken from sea water for industrial use.

Sea water yields about $70 million worth of magnesium each year. Magnesium, weighing one fourth as much as iron, alloys with aluminum to produce a strong metal useful for structures in which light weight is important. The aircraft industry, for example, uses large quantities of magnesium. Finely powdered magnesium and ribbons of magnesium burn fiercely and brightly, which make it useful in flares and other pyrotechnic devices.

Sea water provides 70 per cent of the world's bromine. Small amounts of bromine mixed with gasoline reduce "knocking" in gasoline engines. This is the chief use of bromine.

Red Sea Brines. The Red Sea brines are the most exciting recent mineral discovery in the sea. Three pools of hot brines in deep areas of the Red Sea were found during expeditions in 1964, 1965 and 1966. Mineral concentrations as high as 300,000 parts per million – almost 10 times greater than normal oceanic concentrations – suggest that juvenile gases escaping from the spreading axis down the center of the Red Sea greatly increase mineral concentrations of Red Sea bottom waters. Russian scientists, stimulated by the Red Sea discovery, have found chromite concentrations in sea floor rifts of the Indian Ocean. A United States research program is now searching areas of the Mid-Atlantic Ridge for mineral deposits. If mineral-rich brines occur in sufficient quantities along other spreading axes, they may eventually supply large amounts of mineral wealth.

Minerals From the Sea Floor

The sea floor of the continental shelves consists mainly of unconsolidated deposits of sands, gravels, muds and shells. Locally, heavy mineral concentrates lie hidden along old beaches and in old stream channels. Nodules rich in manganese and phosphorite blanket portions of the sea floor, and processes not yet fully understood gradually add to their volume. Dredging brings these mineral materials to the surface.

Sand and Gravel. Seventy per cent of the surface of the world's continental shelves consists of unconsolidated sediments: gravel, sand, silt, mud and clay. Floating dredges bring up 50 million cubic yards of sand and gravel from shelves each year. Sand and gravel (and 20 million tons of oyster shells) taken from shelf floors fill in shallow water areas for real estate development, help make concrete, are mixed with asphalt to pave highways and contribute to society in an almost limitless number of ways.

Heavy Minerals. The placer process concentrates all heavy minerals, not just gold and diamonds as described in an earlier section. Platinum, diamonds, gold, tin minerals, chromite, native copper, tungsten minerals, magnetite, garnet, rare earth minerals and many others occur both in beach and fluvial (river) placer deposits. Tin minerals, gold and diamonds are currently the most important.

Although they are not usually considered as such, sand and gravel are placer deposits, too. Rivers and streams carry particles of many sizes to the ocean. Ocean waves and currents sort these materials by size and weight. Currents carry lighter clay particles farther out to sea but deposit clean sands and gravels near shore.

Changes in sea level which occurred during the ice ages (see Chap. 4) complicate the search for heavy mineral placer deposits. Three hundred feet of water may cover a beach ridge where 11,000 years ago waves concentrated gold, diamonds or tin.

Manganese and Phosphorite Nodules. At depths between 100 and 1000 feet, widespread deposits of phosphorite nodules blanket the sea floor. Large deposits off the coast of Southern California contain estimated phosphorite reserves of 1.5 billion tons. The low cost of phosphates obtained from mining on land makes economic development of these deposits unnecessary at present, but future generations may extract phosphate from these large deposits.

Manganese oxides and accessory mineral salts precipitate from sea water around tiny nuclei such as shell fragments and sand grains. The nodules average 24 per cent manganese, 14 per cent iron and 1 per cent or less nickel, copper and cobalt. Nickel, copper and cobalt concentrations rather than the manganese concentration—which is significantly less than concentrations in terrestrial ores—attract the most interest.

Nodules usually occur at depths greater than 12,500 feet (the Pacific sea floor averages 31,000 tons per square km), but nodules cover parts of the Blake Plateau at depths as shallow as 1000 feet, and occur at depths of 250 feet in the Great Lakes.

"Fossil" Marine Minerals

Physical processes such as evaporation, and biologic processes such as reef building concentrate many minerals on the sea floor. Sedimentation subsequently covers the mineral accumulations and preserves them as "fossil" deposits. Such "fossil" mineral deposits include salt, limestone, gypsum, potash, sulfur, coal and petroleum. (See Chap. 8 for a detailed discussion of petroleum and coal.)

Salt. Salt deposits usually form in areas of the sea where high evaporation rates and impaired circulation cooperate in raising salinities above maximum solubility levels. In most areas, currents continually exchange local waters with waters from elsewhere. The constant exchange mixes the sea water and maintains an approximately equal concentration of dissolved elements worldwide. (High evaporation in tropic seas slightly elevates salt concentrations, and local biologic activity may affect concentrations of other elements, but worldwide concentrations are remarkably uniform.)

Salt precipitates most frequently in areas where rising sea level or sagging continental platforms allow sea water to invade continental areas and form shallow seas. Sand bars, reefs or tectonic obstacles may form and separate areas of the shallow sea from the main masses of sea water, impairing exchange of waters. Evaporation increases salt concentration in a shallow sea much faster than in a deep basin. In the tropics, evaporation will rapidly (geologically speaking, that is) increase concentrations of dissolved elements if the waters are not frequently exchanged. In a shallow sea, total absence of circulation results in complete evaporation, a small amount of circulation replenishes dissolved elements without excess dilution, and limited circulation probably helped make possible the great salt deposits currently being mined.

Other salt deposits formed as colliding continents gradually approached one another, isolating patches of sea water from the main body of the ocean.

During early stages of drift of North America from Europe and Africa, restricted circulation caused salt precipitation in the long trough which became the Atlantic Ocean and Gulf of Mexico. Salt domes rising from this layer form many of the most productive oil traps of the Gulf Coast. Recent discovery of new salt domes in the northwestern Atlantic has spurred petroleum exploration of the east coast of the United States. A salt horizon flooring much of the Gulf of Mexico also formed in the early days of North American drift.

In the United States, important salt beds underlie four large areas: one comprises parts of New York, Pennsylvania, Ohio and West Virginia; a second includes much of Michigan; the third includes parts of Kansas, Oklahoma, Texas and New Mexico; and a fourth underlies much of the Gulf of Mexico coastal plain and continental shelf. Some of this salt is mined underground, but most is recovered as brine. Water pumped into the beds dissolves the salt; the brine is pumped to the surface and evaporated to recover the salt.

Salt is very mobile and flows readily when squeezed by a few thousand feet of overlying rocks (Fig. 9–20). Squeezed salt frequently punches through overlying sediments, forming salt domes (Chap. 8). In dry climates, salt domes may break through overlying sediments completely, extruding as a mountain of salt. Such a salt mountain (Fig. 9–21) is probably responsible for the biblical story of the transformation of Lot's wife into a pillar of salt.

Figure 9–20. Flow structures in salt. In this picture, the beds are not only tightly folded but also turned upside down.

Figure 9–21. A mountain of salt on the western margin of the Dead Sea. The mobile salt punched its way upward through weak sediments and formed this small mountain.

Limestone. Almost 200 years ago, the great French chemist Antoine Lavoisier carefully evaporated sea water and noted the sequence of compounds precipitating from the briny mixture. Limestone (calcium carbonate) precipitated first, followed by gypsum (calcium sulfate and water), common salt and others. Precipitation from sea water accounts for a substantial percentage of limestone found in rocks today.

Most marine animals build their shells and skeletons of calcium carbonate because it is abundant and easily extracted from sea water. Calcium carbonate skeletons of land animals are a mute reminder of evolutionary forebears who lived in the sea.

Calcium carbonate accumulates on the sea floor as reefs, broken shell material and tiny precipitated crystals. Time and slow chemical reactions convert accumulated calcium carbonate into limestone rock. Limestone is employed as a flux in iron production, crushed limestone is applied in road construction and concrete, and limestone blocks are used in buildings. Limestone heated and squeezed inside the earth turns into marble, a stone widely used for decorative and artistic purposes because of its beauty and the ease with which it can be cut and sculpted.

Gypsum. Gypsum deposits, like salt deposits, result primarily from evaporation of sea water. In

Figure 9–22. Gypsum dune in the White Sands National Park, New Mexico. Quartz sand, the most common variety, tends to have a dirty appearance because of iron staining of quartz grains. Limestone sands are light tannish-yellow in color. Gypsum sands are nearly pure white.

southeastern New Mexico, the wind erodes outcropping gypsum beds and produces the beautiful White Sands (Fig. 9–22). In a wetter climate, water dissolves and carries away the gypsum. Most gypsum is used for concrete fabrication.

Sulfur. Modern technology employs great quantities of certain minerals in such indirect ways that many people are unaware of their use. Sulfur is such a mineral. The United States uses more than 100 pounds per person each year, but few people see any of this sulfur in its pure form. Sulfur goes into fertilizers, textiles, plaster, sulfuric acid and many other components of contemporary technology. The old Anglo-Saxon equivalent of "burn-stone" gave rise to the term "brimstone." Volcanic emissions of sulfur and lava (see Chap. 3) resulted in the term "fire and brimstone" common in medieval and biblical literature.

Volcanically precipitated sulfur deposits are mined in Chile and Japan, but on the whole, volcanic sulfur constitutes only a small fraction of annual sulfur production. The largest reserves are found in sedimentary rocks, where sulfur occurs together with gypsum, salt and limestone—the evaporite sequence. Sulfur occurs in coal and petroleum in concentrations sufficient to cause air pollution but not to warrant profitable extraction. Bacteria are probably responsible for precipitating the sulfur found in rocks, but scientists do not yet completely understand the process. Sulfur also occurs chemically compounded with other elements such as copper and iron. Sulfuric acid is extracted either as the primary product or as a by-product of the metal smelting process.

Because salt domes are often capped by a relatively thin layer of sulfur-rich gypsum, anhydrite and limestone, sulfur early came to the attention of petroleum chemists. In 1890, Herman Frasch, a chemist working for a petroleum company, discovered how to extract sulfur from salt-dome cap rocks. In the Frasch method, superheated water pumped down the outermost of three concentric pipes melts the sulfur, which forms a puddle at the bottom of the well because it is denser than water. Compressed air forced down the central pipe foams the liquid sulfur and lightens it enough for water pressure to push it to the surface through the third pipe (Fig. 9–23).

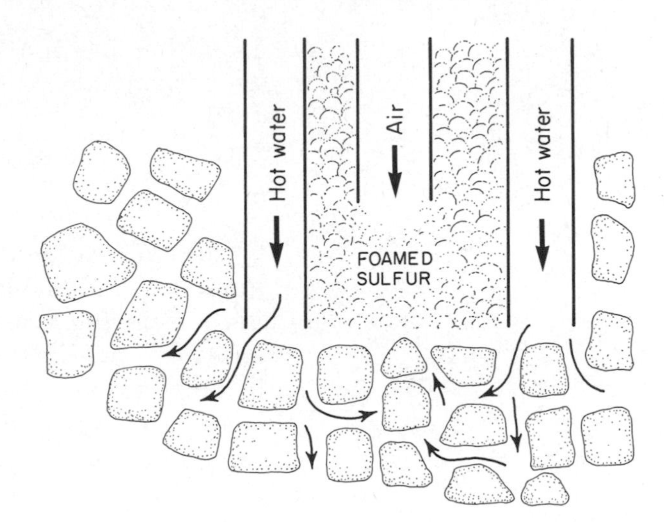

1. Hot water melts sulfur in cracks of rock
2. Air foams and lightens the liquid sulfur
3. Air and water pressure push foamed sulfur up and out of the hole

Figure 9–23. The Frasch process. Superheated water entering the well through the outer pipe melts the sulfur. Water pressure forces the sulfur part way back up the core, where air carried by the central tube foams the sulfur, thereby lightening it sufficiently to allow the water pressure to push the sulfur to the surface. (Adapted from Pratt, C. J.: Sulfur. Sci. Am. *222*:62–77, 1970.)

Eventually almost 90 per cent of all sulfur is converted to sulfuric acid; 60 per cent of the sulfuric acid (or roughly 50 per cent of all sulfur products) is used to manufacture fertilizers. In addition to agricultural uses, sulfur aids in petroleum refining and metallurgy; sulfuric acid dissolves impurities from bauxite and helps produce pure alumina; and sulfuric acid baths remove mill scale from certain types of steel. Sulfur compounds vulcanize rubber, treat illness (sulfa drugs), dye cloth and clean clothes (sulfonated detergents). Elemental sulfur aids in paper manufacture (and contributes to the terrific stink of many paper mills) and is used in the manufacture of rayon, cellophane and insecticides.

Recently, demand for sulfur has grown at a rate of 4 to 5 per cent per year, and consumption rose from 1 million tons in 1900 to about 40 million tons in 1970. Demand outran supply in the early 1950's and middle 1960's. Technologic innovations have since increased sulfur production. Limits on emissions of sulfur dioxide into the air from burning fuels will inadvertently increase sulfur supplies. Manufacturing plants must either use fuels treated to remove sulfur or trap sulfur gases before they escape from the smokestacks. Either method will result in additional sulfur in the marketplace. The present sulfur supply outlook is good.

The Future

The sea and sea floor are the last frontiers of mineral exploration and development. The present yield of minerals from the sea is high, and future production promises to be greater (Fig. 9–24). However, the equilibrium of sea, seashore and sea floor is in delicate balance. Already, dredging has destroyed more than 10 per cent of United States shellfish grounds; hydrocarbons, DDT and radioactive pollution endanger marine and marine-related life. We must plan the use of ocean resources cautiously and thoughtfully.

4 CONCLUSION

Several years ago, an article appeared which hypothesized the evolution of man in the future. This man had an enlarged, egg-shaped head to accommodate his large, superior brain. He had lost his small toes and his legs and arms were weak because he had done little manual labor for centuries. Keen eyesight,

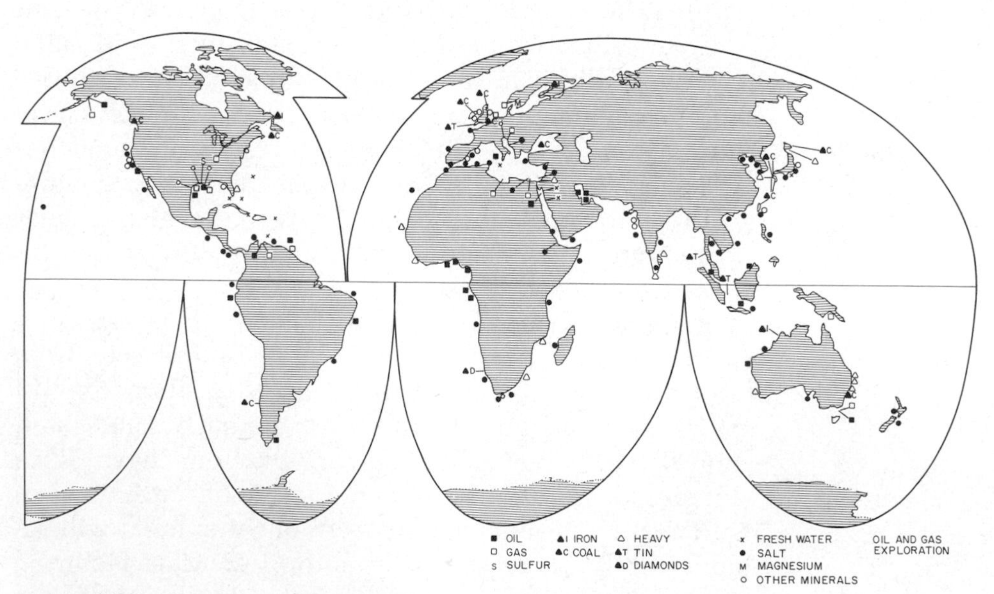

Figure 9–24. Pattern of production of mineral resources from the sea and continental shelves. (Adapted from Wenk, E., Jr.: The physical resources of the ocean. Sci. Am. *222*:167–176, 1969.)

once necessary for hunting food, had deteriorated. He was an evolutionary product suited for life several thousands of years hence. An obvious but tacit assumption in this model of man's evolution was a continued abundance of energy and minerals.

Written today, a similar article might reach different conclusions. A decrease in available mineral and fuel resources would probably cause evolution to regress toward nineteenth-century characteristics. For example, contemporary man in developed countries is too large. He eats too much and thereby requires too much energy and materials to produce his food. To compensate, future man may be smaller so he can survive on less food. He may be stronger and tougher because, like our ancestors a century or more ago, he will have to win his food from the land by his own labor, assisted only by horses or oxen. There may be far fewer men on earth. Large areas will be polluted and unfit for agricultural use. Lacking the steel and fuel to build and operate trains or trucks, he must live closer to the source of his food, which will further restrict population. His health may not be as good. Modern medicine, production of which is based on abundant mineral resources and produced with cheap energy, will become progressively scarcer and more costly. Governmental functions will deteriorate. Copper will not be available for communication lines; postal service and transportation will slow to a crawl. Local governments will reassume their nineteenth (or pre-nineteenth) century functions. Countries may become smaller simply because communications and transportation are inadequate for effective administration of large areas.

Man may cease to exist as a species altogether. The larger reptiles that could not adapt to changing conditions 60 million years ago became extinct. A drastic change in resource availability would severely stress modern society. Fortunately for man, the human race probably has not evolved so far beyond its Cro-Magnon and Neanderthal ancestors that as a species it could not adapt to a more primitive existence.

These are dark thoughts, which may or may not be realistic. Another technologic miracle could save contemporary society. On the other hand, history contains abundant examples of civilizations that flourished, exhausted their mineral and/or energy resources and declined. Will twenty-fifth-century travelers visit our great cities and technologic installations, viewing them with wonder as Renaissance travelers viewed the edifices of ancient Egypt, Athens and Rome?

The answer to this question depends on effective management of the environment. The seriousness of our situation must be recognized; the solution to our problems will require difficult decisions. We *must* use our resources more effectively in the future than we have in the past.

Finite supplies of resources ultimately limit the number of people who can live comfortably on the earth. A substantial body of evidence suggests that this limit has already been exceeded in many areas. Because reallocation of resources involves expenditures of considerable quantities of fuel, it is not a viable solution to the population problem. Population growth rates in developing areas are frequently high, and reallocation of resources would probably do no more than provide temporary relief.

In developed countries such as the United States, inflation (in 1974 at least) is taking an increasing toll in the purchasing power of the average family. Other national problems have obscured the fact that part of the inflationary spiral is due to increased cost of goods resulting from increased costs of energy and raw materials. It seems certain that these costs will continue to rise as mineral and fuel resources are depleted. The people of the United States may not enjoy the "golden years" of the 1950's and 1960's again for many years to come.

There appear to be three factors that will allow man to maintain his present standard of living: decreased population, technologic advances, and better management. Current trends suggest that population growth in developed countries has peaked; unfortunately, developing countries, which have the greatest need for resources, are still in the grip of high population growth. Major technologic advances in the twentieth century have lulled many into the belief that this trend will continue. History, on the other hand, suggests that technologic innovations are sporadic. Furthermore, these inventions and discoveries have produced as many environmental problems as they have solved. Pollution levels have risen, hazards have increased and demands on resources are at an all-time high. It seems ill-advised, therefore, to rely on new technology to solve our problems. Of the three ways of maintaining our present standard of living, better management is the most promising and the least expensive.

Because resources are diminishing, we must make better use of those that remain. Broad interdisciplinary cooperation, planning and action are imperative if we are to avoid the alternatives of drastically decreased standards of living, food shortages, famines in some areas of the world and possibly devastating wars. The future of civilization may truly be in our hands.

10

SOILS

Lateritic soil. Although the original grain shapes are preserved, solution has removed so much material from this rock that only a crumbly soil remains. (Pen indicates scale.)

The Sahelian region lies along the southern margin of the Sahara, extending east-west across Africa (Fig. 10–1). The land is a semi-arid desert with enough rainfall to support some marginal agriculture and a native grassland. The Sahelian people had arrived at a way of life that fitted into the ecologic balance and allowed them to live within the limitations of their environment. They and their herds moved as far south as possible during the dry season. When the rains began, the nomads and their herds moved northward behind the falling rain. When they reached the southern margin of the Sahara, they turned back southward, grazing on the grass that had grown behind them as they moved north. Migration routes and rules were carefully observed so that overgrazing and conflict among tribes were avoided.

The farmers of the region used agricultural methods adapted to the environment, allowing land to lie fallow for as much as 20 years and cultivating crops according to growing season. Thus, both nomad and farmer had worked out a way of life within the limits of the ecologic balance of the region.

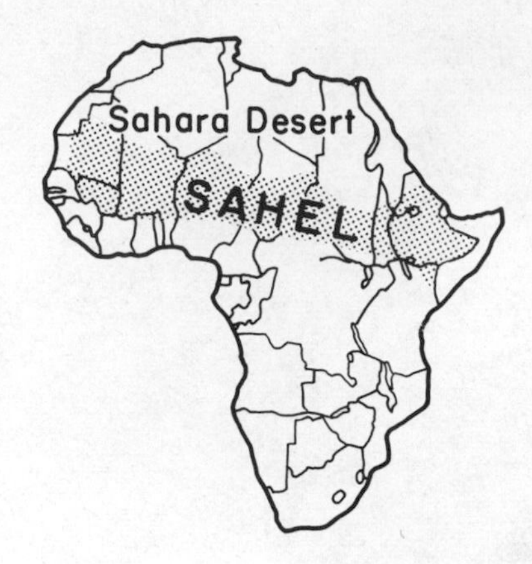

Figure 10–1. The Sahelian region of Africa.

Then a combination of events disturbed the system: overpopulation, climatic change and the influence of Western civilization. The introduction of Western medicine and medical practices helped increase the rate of population growth. With the increase in human population, there was also an increase in nomadic herd animals. Forage sufficient for the previous animal population was no longer adequate. The need for more food for the human population and the desire for cash crops (another Western innovation) resulted in a change in farming practices. Land was farmed without allowing it to lie fallow, or with only a short fallow time, so that overuse of the farmland occurred. Moreover, marginal farmlands were withdrawn from forage use and put under cultivation. Cash crops, not suitable for the type of soil, depleted the soil as a resource. The resultant overfarming and overgrazing left the soil exposed and vulnerable to erosion by wind and water. The final push toward desertization came from a drought. The combination of all these events caused starvation of both herds and people. At least 100,000 Sahelians are estimated to have died of starvation in 1973, and the situation was equally serious in 1974.

This story illustrates all too tragically a recurrent theme of this work: man exists under the control of nature. He cannot escape the consequences of upsetting the ecologic balance. In this chapter, we will discuss the origin of soil and our use or misuse of this vital resource. Changing land-soil use is an important issue because on a finite earth with limited space only certain soils (and climates) are suitable for agriculture (Fig. 10–2).

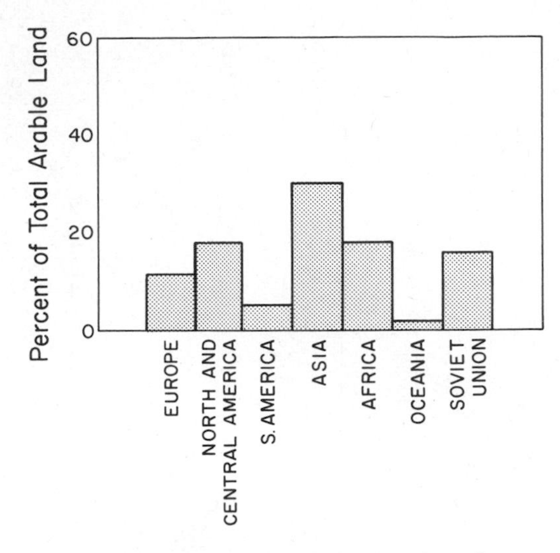

Figure 10–2. Distribution of total arable land. Cultivation of much of the soil of the world is limited by climate and topography.

Most of our food comes from the soil, either directly or indirectly. The soil is a versatile resource, serving other uses such as mining, forestry, transport, industry, urbanization and recreation. The current conflict between use of land for purposes other than farming is important in considering soil as a resource, because when land is taken out of cultivation it is lost to agricultural exploitation, and generally the loss is irrevocable. For example, extensive use of the automobile has resulted in the removal of 50 acres of land from cultivation for each mile of two-lane highway. Superhighways cover much of the United States where fields of corn, wheat and vegetables once grew. In many areas the most fertile agricultural soils lie on flood plains, which are level, well drained and suitable for construction. These regions have been preempted by builders of highways and suburban housing developments. Thus land is being taken from cultivation just when the world is in need of more food. The fertile soil, perhaps our most valuable resource, is dwindling away (Fig. 10–3).

Conflicting use of the land is not the only problem in regard to soil resources. Much farmland has been abused and overused in the United States and other parts of the world, just as in the Sahelian region. The result is badly gullied and eroded land, a "dust bowl" on its way to desertization. The President's Council on Environment estimated that in 1967, two thirds of the

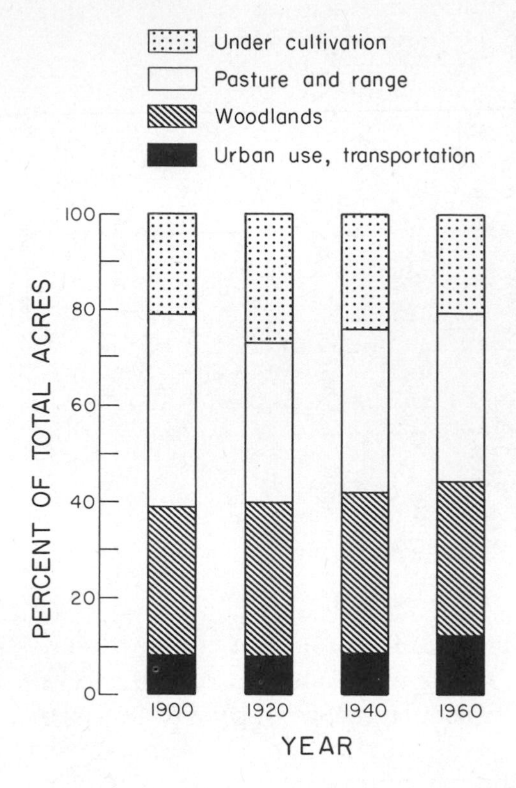

Figure 10–3. Changing land use in the United States, 1900–1960. Data from U.S. Department of Agriculture.

land used for crops, timber or grazing was in need of conservation treatment (Fig. 10–4).

Soil is a precious resource created by geologic forces working over long periods of time. An exception is Hawaiian basaltic lava poured forth by the volcano Kilauea in a 1959 eruption (see Chap. 2), which weathered so rapidly in the tropical climate that papaya trees grew on the lava only a few years later. However, soil formation usually proceeds very slowly. Basaltic lavas cover parts of Idaho and other arid regions of the western United States in which no soil has

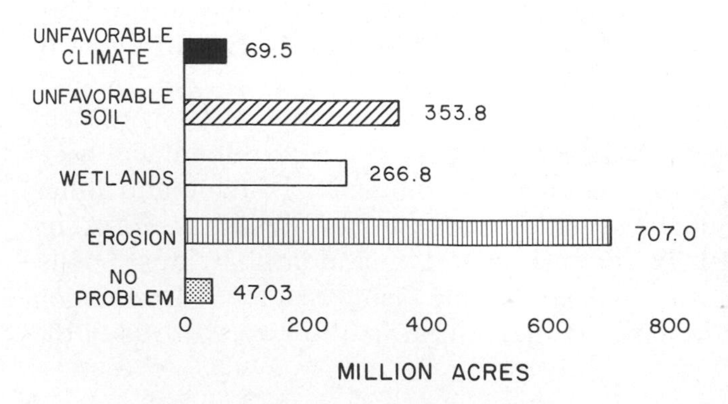

Figure 10–4. Soil problems in the United States on non-federal rural land, 1967. (Adapted from Schwab, G. O., et al.: Elementary Soil and Water Engineering. New York, John Wiley & Sons, 1971.)

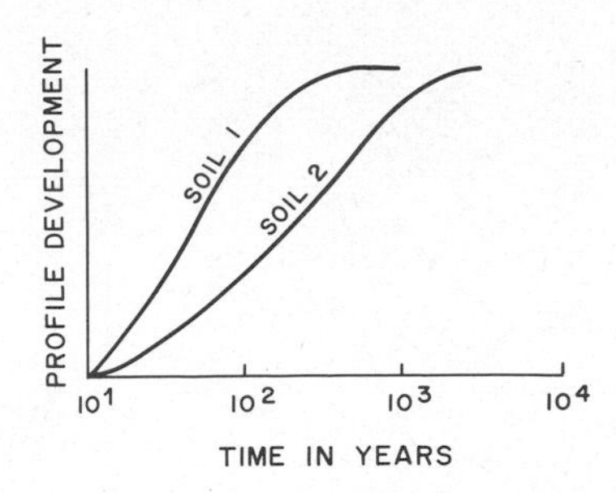

Figure 10–5. Time required for development of an A horizon. (Schematic after Birkeland.)

developed in the thousands of years since extrusion of the lava. Soils develop at different rates depending upon environmental conditions such as moisture, temperature, vegetation, parent rock type and topography (Fig. 10–5).

Soil results from the weathering of rock or other parent material such as alluvium, glacial till or landslide debris. Soil is a dynamic, changing medium, which has characteristic properties derived from the interaction of parent material, climatic factors, biota, topography and time.

1 WEATHERING

Physical Weathering

Rocks are the basic building material of the soil. On exposure to the atmosphere and hydrosphere, solid rock is gradually broken down physically and chemically by the processes of weathering. *Physical weathering* refers to the mechanical disintegration of rock. This is accomplished by stresses from freezing water, expansion due to heat and moisture and pressure release. Frost action, the freezing and thawing of water, is the most important physical weathering agent. When water freezes it expands. If it freezes under confinement, extreme pressures are exerted on the surroundings. Freezing water can enlarge cracks and pores of the rock in which it is enclosed and eventually break it up. Such frost action is most effective if freezing and thawing take place repeatedly.

Another way in which pressure may be exerted to break rock is by *temperature change*. Most materials, unlike water, expand when heated and contract when cooled. Intense heat from forest fires has been known

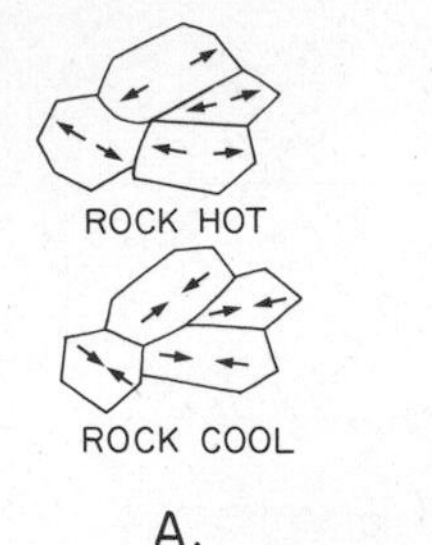

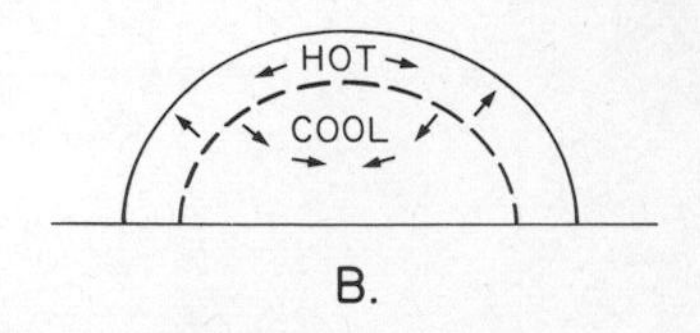

Figure 10–6. Two theories to explain rock breakage by heating. *A*, Heat from the sun causes differential expansion of minerals in the rock. This causes stresses between the minerals, which break apart. *B*, Solar radiation heats up the outside of the rock, while the inside remains cool. The temperature difference (or gradient) results in stress.

to burst rocks, but because such heat is uncommon, direct heat is not an effective breaking agent on earth. (Lunar rocks apparently do weather in this way owing to temperature extremes.) It is thought that heat from the sun, which causes differential expansion of various minerals, sets up minute stresses in the rock. After a long period of time, breakage results (Fig. 10–6A). However, recent studies indicate that thermal stresses in rocks may be caused by temperature gradients rather than by differential expansion of minerals (Fig. 10–6B). Temperature gradients result when the outer part of the rock becomes hot while the inside remains cool: the outer, hot rock expands while the cool inner rock is contracted. This thermal difference causes a stress, which will in time break the rock.

Minerals such as the micas and clays have an internal lattice composed of sheets of elements stacked on one another and held together by *cations* (positive ions). When these minerals are wetted, water is absorbed between the sheets or layers (Fig. 10–7), pushing them apart and expanding the crystal lattice. When the water is removed by drying, the lattice contracts. (This mechanism causes mud to swell when wet and to crack when dried out.) Clay minerals or the micas may thus create pressure by swelling when they become wet and cause the rock to break.

Silica Tetrahedra
Alumina Tetrahedra
Silica Tetrahedra
$\pm H_2O$ — spacing variable
Silica Tetrahedra
Alumina Tetrahedra — 9.6Å
Silica Tetrahedra

Figure 10–7. Schematic diagram of the crystal structure of montmorillonite, a hydrous clay that expands when water is added between sets of sheets or layers. (Adapted from Russell, M. B.: Physical properties in soil. *In* 1957 Yearbook of Agriculture. Washington, D.C., Government Printing Office, 1957.)

Figure 10–8. Exfoliation of granite in the Laramie Range, Wyoming.

Unloading pressures occur when igneous or metamorphic rocks, formed under confining pressures deep within the earth, are uplifted and then exposed by erosion. Release of the confining pressure causes a relaxation in the rock and an outward expansion. Such release occurs parallel to the rock surface and hence results in a breaking away of sheets of rock in a process called *exfoliation* (Fig. 10–8).

Chemical Weathering

Physical weathering is the breakdown of rock into smaller pieces in which the original mineral composition is retained. Alteration of the mineralogic components of a rock occurs in *chemical weathering.* Various weathering processes affect the minerals of a rock, changing them into other minerals. Some of the constituent elements may be lost, and the entire chemical composition of the weathered rock differs from that of the original rock.

Whereas water plays an important part in the physical weathering of rocks, it is even more important in the chemical weathering processes. Solution is a chemical process especially effective in the development of soils. Many minerals, such as calcium carbonate, potash and nitrates, are soluble in weak acids. When they are dissolved out and carried away, this loss not only alters the chemical composition of the rock but may also cause it to break up if the dissolved material was cementing the grains. Removal of elements by *leaching* (solution) is critical in the formation of soil.

As discussed in the section on physical weathering, many minerals (particularly the clays) react with water not by dissolving in it but by absorbing it into their molecular structure. The reaction is reversible; that is, the mineral may also lose the water. Such minerals are said to be *hydrous*. The clays, micas and gypsum are common hydrous minerals. When water is absorbed into the crystal lattice, swelling takes place, whereas contraction occurs when water is lost. The chemical reaction in hydrous minerals causes a physical response (swelling and shrinking), so chemical and physical weathering work together as one process.

Hydrolysis is the ionization of water into positive hydrogen (H^+) and negative hydroxyl (OH^-) ions. Hydrolysis is extremely important in soil formation and nutrient supply because the hydrogen ion can readily replace other cations such as sodium, calcium and magnesium in minerals. This process, called *base exchange* (or *cation exchange*), occurs commonly in clays and organic compounds. The hydrogen ion, by replacing some other cation in the mineral, releases it for use by plants.

Carbon dioxide from the air may dissolve in soil water and react with minerals to form soluble carbonates. In this process, called *carbonation*, a weak acid is formed. This is very effective in chemical reactions because hydrogen ions are made available for cation exchange.

H_2O +	CO_2 ⇄	H_2CO_3 ⇄	$2H^+$ +	$CO_3^{=}$
water	carbon dioxide	carbonic acid	hydrogen ion	carbonate ion

In fact, any acid in the soil facilitates chemical reactions by providing hydrogen ions (H^+).

Oxidation-reduction is another chemical process that takes place during weathering and soil formation. An example is the weathering of olivine, an iron-rich mineral. Ferrous oxide is formed which is then oxidized to a hydrous ferric oxide. Other iron-rich minerals such as the amphiboles and pyroxenes are also easily weathered by oxidation. The process is readily recognized by the reddish color of the oxidized (ferric) iron minerals, the familiar iron rust. Because the oxidized iron oxide is hydrous, the physical effect of expansion and contraction of the mineral also takes place.

Chemical weathering often results in physical stresses, as we have shown. Weathering works to create more stable types of minerals and disintegration of rock. When organic material is added, the broken and altered rock becomes soil.

Biochemical Weathering

The soil-forming processes are accelerated by changes brought about by living organisms. Biologic agents play a minor part in the physical break-up of rock and soil by the prying and disturbing action of roots (Fig. 10–9) and by the turnover of soil by worms, ants, rodents and man. However, the primary biologic processes are chemical. Fungi, algae and lichen are significant in bringing about decomposition of organic and rock material and releasing elements for use by other plants.

Humification is the decay of organic matter. It is carried out by various microorganisms at rates deter-

Figure 10–9. Roots enlarge cracks or joints, as shown by this tree growing in sandstone. Near Canyonlands National Park, Utah.

mined by temperature, oxygen supply and drainage and results in the release of minerals from dead plants. These minerals are reused as nutrients for living plants.

The microbiota are important agents of decay, facilitating the recycling of mineral nutrients, setting carbon free as carbon dioxide and fixing and freeing nitrogen. The nitrogen-fixing bacteria perform the essential function of altering nitrogen and nitrogen compounds so that they are made available to plants. Many other element-fixing bacteria are vital to a productive and fertile soil.

Plants are able to extract mineral nutrients, primarily metals, from the soil by a process called *chelation*. This involves the exchange of hydrogen ion for a metallic ion, generally in the root region.

2 SOIL HORIZONS

Soils are derived from bedrock that has undergone the following series of changes: physical and chemical weathering; humification of the mineral debris by decay of organic matter; and transposition of mineral and organic matter by leaching and deposition. As soils are formed, horizontal subdivisions called *horizons* develop from the bedrock (Fig. 10–10). These contrasting layers of the soil profile are differentiated by the redistribution of weathered and organic materials. The upper, or A horizon, is the surface layer, containing organic matter mixed with loose material. When well developed, the lower part of the A horizon is leached of soluble salts. Eventually even the more durable iron oxides and clay minerals are removed.

Material leached from the upper layers is rede-

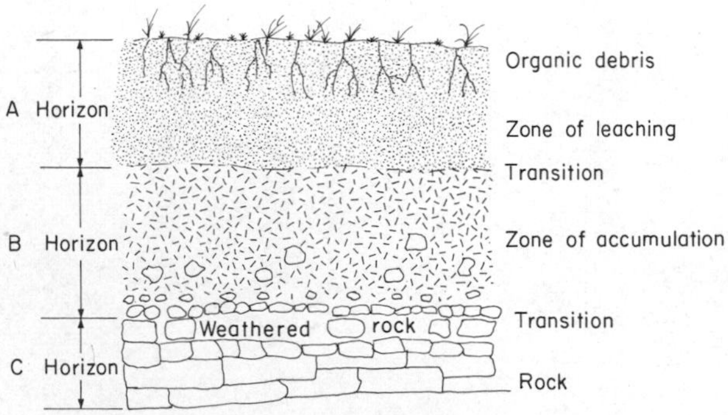

Figure 10–10. Soil horizons.

posited below by percolating waters to form another layer, the B horizon. Thus the B layer is enriched with deposits of humus, calcium carbonate, oxides of iron and aluminum and clay minerals. Sometimes it is so enriched that the *illuvial* (or depositional) material forms a hard layer called a *hardpan*. Hardpans may be further categorized as clay pans or iron pans, depending on the type of material present. In the tropics, the well-developed B horizon is greatly enriched with the compounds of iron and aluminum, forming a *laterite*. In arid and semi-arid regions, a layer of calcium carbonate, or *caliche*, is often formed.

At depth the B horizon grades into broken bedrock, that is less and less weathered, until it merges with the solid bedrock, or C horizon.

Several other processes take place in the soil, including the gradual acidification by the replacement of exchangeable cations in the soil water with hydrogen. This results in increasing hydrogen ion concentration, and the cations removed may be completely lost to the soil system. Another process that occurs over a long period of weathering is the breakdown of the silicate mineral structure. The crystal lattice of the feldspars, micas and some clays are reconstituted as kaolinite clay with the loss of some silica. In addition to silica loss (desilicification), the break-up of the silicates releases iron and alumina to the B horizon. This accumulation of compounds of iron and aluminum is called *podsolization* and is the sign of a mature soil.

3 SOIL CLASSIFICATION

Many different classifications have been devised for distinguishing soil types. One of the more modern and widely accepted classifications was prepared by the United States Department of Agriculture. The stage of development of the soil and its various horizons are used as the basis. A simplified version of the classification is as follows:

Entisols. Shallow mineral soils with some organic litter. No B horizon. Soil just beginning to form.

Inseptisols. Young soils with a weak horizon development. Some leaching in the A horizon but no well-developed B horizon.

Alfisols. Illuviation (deposition) in the B horizon. Strong development of decayed organic matter. Leached and acid A horizon. High percentage of clays in the B.

Spodosols. Highly organic surface horizon, also acidic. Accumulation of compounds of iron and aluminum in the B horizon, with an iron pan.

Ultisols. Intense chemical weathering and much leaching. Soils are deep and acidic. Clay-rich B horizon, with compounds of iron and aluminum. Formed in subhumid, warm climates.

Oxisols. Humid tropical red soils. Deep B illuvial horizon below a well-leached and acidic A. Lateritic (clays and compounds of iron and aluminum).

Mollisols. Grassland soils, with a deep, dark, humic surface horizon (A), rich in calcium and magnesium cations. Clay in the B horizon.

Aridisols. Soils of arid regions. Essentially only minerals with little organic matter. Very little leaching with a cemented pan of calcium carbonate (caliche).

Vertisols. Clayey soils which swell and contract. Very little profile development.

Histosols. Organic soils, generally waterlogged so that humus cannot decay. Very high carbon/nitrogen ratio.

Many variations exist within each class of soil, depending upon the parent rock composition, climate, biota, topography and time. We have seen that the parent rock supplies the basic minerals composing the soil and determines many of its characteristics. The mineral composition tells us the elements present and the chemical nature of the soil, which are indicators of fertility. The climate determines weathering processes. Rainfall provides water for leaching and other physical and chemical reactions. Temperature affects the rate of chemical reactions, the freeze and thaw cycles and the type of biota present. Chemical reactions in the soil proceed very slowly at low temperatures, whereas such reactions occur rapidly at high temperatures. Weathering takes place three times as fast in humid tropical zones as in temperate climates. Thus we find thick well-leached (but unproductive) soils in wet tropical regions. In cold regions the soil is not well developed and often contains undecayed organic matter such as peat. Desert soils are usually lime-rich with very little leaching and almost no organic matter. The most fertile soils are those developed under subhumid to semi-arid climates that are either tropical, temperate or cool. Under such conditions chemical, physical and biologic reactions combine to cause formation of a rich, productive soil.

Quantity and type of vegetation and animals present are important in soil formation because organic materials supply many nutrients and often determine decay rates. If organic decay is equal to supply of organic material, a steady state is reached and a healthy soil results. If the supply of organic matter is greater than can be decayed, an unproductive peat bog forms.

Topography influences drainage and rate of percolation or run-off. Leaching is a function of angle of slope. Steep slopes allow little development of soil because material loose on the surface is lost as it moves downhill and water does not penetrate the surface. On the other hand, level areas may be poorly drained and, again, peat bogs may be produced. And finally, time is an important factor because conditions change as soil development and weathering proceed. In fact, many *pedologists* (pedo = Greek *soil*) consider time to be the primary determining agent in soil development. They believe that all soils would ultimately be the same if given enough time. The fact that many different soils exist is the result of the interaction of all the above variables within a limited time frame. The soil is a dynamic system with changing properties depending upon the environment.

4 SOIL FERTILITY

Fertility refers to the ability of soil to grow crops, supplying all the nutrients needed. Because world population is currently increasing faster than food supply, fertility of the soil must be maintained at a high level. Previously, especially in the United States, soil was used and then abandoned when it was worn out. Thousands of acres are barren because of bad cropping, overuse or gullying (Fig. 10–11), and thousands more have been used for highways and urbanization rather than cultivation.

The ability of the soil to produce crops depends upon *soil texture, structure* and *nutrients*. Soil texture refers to the grain size of the mineral components and to the distribution of the sizes. Size of the soil particles varies from boulders and gravel to very fine clay. According to Department of Agriculture definitions, gravel consists of particles 2 to 100 mm in diameter, sand ranges from .05 to .2 mm, silt is .002 to .05 mm and clays are less than .002 mm in size. Soil textures are classified according to size distribution as shown in the textural diagram, Fig. 10–12. Texture is very closely related to mineral composition. For example, the clay minerals are very small, and silts are most often quartz-rich with some micas and clays. In general, sands are composed predominantly of quartz, and gravels are likely to be heterogeneous in composi-

Figure 10–11. Gullying resulting from overuse of the soil. (From USDA, SCS Mo 1–852.)

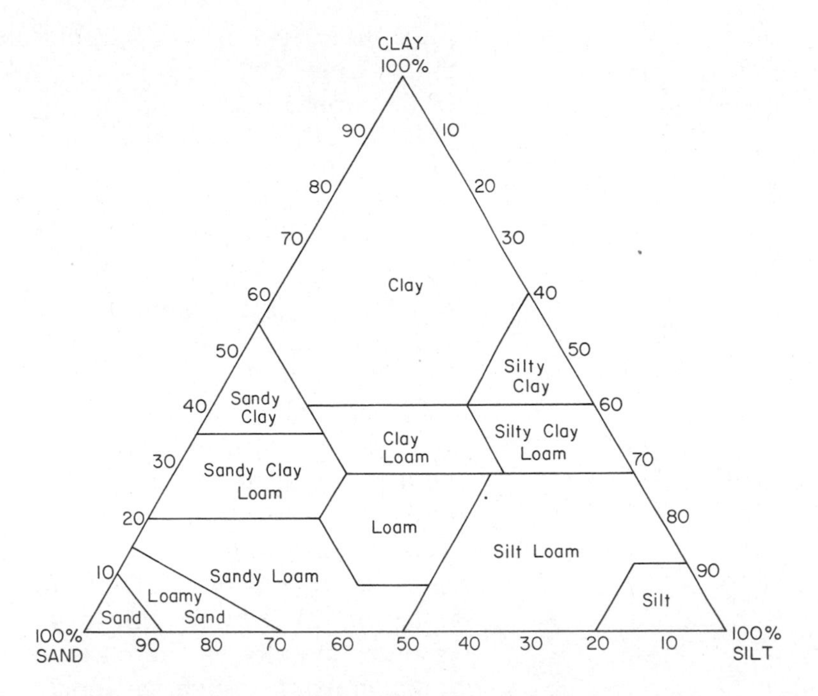

Figure 10-12. Textural diagram showing the size ranges of types of soil.

tion. Soil texture influences the drainage behavior of soil. Gravels and sands transmit water very easily and cannot hold moisture. They are also noncohesive. Silts are more cohesive, less drainable and can hold moisture. Clays are cohesive and retain moisture, with high water absorption. More importantly, most clays have a high cation exchange capacity and thus can supply nutrients by substitution with hydrogen.

Soil structure refers to the way in which soil particles are put together (Fig. 10-13). Single mineral grains or aggregates of fine grains in a *crumb structure* are best suited for agricultural purposes. However, over a period of time, cultivation tends to destroy the structure by compaction or crusting. Soils rich in clays tend to have crumb structures that may be retained for a long time owing to the plasticity and adhesiveness of the clay. Organic *colloids* (materials less than 10^{-5} cm in size) also aid in stabilizing the crumb structure because they possess the same cohesive and plastic properties as clay. By determining porosity and permeability (see Chap. 11), structure also influences the drainage properties of soil.

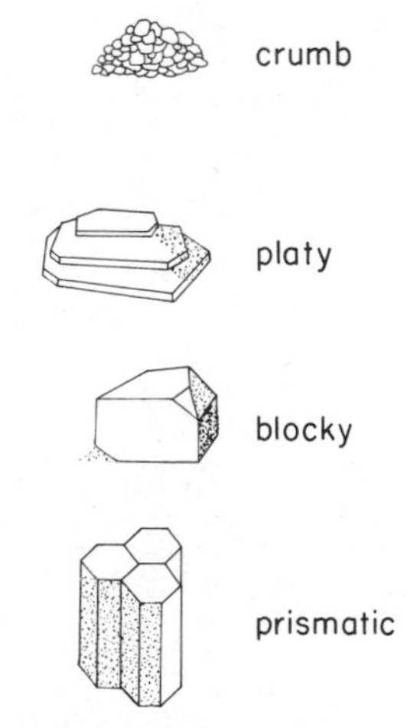

Figure 10-13. Some kinds of soil structures.

Nutrients necessary for plant growth are derived primarily from the minerals and organic matter present in the soil. Elements vital for growth include carbon (C), hydrogen (H), oxygen (O), nitrogen (N), phosphorus (P), sulfur (S), potassium (K), calcium (Ca), magnesium (Mg), Iron (Fe), manganese (Mn), zinc (Zn), copper (Cu), molybdenum (Mo), boron (B) and chlorine (Cl). These elements are available to plants only through the soil water, which contains them as ions. They can be released to the soil water by cation exchange with hydrogen ion. Therefore, the amount of hydrogen ion present in the soil water is a factor in soil fertility. Hydrogen is measured by the *pH* level, which varies on a logarithmic scale. The neutral level is 7; less than 7 means the soil is acidic.

Cation exchange takes place under three circumstances: 1) Hydrogen ion can substitute for some cation held in the lattice of clay minerals or the micas. These include such cations as sodium, potassium, calcium, magnesium, aluminum and iron, all of which can be released and made available for leaching or for plant nutrition. 2) Organic colloids have the same cation exchange capacity as the clay minerals, releasing additional nutrients such as phosphorus, boron, zinc and copper, under favorable conditions. The cation exchange capacity of organic colloids generally increases with increasing pH. 3) The surfaces of roots carry a negative charge and thus retain hydrogen ions, which can be exchanged for nutrients by the process of chelation. Chelation is known to occur under both acidic and alkaline conditions.

Soil is *limed* primarily to encourage cation exchange (Fig. 10–14). Addition of lime (CaO) allows the cal-

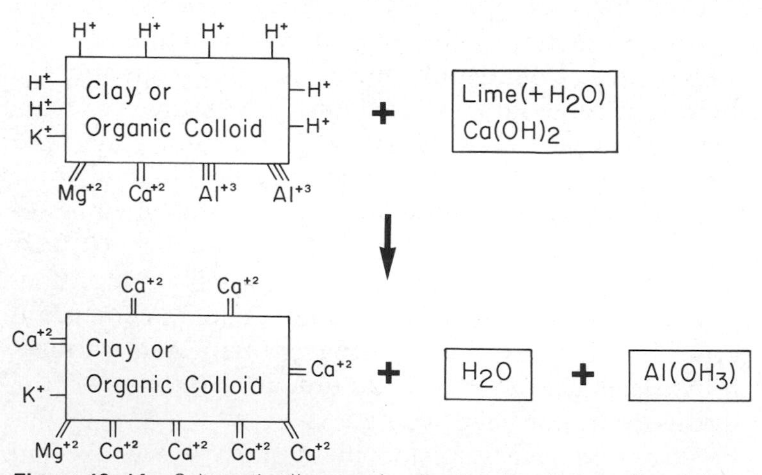

Figure 10–14. Schematic diagram showing the chemical reactions taking place with liming of soil. (Adapted from Allaway, W. H.: pH, soil acidity and plant growth in soil. *In* 1957 Yearbook of Agriculture. Washington, D. C., Government Printing Office, 1957.)

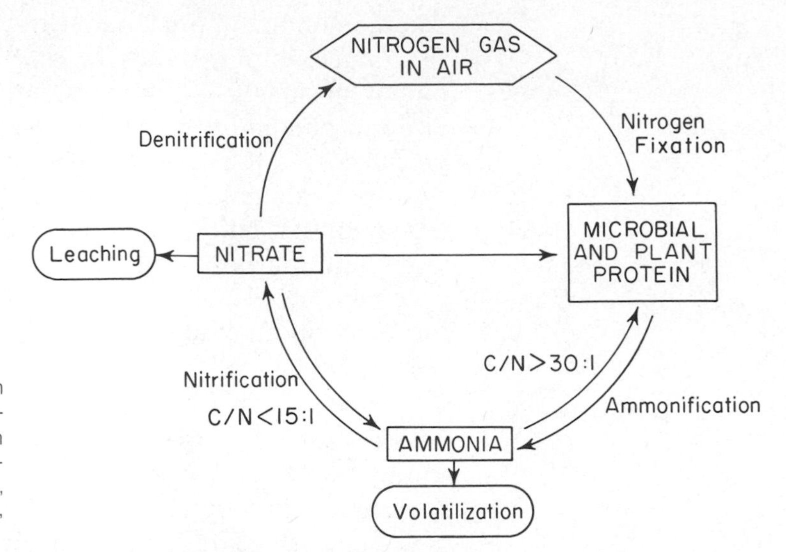

Figure 10–15. The nitrogen cycle. (Adapted from Broadbent, F. E.: Organic matter in soil. *In* 1957 Yearbook of Agriculture. Washington, D.C., Government Printing Office, 1957.)

cium to replace the hydrogen in clays and organic colloids, thus restoring the calcium to the soil. Lime also acts to neutralize the soil so that bacteria and other microorganisms can function better. However, the amount of lime to be added must be carefully calculated, because a soil that is too limey or alkaline is not fertile.

Many soil nutrients undergo a mineral-plant-mineral (or soil) cycle and may be lost along the way by leaching or removal in crops. The nitrogen cycle illustrates this mechanism (Fig. 10–15). Nitrogen is removed from the air and fixed by bacteria as animal or plant protein. When the animal or plant dies, microorganisms decompose the organic matter to ammonia (NH_4). Ammonia, in turn, can be assimilated by plants and reconverted into protein or it can be further altered by microorganisms to nitrate. The fate of the ammonia is determined by the carbon/nitrogen ratio. If the ratio is high, ammonia is reconverted into plant protein. If the ratio is low, the excess nitrogen is oxidized to nitrate, which is soluble and can be dissolved away or supplied to plants from the soil water. The cycle can be completed if microorganisms attack the nitrates and free the nitrogen back into the air.

At every step of the cycle, nitrogen may be lost to the system. If the soil is being cultivated, nitrogen and other nutrients are consumed by crops and must be replaced. Nutrients are stored in the soil over long periods of geologic time by natural processes of weathering and annual flooding. River valleys such as the Tigris-Euphrates, Nile and Mississippi became fertile because the rivers flooded their banks annually and replenished the nutrients removed by the previous year's crops. If a river is prevented from flowing over the flood plain, the soil will be impoverished as it is tilled. In 1964 the Nile was dammed at Aswan. Ten years later, farms along the lower Nile, deprived of the sediment that brought fertility, require artificial fertilization at great labor and expense.

5 PROBLEMS IN SOIL USE

Dangers in exploitation of soil resources include misuse, overgrazing, drainage, irrigation and erosion. Clearing of forests in order to till the land is an example of soil misuse. Generally, forest soils are not fertile or suitable for growing crops, so nutrients in the cleared land are soon depleted. In the New England and Ozark regions, many abandoned farms testify to this type of misuse. The so-called Desert of Maine was once a lush forest. Settlers moved in, cleared the land and worked it as farmland for a while. But the soil under the forest was thin and poorly developed and covered a layer of sand. Soon spots of the underlying sand appeared as the soil was tilled season after season. These spots of sand eventually grew so large that the farms had to be abandoned. Now the sand has completely taken over, and the farms are a desert of sand dunes.

Overgrazing reduces the vegetation cover that protects the soil. Generally, grazed lands have reached an equilibrium with the environment and if treated properly will continue to support native vegetal cover for forage. But with overgrazing, little vegetation is left to decay and fertilize the soil with recycled nutrients. Soil conditions degenerate and vegetation decreases. The final result is gullying and accelerated erosion by wind and water. Downstream of the gullies, siltation occurs with far-reaching effects on flooding, recreation, land use and wildlife. This type of soil misuse has caused difficulty in the semi-arid West. Depletion of plant cover by overgrazing on public lands in northern Utah resulted in gullying and concomitant siltation downstream (Fig. 10–16). Only the overgrazed areas

Figure 10–16. A view of a slope almost denuded of vegetation by overgrazing. At the head of Ford Canyon near Centerville, Utah. The destruction of the plant cover and packing of the soil by the animals provided a perfect setting for torrential rains to wash over the surface, gullying it and flooding the area below. (Photo from USDA F253169, USDA Misc. Pub. 196, Fig. 8.)

suffered these ill effects; nearby watersheds, carefully grazed and still covered with native vegetation, were not gullied and had no excessive flooding.

Drainage of wetlands has been a popular method of increasing the amount of land under cultivation. Although suitable land for cropping may be obtained in this manner, indiscriminate drainage without previous analysis of soil fertility may have disastrous effects. Many acres of wetland in Minnesota and Florida have been drained, leaving only infertile peat bogs.

Irrigation of fertile soils in dry regions can add tremendously to our food supply. However, this too must be done with care and with a full realization of the inherent dangers. In Yuma, Arizona, overirrigation has created a swampland. Incessant irrigation in a dry region causes dissolved salts to be carried upward as the soil moisture is drawn up by evaporation. When the soil water evaporates, salts are deposited, building up the mineral content to the point where the soil becomes infertile.

In the Salt River valley of Arizona, three years of irrigation resulted in a twofold increase in salinity of the soil at the surface. After years of irrigation in the Imperial Valley of California, the soil has become so salty that it has to be flushed out in order to retain its ability to grow crops.

Drainage of water through the soil during times of

Figure 10–17. Scene from the dust bowl in the United States. (From the USDA.)

intense irrigation carries away some of the salts, but it also engenders a new problem: the salt content of the water is increased with each use. The final result is a river so salty that it cannot be used for further irrigation or even for domestic purposes. This problem is exemplified in the controversy between the United States and Mexico over the waters in the Rio Grande and the Colorado River. These waters have been used so many times for irrigation in the United States that by the time they reach Mexico they are too saline for any use. This caused a tense international situation, which was relieved only by the expenditure of millions of dollars by the United States to desalinate the rivers and construct aqueducts, which provide usable water to Mexico.

Perhaps the classic example of the effects of soil misuse was in the midwestern section of the United States. The once fertile and lush prairie was converted by overuse and bad land management practices into a region of open loose soil exposed to the geologic forces of erosion. In the early 1930's, the area from the Dakotas southward became a large dustbowl (Fig. 10–17), losing tons of fine-grained soil to the winds that blew over it. In addition, denudation of the forests in the Mississippi River basin and resulting fluvial erosion led to the loss of much topsoil in the Midwest. Some agriculturalists fear we are beginning another dry cycle that may lead to dustbowl conditions in the American Midwest.

6 SOIL MANAGEMENT

Significant depletion of agricultural soil resources during the 1930's caused people to realize that human life depends upon healthy soil. Since that time, land management practices have been devised and implemented to improve care of soil. These include a variety of methods, the most prominent of which is fertilization. The older practices of liming and manuring have been partially replaced by use of industrial fertilizers, which are applied only after soil has been analyzed to determine its nutrient needs. One relatively modern technique (though it has been practiced in a different form for centuries) is the spraying of agricultural land with sewage. This accomplishes the double purpose of fertilization and sewage disposal. Allowing the land to lie vacant, or fallow, in order to renew itself is a beneficial practice based on natural soil dynamics. The same purpose can be accomplished by *crop rotation* or *strip cropping*. Crop rotation seeks to renew the soil by using different crops in alternate years so that certain nutrients are depleted while others accumulate. In strip cropping (Fig. 10–18), different crops are planted at the same time but in strips. Thus the soil

Figure 10–18. Strip cropping. (From USDA, SCS ND-589.)

can be kept nutrient-rich by alternating crops in the strips from year to year. Corn, soybeans and cotton are the most soil-depleting crops. Those most helpful in restoring fertility are legume crops such as clover and alfalfa.

Erosion on bare cultivated fields can be ameliorated by mulching, terracing and the use of waterways, diversion ditches and other structural controls. A *mulch* not only protects the soil from being washed washed away but also retains soil moisture. *Terracing* tends to delay run-off and decrease its velocity (and hence its energy) so the soil will not be carried off. Two major types of terracing are used (Fig. 10–19). Gullies already present can be smoothed, banks reshaped and grass planted on bed and banks to provide a grassed waterway through which run-off can be diverted from a field. *Diversion ditches,* leading water away from gully heads, can prevent the spread of the gully. The main purpose is to break up any concentrated sheet flow overland. Sheet flow can also be prevented by plowing along the contours of a hill.

Structural controls of erosion include small check dams, drop inlets, drop spillways and chutes (Fig. 10–20), all of which serve in various ways to check the velocity and/or decrease the amount of water flowing, thus reducing the energy of the water and lessening its erosive power (see Chap. 5).

Figure 10–19. Terracing of farmland. (From USDA, SCS la-2707)

Figure 10–20. Structural control of gullies. (From USDA, SCS Mo-1939 and Conn-97.)

The United States Department of Agriculture's Soil Conservation Service attempts to bring these problems and their remedies to the attention of farmers and to persuade them to apply sound management techniques for the betterment of the soil and crop production. One such method involves testing of soils and making soil maps. The soils of a region are studied, classified and categorized. The categories indicate the uses to which a soil can be put and the management practices required to maintain productivity (Table 10–1).

TABLE 10-1. AGRICULTURAL CAPABILITY UNITS*

Class	Description
I	Wide range of uses. Can be cultivated without damage to soil. Generally on level ground, well drained and easy to work. Very productive and can be used continuously without damage or extra care.
II	Can be cultivated regularly but has less range of crop suitability than class I. May be on gently sloping land with some danger of erosion under cultivation and is slightly less well drained.
III	May undergo regular cultivation but needs care in preventing erosion and in management practices. Crop versatility is less than class I or II.
IV	Needs careful management and can be cultivated only occasionally.
V	Wet and has low fertility. Relatively unsuited for agriculture; should be used for pasture or forest.
VI	Completely unsuitable for agriculture. Should be used only for forage or forest.
VII	May be used for pasture or forest but only under careful management.
VIII	Of no use except as scenery or for recreation.

*After U.S. Department of Agriculture.

The expanding population of the world will require a vast increase in world food supply. This means both more intensive use of land already under cultivation as well as the use of much land not now used for agriculture. Most of the presently unarable land lies in tropical regions, where the highly leached soils are unproductive. Careful management practices are needed to bring such poor-quality soil into food production, but it can be done. Still another source of potential arable lands is in the arid regions, where soils are highly alkaline and poor in nutrients. Irrigation, wise management and fertilization may allow these soils also to be utilized for food production.

The fertile soils of the world are already under cultivation and must be carefully preserved. Current attitudes toward the soil and use of agricultural land must change. Soil is a resource not to be wastefully exploited or destroyed. Prudent use of soil is necessary in order not only to maintain a high standard of living but simply to live. Like air and water, the soil is indispensable to human life.

11

WATER

Soil flies into the air and seismic waves ripple through the ground in the Arizona desert. Seismic wave velocity differences indicate location, depth and thickness of water-bearing rock in the area.

1 INTRODUCTION

Water plays a major role in human life. A vital necessity, it can also be a destructive force. In this chapter we will concentrate on the occurrence of water, its physical and chemical properties and particularly on the fresh water, streams, lakes, and so forth, needed by man.

2 ORIGIN OF WATER

During the formation of the earth its interior was molten. The surface may or may not have been molten, but conditions were such that water and other volatiles were not driven off into space. The water was dissolved or trapped in the near-surface interior and existed as steam in the primitive atmosphere. Eventually the surface cooled enough for the steam to condense, forming the early oceans and atmosphere. The water trapped in the interior, *juvenile water*, is released by igneous activity.

The early cooling of the earth's surface that allowed

the formation of surface water occurred more than 4 billion years ago; the release of juvenile water from the interior continues to this day. However, the amount of juvenile water supplied to the surface is insignificant relative to the available water supplies. The study of the origin and history of water is of great scientific interest, but our immediate concern is with its distribution and occurrence. The fact is that the earth has plenty of water, but where is it?

3 DISTRIBUTION OF WATER

Table 11–1 lists the estimates made by the U.S. Geological Survey regarding distribution of water. The total volume of water is 326 million cubic miles, but 97.2 per cent of this is salt water in the oceans, which is not directly usable. The oceans are an invaluable resource nevertheless (see Chap. 4). Of the remaining water, while almost all of it is fresh, most of it is frozen in the icecaps of Greenland and the Antarctic. In summary, less than 1 per cent of all the water on the earth is readily available as fresh water.

Surface water (streams and lakes) and water that fills the voids and openings below the surface, or *subsurface water*, comprise our present available water supply. In the future, development of large-scale desalina-

TABLE 11–1. DISTRIBUTION OF WORLD'S ESTIMATED WATER SUPPLY*

Location	Surface Area (square miles)	Water Volume (cubic miles)**	Percentage of Total Water
Surface Water			
Fresh-water lakes	330,000	30,000	.009
Saline lakes and inland seas	270,000	25,000	.008
Average in stream channels		300	.0001
Subsurface Water			
Vadose water (includes soil moisture)		16,000	.005
Ground water within depth of 1/2 mile	50,000,000	1,000,000	.31
Ground water–deep lying		1,000,000	.31
Other Water Locations			
Icecaps and glaciers	6,900,000	7,000,000	2.15
Atmosphere (at sea level)	197,000,000	3,100	.001
World ocean	139,500,000	317,000,000	97.2
Totals (rounded)		326,000,000	100

*From Department of the Interior news release, August 13, 1972.
**One cubic mile of water equals 1.1 trillion gallons.

tion plants and ways to use the ice of Greenland and the Antarctic could add to water supplies. Because water is a renewable resource (see Chap. 7), this problem is one of availability and suitability in a particular area rather than of overall depletion as with minerals.

4 THE HYDROLOGIC CYCLE

The unique properties of water are beneficial in many ways. For example, its solid form, ice, has a lower density than the liquid, water. This is fortunate because if ice were denser than water, it would sink rather than float, and lakes and rivers would fill with ice during the winter. (Philosophers during the Middle Ages pointed out that it was logical and necessary for ice to be less dense than water; otherwise the lakes of Europe would fill with ice, and God could not provide mankind with fish during the winter!) Such natural phenomena form the basis of the delicate physiochemical conditions that allow such a large portion of the earth to be habitable.

Water also occurs as a gas, water vapor. In addition to evaporation, water vapor is also formed by *transpiration*, which occurs when water absorbed by plants is released to the atmosphere from stomata of the plant tissue.

In nature, the various forms of water occur depending on temperature and geographic location. The *hydrologic cycle* (Fig. 11–1) relates the processes of *evaporation* and *condensation* by which rainfall is produced from water vapor. This requires a climate warm enough to cause water vapor to form from the surface of large bodies of water, principally the oceans, and allows for the condensation and precipitation of the water vapor as a liquid (rain). A temperate coastal region such as the eastern United States fulfills these requirements. The hydrologic cycle is altered during the winter, when the precipitation may occur as a solid, in the form of sleet or snow. The presence of a solid leads to *sublimation*, which is the direct production of water vapor from the solid, without the intervening step of melting.

What happens to the water that falls on the earth's surface? Some of it stays on the surface as *run-off* (see Chap. 5); the rest infiltrates the ground to become part of the *subsurface water system*.

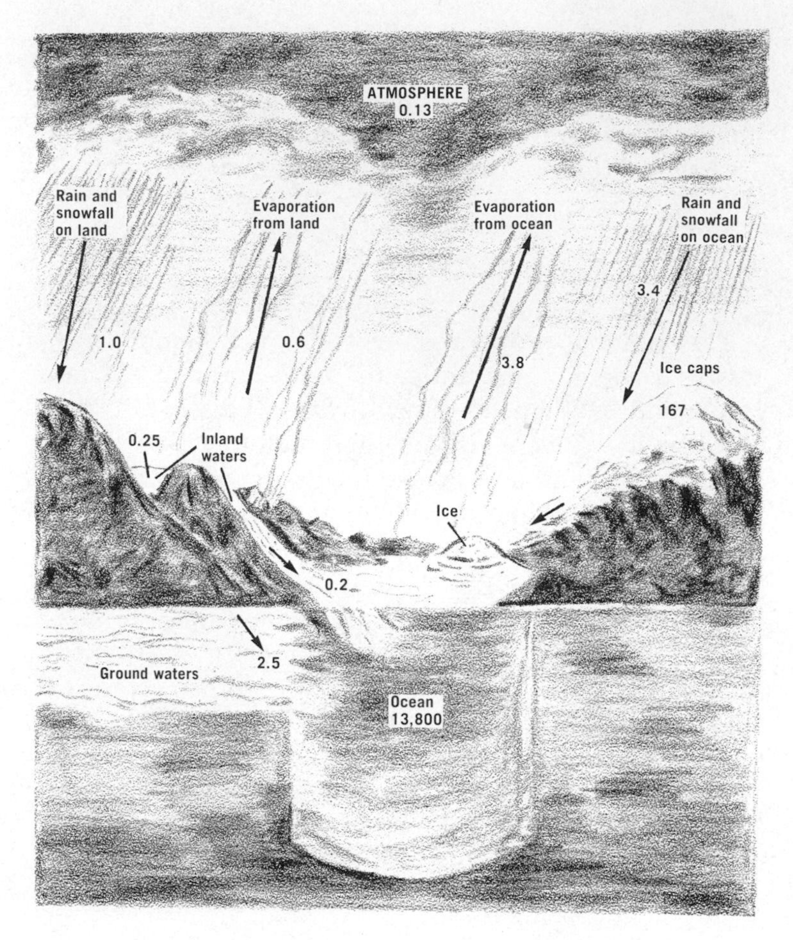

Figure 11–1. The hydrologic cycle. (From Turk, A., et al.: Environmental Science. Philadelphia, W. B. Saunders Co., 1974.)

5 SUBSURFACE WATER

The diagrammatic cross-section in Figure 11–2 illustrates the relationship between the *zone of undersaturation* or *zone of aeration* and the *zone of saturation*. The zone of aeration contains *vadose* water; the zone of saturation is composed of ground water. The zones are separated by the *water table*, which is the surface of the saturated or ground water zone.

Vadose water is the moisture of the soil and other loose material near the surface. Openings and voids in these near-surface materials are only partially filled with water. Voids and openings below the water table are completely filled with water.

The volume of openings and voids in a rock or other material is known as the *porosity*. More specifically, porosity is defined as the ratio of the volume of openings or voids to the total volume. Unconsolidated or

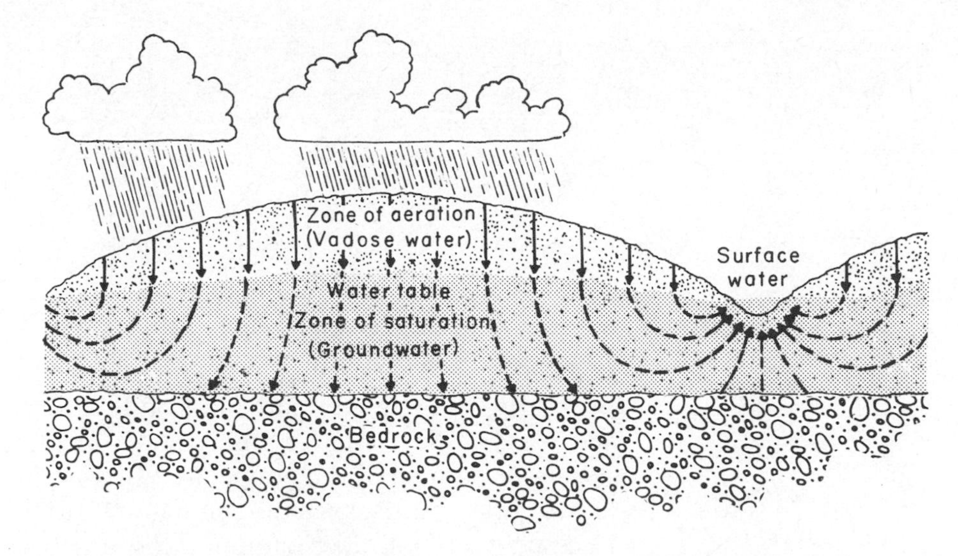

Figure 11–2. Diagrammatic representation showing water table, ground water and surface water. (From Donn, W. L.: The Earth. New York, John Wiley & Sons, 1972.)

loose materials such as gravel or sand (Fig. 11–3 *A*, *B*, *C* and *D*) with many openings between the grains generally have 20 to 40 per cent voids or porosity. Factors such as packing, sorting and grain shape affect porosity in sediments. In a poorly sorted sediment, Figure 11–3 *D* (one with a variety of grain sizes), the porosity is reduced because the smaller grains fill in part of the void space around the large grains. Porosity is reduced by closer packing of spherical grains, as shown in Figure 11–3 *A* and *B*. Irregular grain shapes (Fig. 11–3 *C*) can also pack closer to reduce porosity in a sediment.

Sediments that have undergone compaction and cementation (or *lithification,* the formation of rock

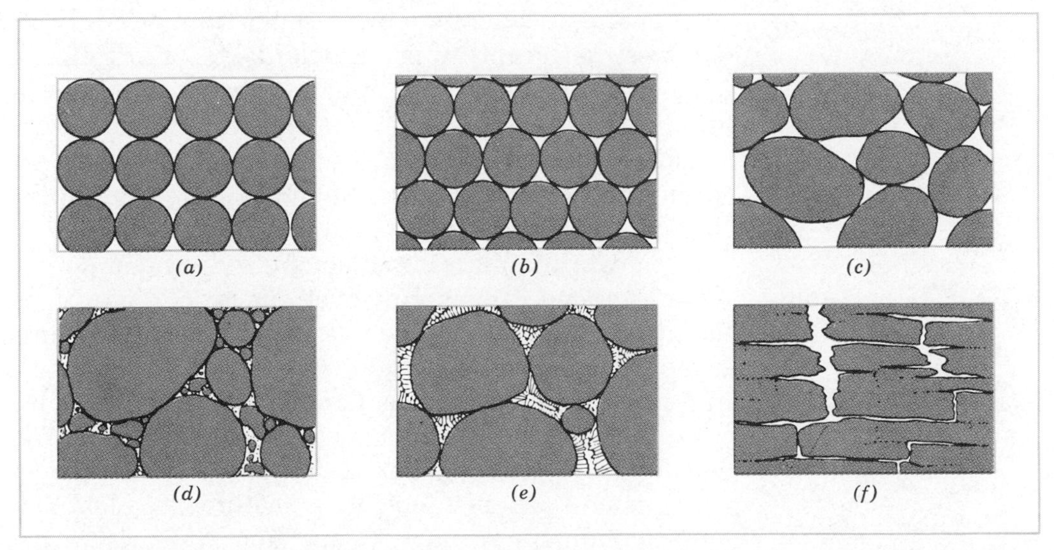

Figure 11–3. Diagrammatic representation of porosity in rocks. (From Donn, W. L.: The Earth, New York, John Wiley & Sons, 1972.)

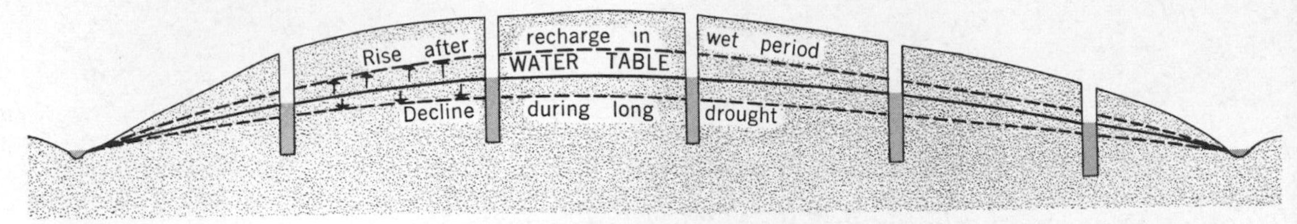

Figure 11–4. Relationship of water table and rainfall. (From Strahler, A. N.: Introduction to Physical Geography. New York, John Wiley & Sons, 1973.)

from sediment) are called consolidated sediments and generally have lower porosities than unconsolidated sediments. Cementation is usually initiated from points of contact of the grains, and incomplete cementation leaves voids, as in the cemented sand (sandstone) in Figure 11–3 *E*. Limestones often are partially dissolved, thus increasing porosity (Fig. 11–3 *F*).

However, packing, sorting, grain shape and cementation are not the only considerations in understanding porosity. In well-crystallized limestones, shales and metamorphic and igneous rocks, which have very low porosities because the grains are tightly interlocked, porosity is related to fracture and breaks rather than intergranular voids. Various rock types exhibit a wide range of porosity. Real porosity values are therefore empirically determined because of the complexity of controlling factors.

Permeability is the ability of a rock or sediment to transmit fluid. (Discussions of porosity and permeability apply equally to fluids such as gas and oil as to water.) In terms of groundwater movement or the practical problem of attaining water from groundwater sources, permeability is a crucial factor. For example, clay sediments, although highly porous, are *impermeable* and therefore will not yield or transmit water readily. This is because the envelope of water surrounding the clay grains is bound to the grain, inhibiting movement of the water.

Permeability implies that ground water moves, which indeed it does. However, the type of movement is different from surface water movement. Groundwater flow is nonturbulent (*laminar*) as opposed to the turbulent flow of a stream. Ground water moves downward but eventually exits back into surface run-off via springs or groundwater intersection with rivers, streams and other surface water features (see Fig. 11–2 and section on surface water). The water table generally follows the topography but is flatter and mobile. Figure 11–4 illustrates the relationship of the water table to rainfall.

The rate of groundwater movement, which is measured in units of feet per year, is slow compared to the rate of flow of a stream, which is measured in feet per second. Groundwater velocity ranges from a few feet per day to a few feet per year, and generally is about 50 feet per year.

Wells

A well is a drill hole that penetrates below the water table (Fig. 11–5), allowing water to seep in and be available for pumping. If the well bottom is located in an *aquifer*, a rock or sediment unit that is porous and permeable and thus readily transmits water, a significant yield of water is obtained. Aquifers can be local or regional in extent, consisting of rocks or sediment such as sands, gravels, sandstones or limestones that have solution voids or fractures. Rarely, highly fractured igneous or metamorphic rocks are aquifers (Fig. 11–6). An example of a local aquifer (discussed at the end of this section) is the Castle Hayne limestone, which underlies the coastal region of eastern North Carolina.

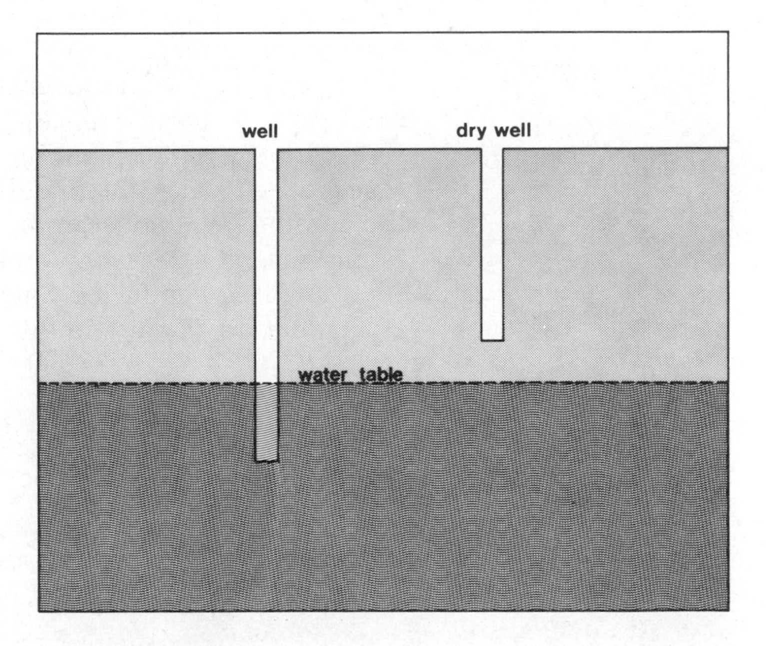

Figure 11–5. Relationship of water table and wells.

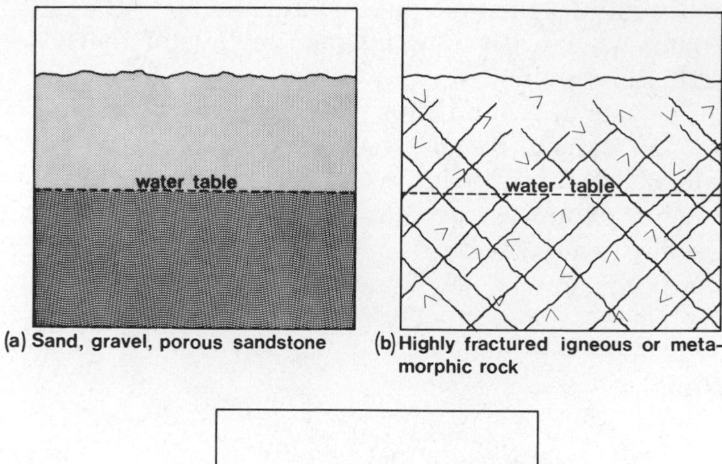

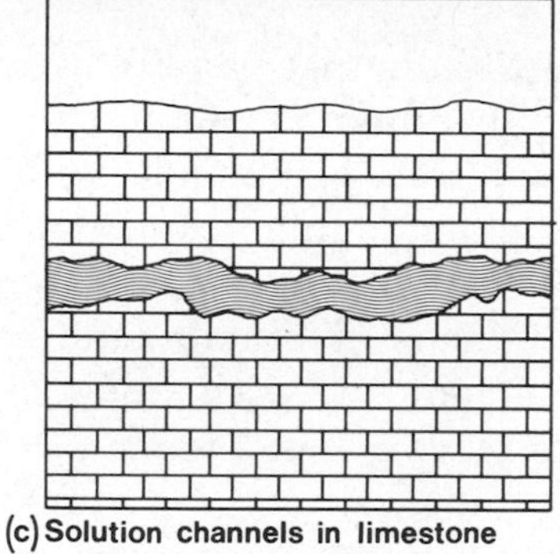

Figure 11–6. Potential aquifer materials.

In areas where aquifers are regional in extent, such as the Dakota sandstone, the St. Peter sandstone in Wisconsin and the aquifer under the Sahara (Figs. 11–7 and 11–8), an impermeable layer is necessary to contain the water in the aquifer (Fig. 11–8). The conditions required for occurrence of a regional aquifer can also lead to an *artesian* type of well, where the water flows without pumping. (A stricter definition of artesian well states only that the water must rise above the confined aquifer.) Confined water of this type explains several interesting and important phenomena, such as the oases in the Sahara (Fig. 11–9). There is an abundance of water in this region, and it can be ob-

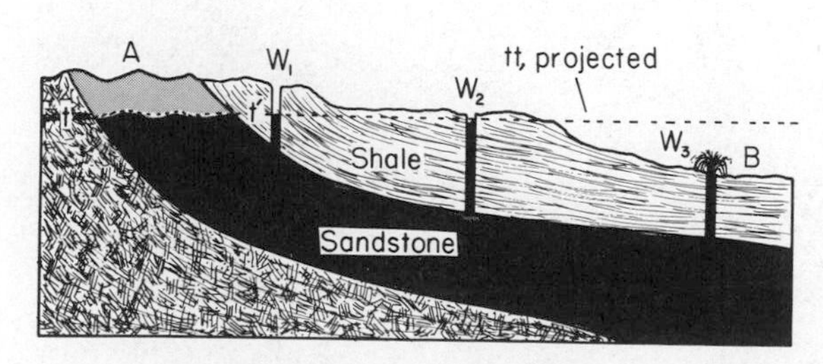

Figure 11–7. Cross section showing a series of wells penetrating a confined aquifer. The water table in the recharge area is t –t′. (From Gilluly, J, et al.: Principles of Geology, 3rd ed. San Francisco, W. H. Freeman & Co., 1968.)

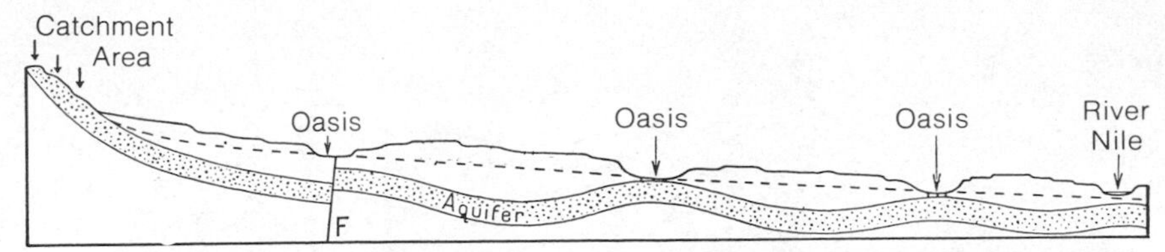

Figure 11–8. Schematic section across the Sahara to illustrate conditions favorable to the development of oases. (From Holmes, A.: Principles of Physical Geology, 2nd ed. New York, Ronald Press, 1965. Copyright © 1965 by Arthur Holmes. Reprinted with permission of Ronald Press Co., New York, and Thomas Nelson and Sons Ltd., London.)

tained if deep wells are drilled to tap these deep aquifers.

When a well is pumped, a local cone of depression forms around the well (Fig. 11–10). If water is removed faster than the ground water can replenish it, the well will go dry. If there are too many wells in an aquifer or if a drought occurs, the water table will fall.

Figures 11–11 and 11–12, taken from a report on the effects of high-volume pumping by a mining operation in the coastal plain of North Carolina (Fig. 11–13 shows

Figure 11–9. Part of the great oasis of Colomb Bechar in the northwest of the Algerian Sahara, as it was about 1930. Since then the abundant supplies of ground water have been conserved by underground dams and reservoirs; coal has been discovered and, not far away, ores of iron and manganese. Colomb Bechar has now become the center of a rapidly expanding industrial area. (From Holmes, A.: Principles of Physical Geology, 2nd ed. New York, Ronald Press, 1965. Courtesy of Paul Popper Ltd., London.)

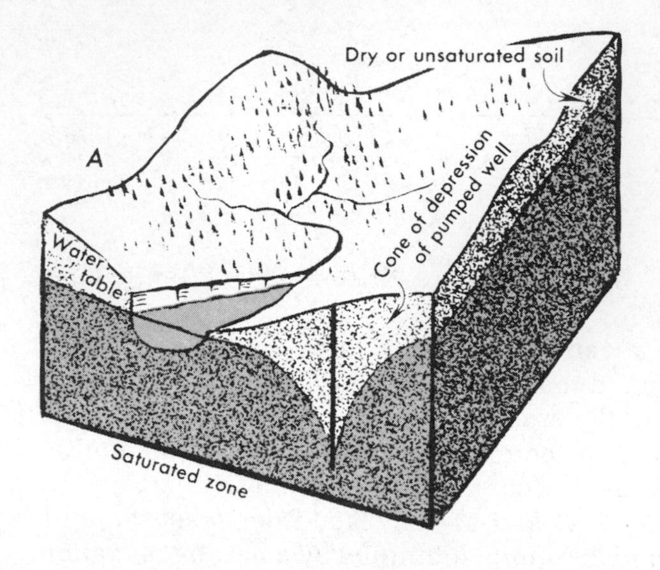

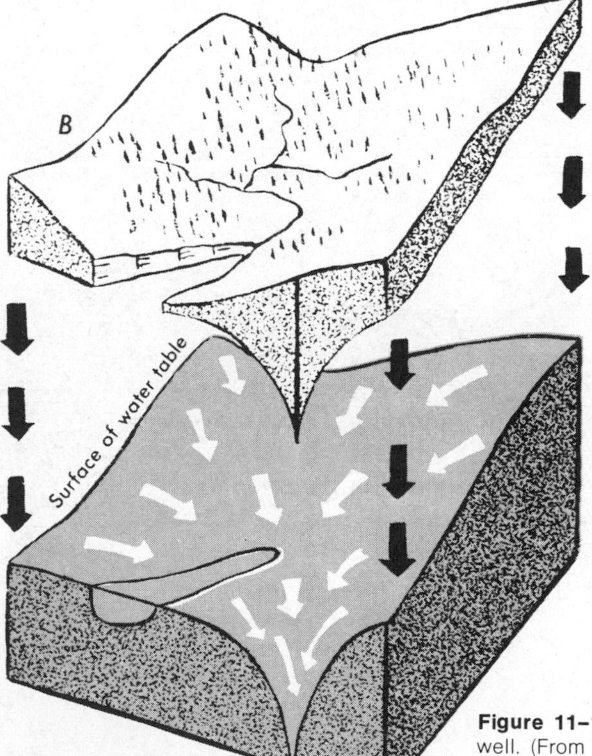

Figure 11–10. Effect on local water table of pumping a well. (From Leopold, L. B., and Langbein, W. B.: A Primer on Water. Washington, D.C., United States Department of the Interior Survey, 1960.)

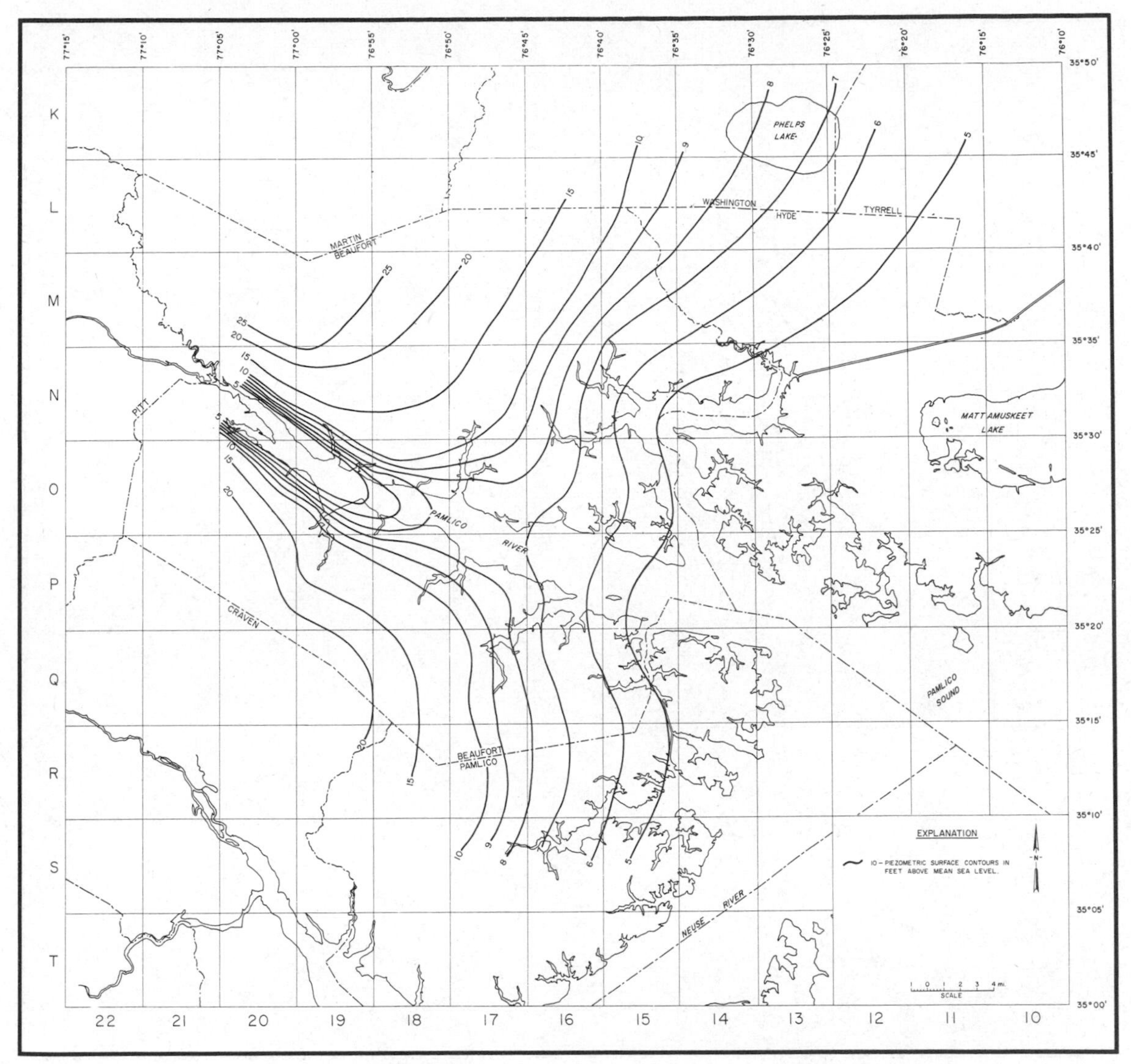

Figure 11–11. Piezometric surface prior to 1965, Beaufort County, North Carolina.

location), provide a dramatic example of the effect of pumping on a water table. The piezometric surface indicates the behavior of the water table, which is a particular piezometric (now called potentiometric) surface.

Underlying Beaufort County in eastern North Carolina is one of the largest phosphate deposits in the world, containing an estimated 10 billion tons of phosphate, which is a necessity in modern chemical agriculture. The mining began in 1965 at the Lee Creek Mine of the Texas Gulf Sulphur Company. The open pit mine penetrates well below the water table and into

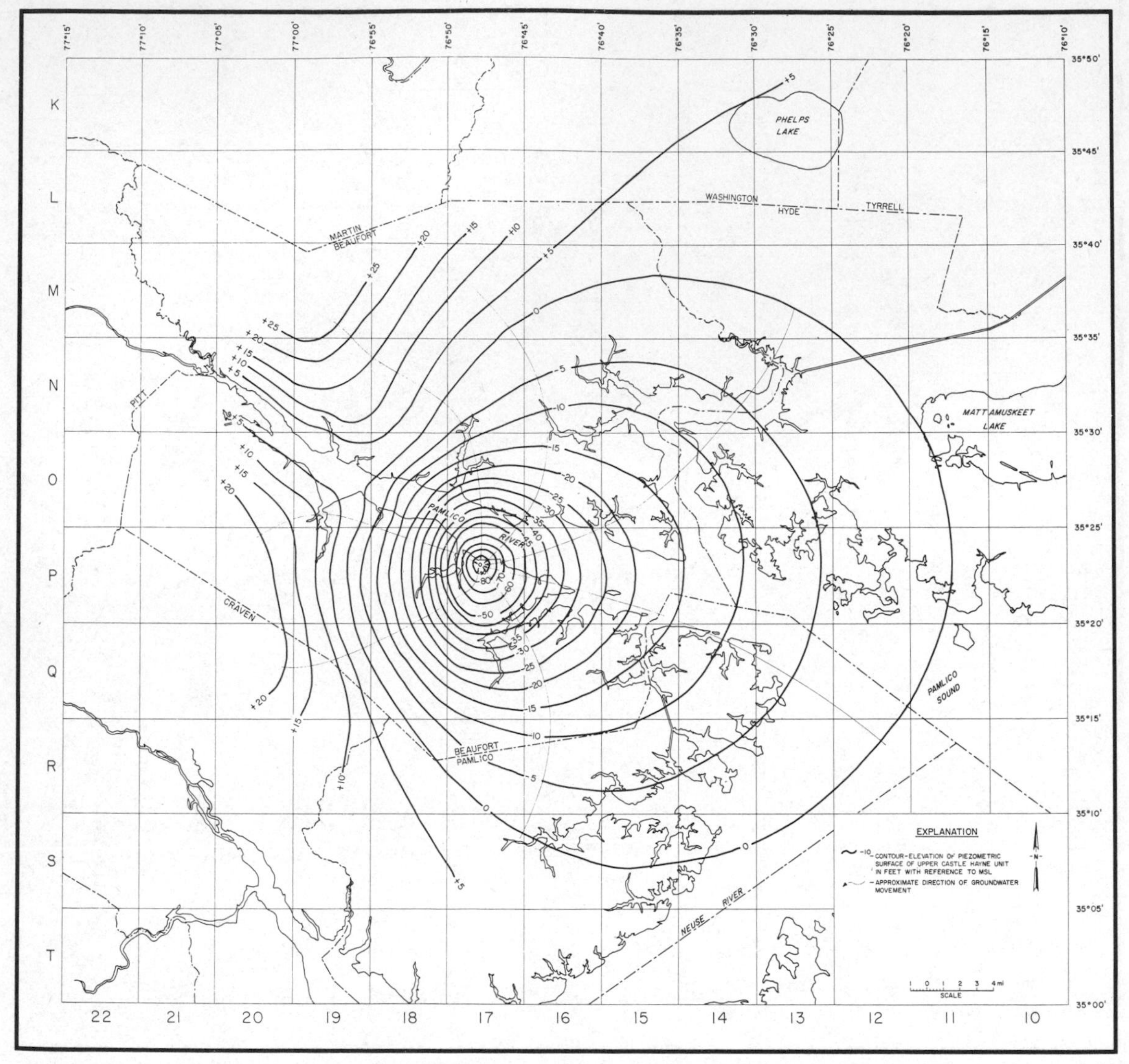

Figure 11–12. Piezometric surface mid-1970, Beaufort County.

the local aquifer, the Castle Hayne limestone. The Castle Hayne limestone is a solution cavity aquifer, where the fossil shells have dissolved out, leaving a highly porous and permeable rock.

One of the problems of overexploitation of an aquifer of this type is loss of water for wells in the region. Figure 11–14 shows the location of wells in the Lee Creek mine section of the Castle Hayne. The numerous small wells had little effect on the water table prior to 1965 (see Fig. 11–11), but the open pit mine caused the water table to drop from approximately 8 feet above sea level to more than 90 feet below sea level, and the

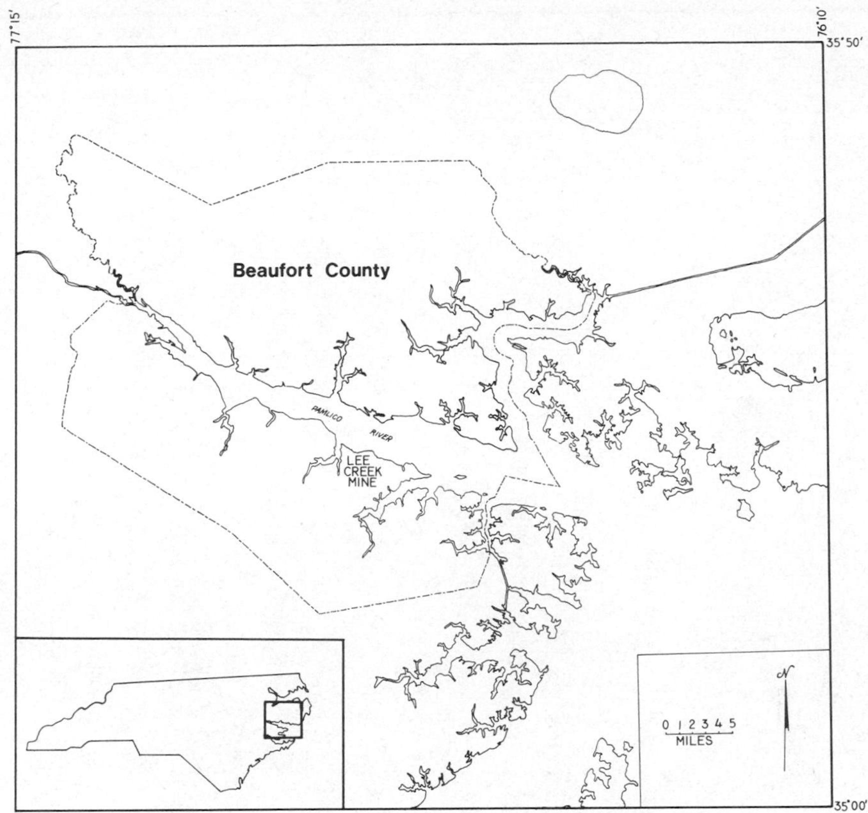

Figure 11–13. Location map, Beaufort County, North Carolina.

shallow wells in the vicinity had to go dry or be drilled deeper.

However, a more serious problem than lowering of the water table exists. In a coastal region, saltwater encroachment into the fresh ground water presents a significant danger.

Figure 11–15 illustrates the relationship between fresh and salt water in a coastal region. Because the fresh water is less dense than the salt water, it "floats" on top of the salt water. The boundary is not precise, however; some mixing occurs through diffusion and tidal changes. If the fresh ground water is removed more rapidly than it is replenished, the boundary will move inland, with a consequent contamination of wells. Contamination is measured by the chloride content of the water because the principal salt of sea water is sodium chloride.

At the Lee Creek mine, the chloride content increased by a factor of 2.5 in a period of five years, from

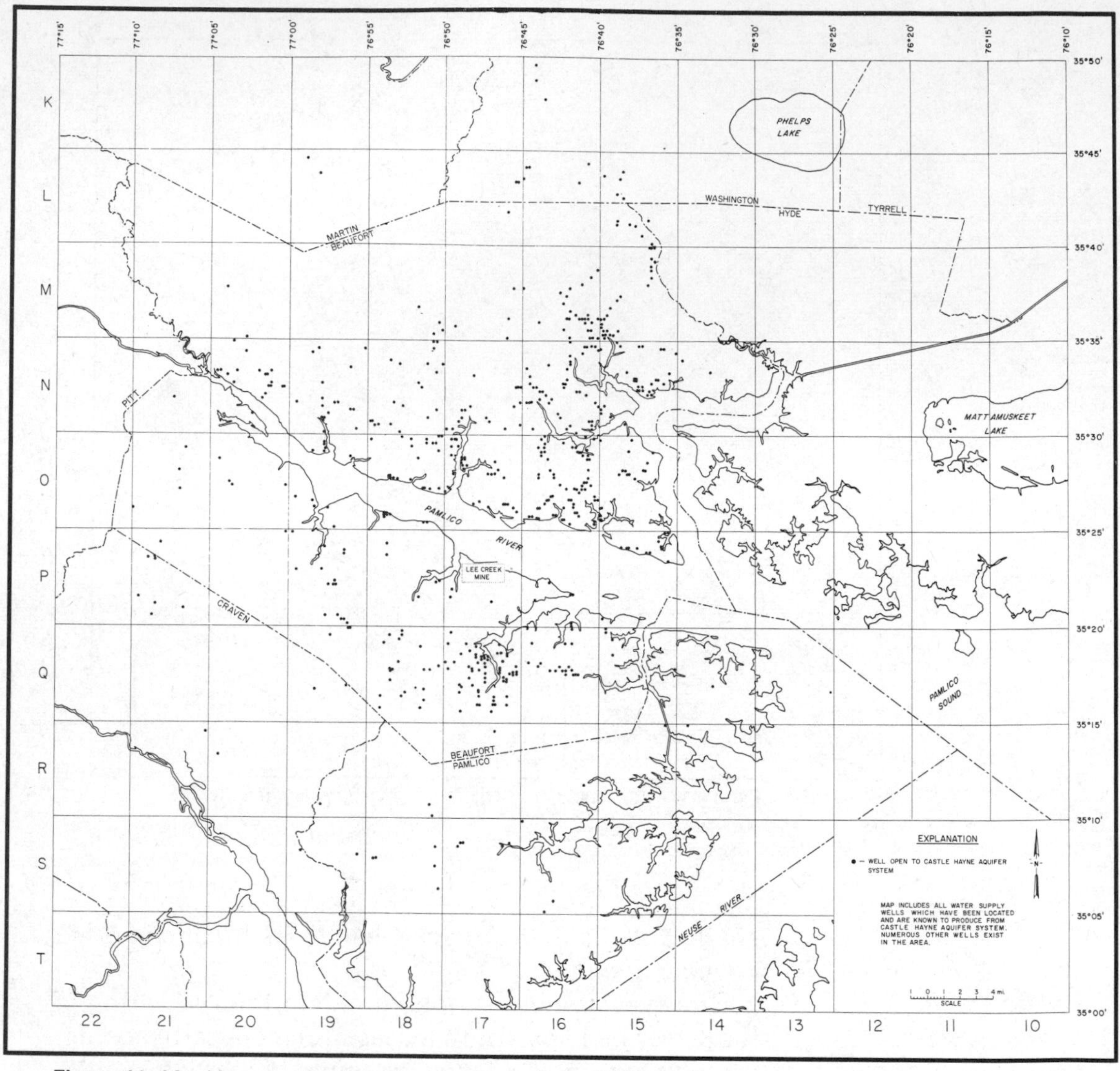

Figure 11–14. Map showing location of water wells in vicinity of the Lee Creek mine, Beaufort County.

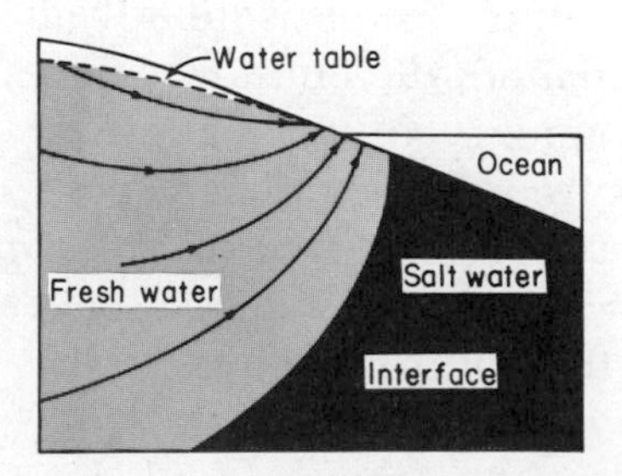

Figure 11–15. Cross section showing fresh-water flow lines in relation to the contact with underground salt water. (From Gilluly, J., et al.: Principles of Geology, 3rd ed. San Francisco, W. H. Freeman & Co., 1968.)

approximately 50 ppm chloride (1 ppm or part per million equals 0.0001 per cent) to 120 ppm chloride (Figs. 11–16 and 11–17). While this chloride ion concentration level is not disastrous (federal government standards for drinkable water allow up to 250 ppm chloride ion), if the chloride ion content increased 2.5 times in five years, what does the future hold? According to Texas Gulf Sulphur and North Carolina Phosphate Company in a joint report with the North Carolina Department of Water and Air Resources, there is no danger of serious saltwater incursions for at least 100 years (Fig. 11–18).

Figure 11–16. Chloride ion content prior to 1967, Beaufort County.

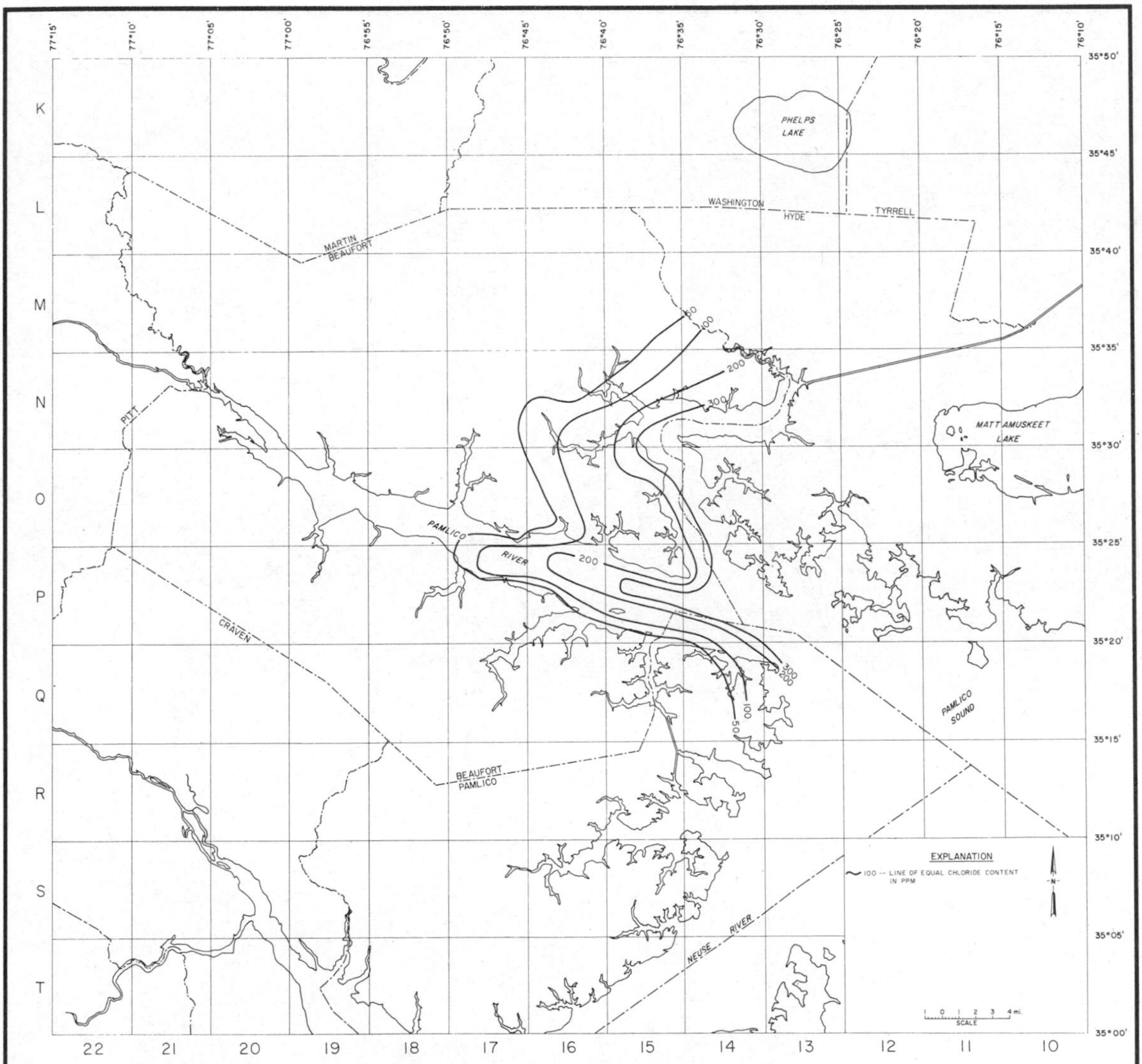

At the Lee Creek mine, the chloride content is projected to go up to 180 ppm. Figures 11–19 and 11–20 show the projected effects of adding another producer, the North Carolina Phosphate Company, south of Lee Creek mine. The water table is lowered to more than 110 feet below sea level, and the chloride ion content is increased to 150 ppm (still not undrinkable, according to federal standards).

The question we must ask ourselves is whether these rosy predictions are accurate. Clearly, they are advantageous to the companies and the State Bureau. The companies need long-term permits in order to

Figure 11–17. Chloride ion content, 1970, Beaufort County.

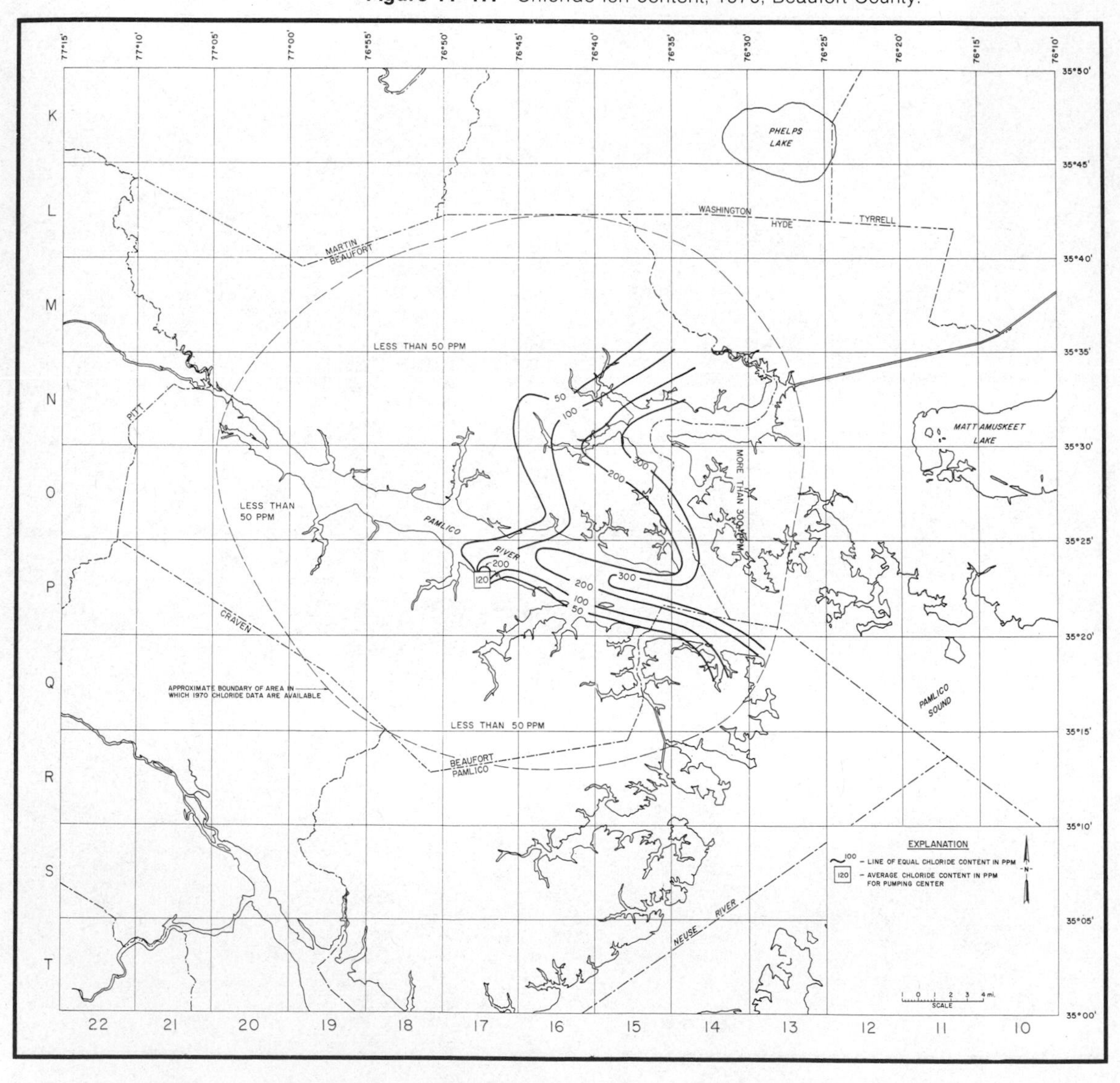

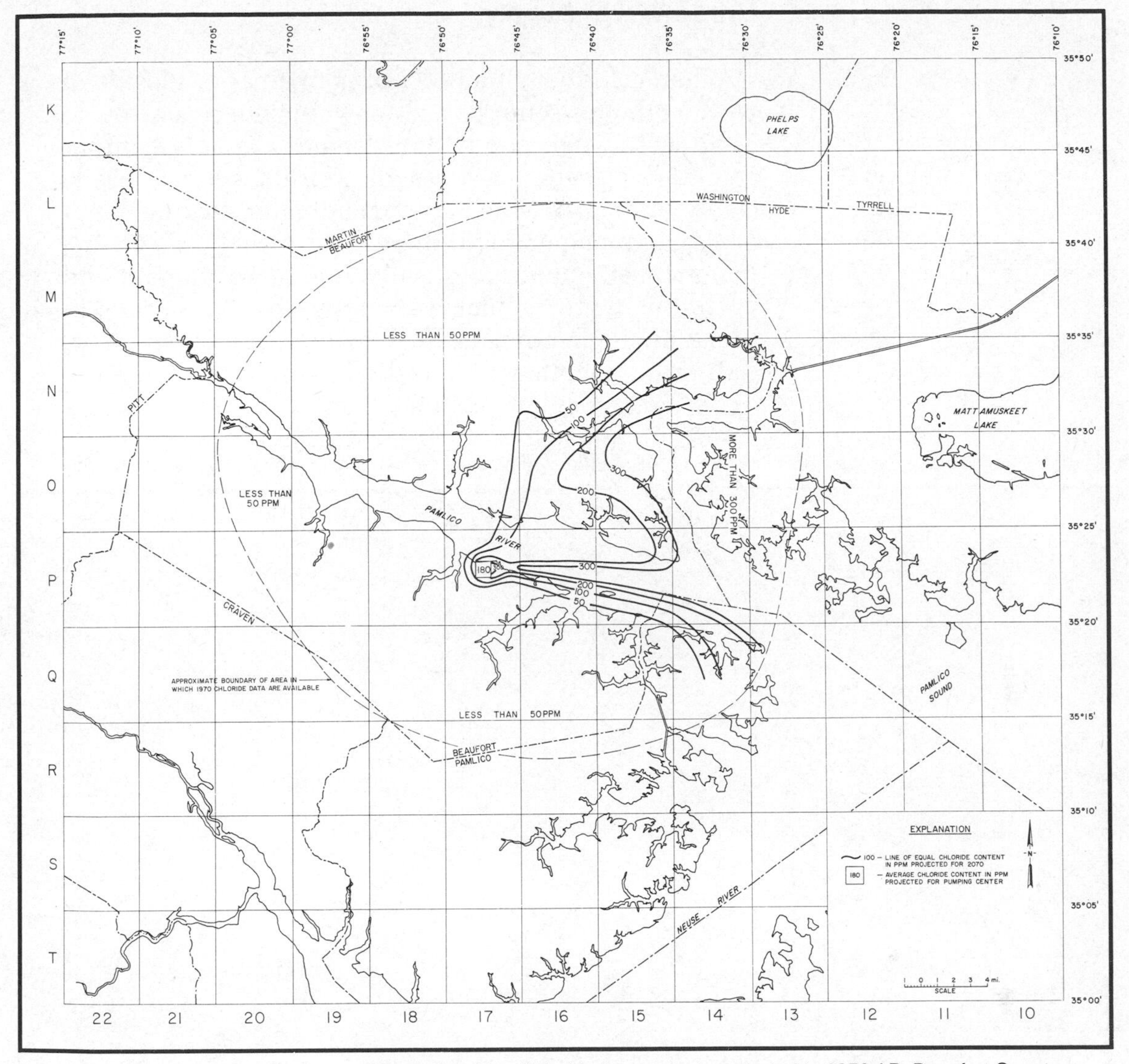

Figure 11–18. Chloride ion content, projected to 2070 AD, Beaufort County.

operate profitably and amortize equipment (the reserves are estimated to last for 130 years at present rates of production), and we, the society, need the phosphate—but we also need water. Thus we are dealing with an environmental trade-off. In this case not even considering the esthetic and land damage, we can pose a straightforward question of phosphate versus water. In this area of low population density and essentially no manufacturing, water needs are relatively easy to meet. The trade-off is fairly simple, although the political viability of this solution may be more complex.

Urban Water Supply

When a similar situation arises in areas of Florida or New York, for example, the answer is not so simple. In both cases, high population densities greatly complicate the picture. In areas of Florida such as Dade County (Miami), which are currently undergoing population explosions, withdrawal of water is causing lowering of water tables, saltwater encroachment and pollution. Such problems are not unique to Florida. In the arid, high population density areas of Southern California, which are naturally low in water, the prob-

Figure 11–19. Projected piezometric surface for two mines, Beaufort County.

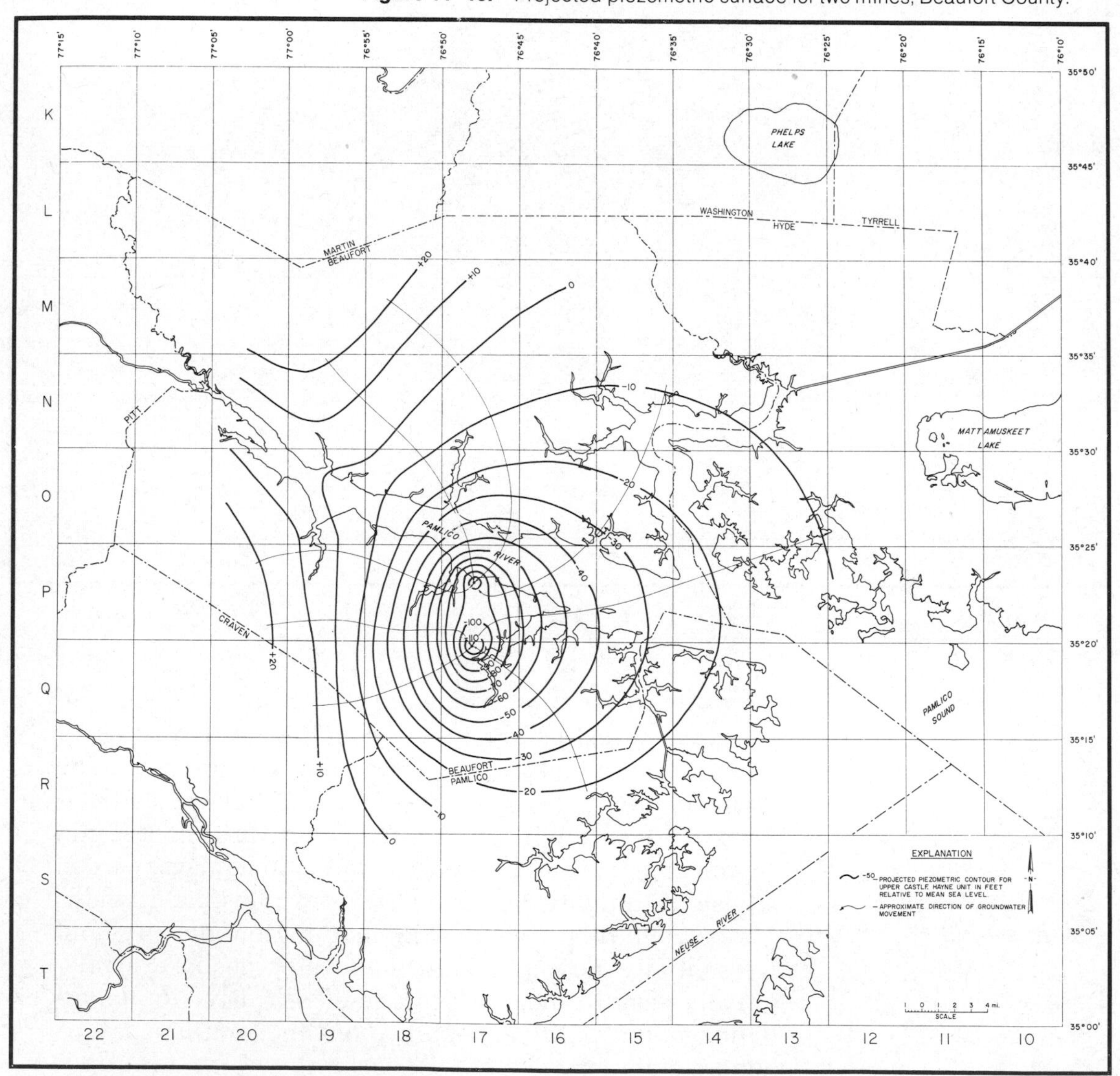

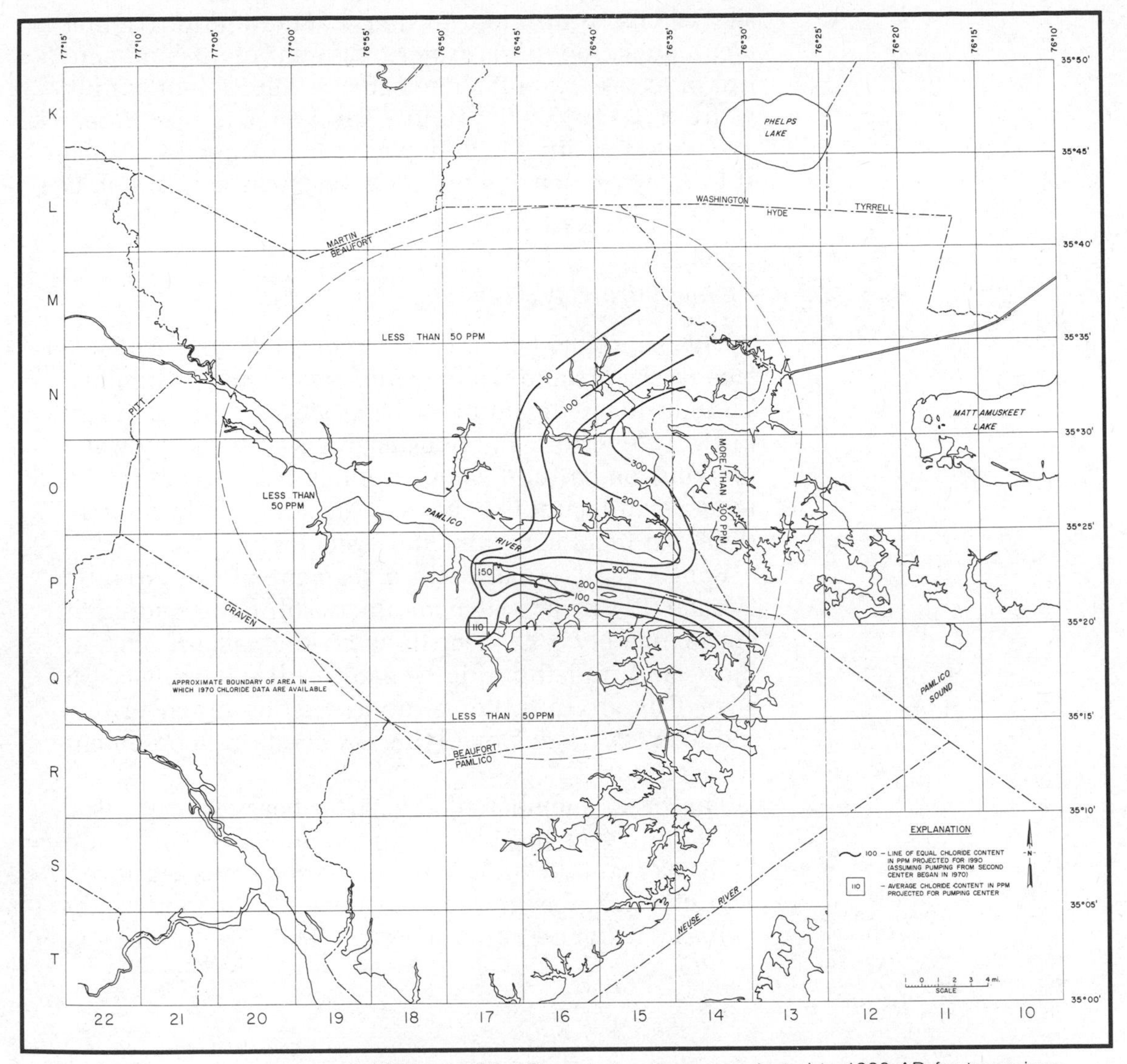

Figure 11–20. Chloride ion content, projected to 1990 AD for two mines, Beaufort County.

lem has been to find ways to transport water into the area. This is a political as well as an economic dilemma because, in essence, Southern California is competing with neighboring states for all water in the area.

Southern Florida is a well-studied area because there has been an awareness of water problems since the first wells were drilled into the Biscayne aquifer in 1896. By 1925 the original wells had to be abandoned because of saltwater contamination. New wells were drilled farther inland, but within a few years these too were in danger of saltwater encroachment. However, remedial techniques such as saltwater barriers in drainage canals prevented further contamination. Other techniques, including water treatment plants and control of excess run-off should insure a sufficient supply of fresh water for the future. But water is not "free," and detailed studies and planning are needed. Such studies are expensive but necessary where so much is at stake.

New York City Area

The New York City area has a complex system of surface reservoirs and wells in unconsolidated sediments. The city itself is supplied from surface reservoirs in upstate New York, but Long Island is supplied by wells drawing on groundwater supplies (Fig. 11–21). The Long Island supplies have been studied by federal, state and local governmental agencies.

Long Island is basically a segment of the Atlantic Coastal Plain, a wedge of unconsolidated sediment approximately 2000 feet thick on the eastern edge of the island and thinning to zero on the northwestern edge. The groundwater reserves are contained in this sedimentary wedge overlying the crystalline basement (Fig. 11–22).

The basal member of the water-bearing units is a sand known as the Lloyd sand member. It is overlain by a clay unit, which acts as a semipermeable layer. The clay member retards but does not prevent water movement in and out of the sand. These water-bearing

Figure 11–21. New York City area. (From Heath, R. C. et al.: The changing pattern of groundwater development on Long Island, New York. Washington, D.C., U.S. Geological Survey Circular 524, 1966.)

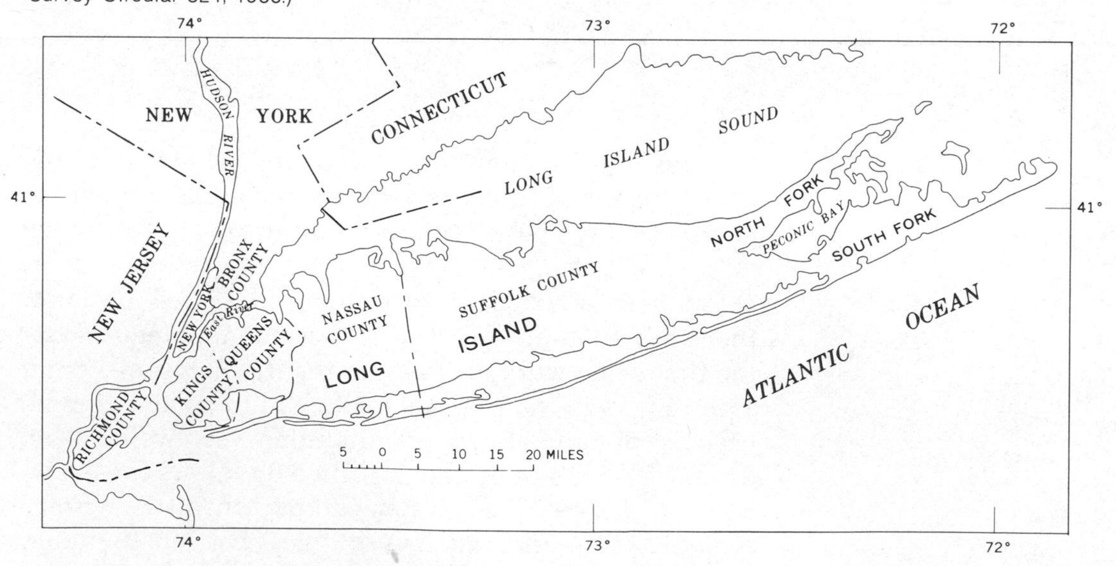

is discharged as surface run-off through drainage pipes and thus is lost to the groundwater supplies. By 1947, essentially all groundwater pumping in Brooklyn was discontinued except for water for air-conditioning, and even that is permitted only if the water is returned to the ground by injection wells.

The eastern part of the island, Nassau and Suffolk counties, will continue to rely for at least several decades on groundwater reserves for their water supplies. In Suffolk County the population is still low enough that saltwater intrusion has not yet occurred, and water supply is sufficient for the near future. Nassau County, however, has real and immediate problems. The population is presently two million, and the groundwater withdrawal rate is much higher than the recharge rate. Salt water is encroaching on the fresh ground water. A comparison of Figures 11–25 and 11–23 illustrates the change in position of the

Figure 11–23. Diagrammatic section showing predevelopment (phase 1) generalized groundwater conditions. (See also Fig. 11–26.) (From Heath, R. C., et al., 1966.)

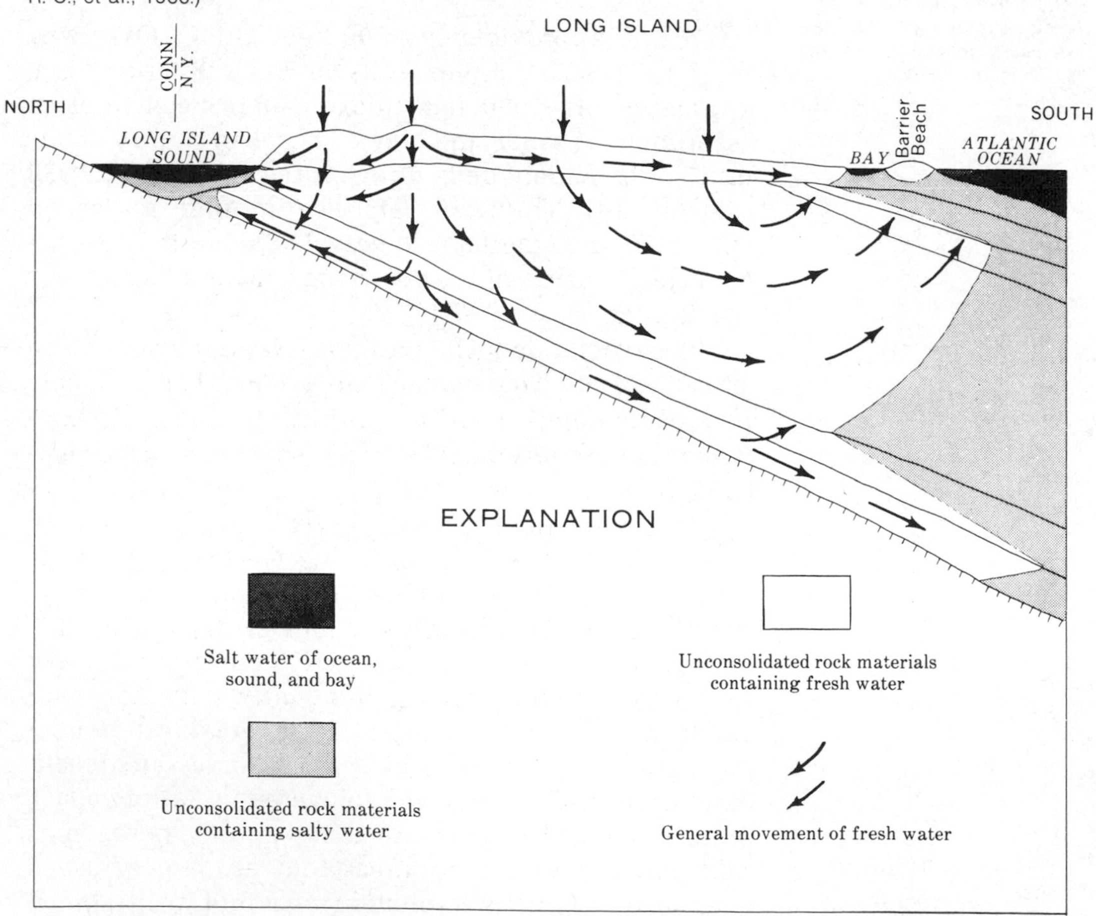

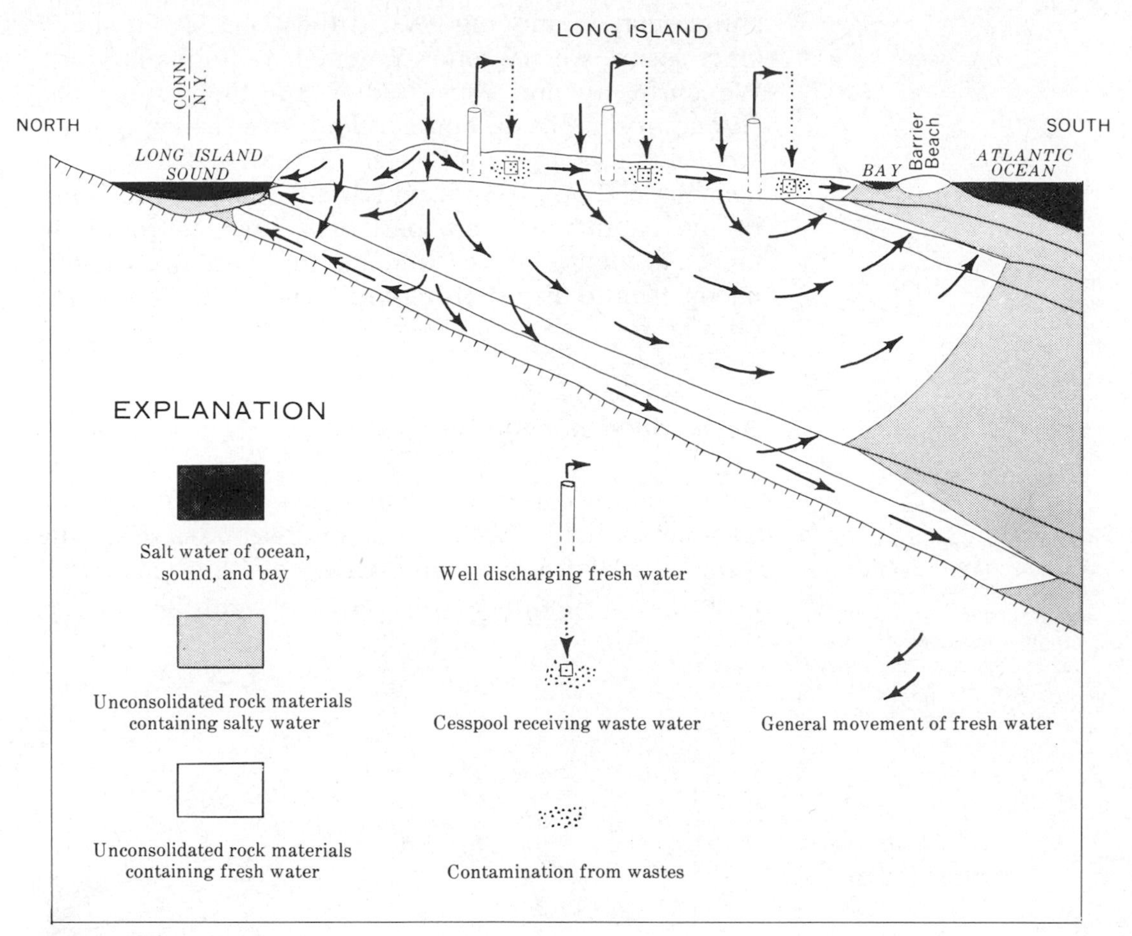

Figure 11–24. Diagrammatic section showing generalized groundwater conditions during phase 2 of groundwater development (shallow supply wells and waste disposal through cesspools). (See also Fig. 11–26.) (From Heath, R. C., et al., 1966.)

saltwater–fresh groundwater interface. Figure 11–26 is a summary of the current groundwater situation for the island.

It is clear that population growth and a probable acceleration of per capita consumption of water will result in continued overexploitation of groundwater reserves on Long Island. Additional methods of obtaining water will have to be used, such as desalination and water reuse through sewage treatment plants, new surface water supplies and so forth. However, the immediate and most practical technique, sewage treatment, involves the problem of getting rid of the sludge (solid waste product of sewage treatment). New York City, which has been rapidly increasing its sewage treatment facilities, is now faced with an increasing problem of sludge disposal. Traditionally, the sludge has been dumped about ten miles offshore in the New York Bight area. Recent studies have suggested that the sludge may be moving toward the Long Island shore. Once again, we see environmental trade-offs at work. We clean up and reuse water but then dump the sludge in the ocean, and suddenly we realize that we are endangering a priceless natural resource, the beaches of Long Island, which are some of the finest in the country and are part of a large metropolitan area. As mentioned earlier, environmental trade-offs become more and more complicated with increasing population.

Bermuda Water Supply

An example of an area where ground water is not a feasible source of water is in the Bermudas. Here the thin layer of fresh water floating on the denser salt water is polluted by sewage effluent, and the only re-

Figure 11–25. Diagrammatic section showing generalized groundwater conditions during phase 3 of groundwater development (deep supply wells and waste disposal through cesspools). (See also Fig. 11–26.) (From Heath, R. C., et al., 1966.)

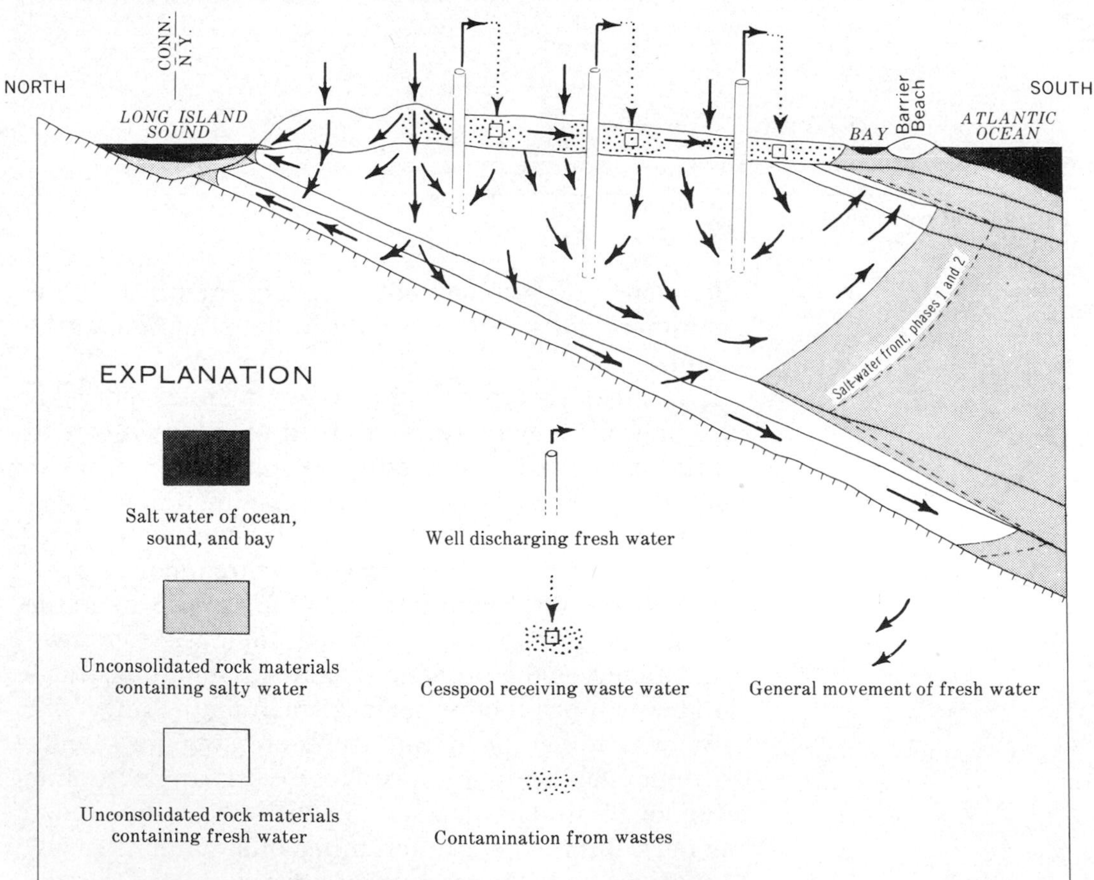

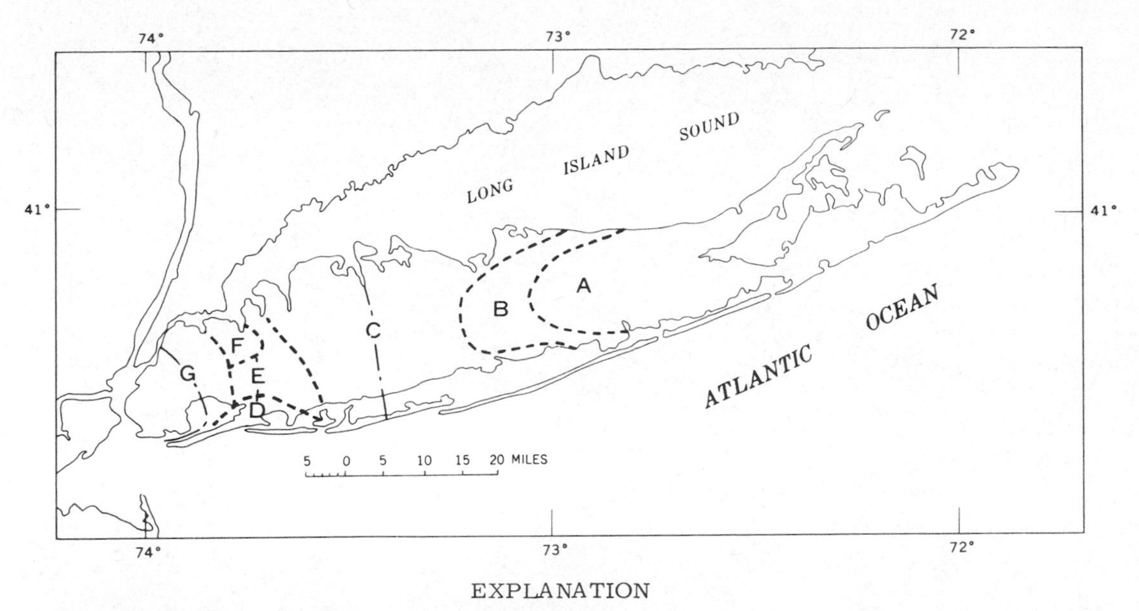

EXPLANATION

Figure 11–26. Water-development subareas in 1965. (From Heath, R. C., et al., 1966.)

Subarea Characteristics

A—Phase 2 of development. Pumpage mainly from shallow privately owned wells. Waste water returned to shallow glacial deposits through cesspools; local contamination of glacial deposits by cesspool effluent. System virtually in balance; positions of salt water fronts unchanged.

B—Transition between phase 2 and 3. Pumpage from privately owned and public-supply wells. Waste water returned to shallow glacial deposits by way of cesspools; areas of cesspool-effluent contamination spreading. System virtually in balance.

C—Phase 3 of development. Pumpage mainly from deep public-supply wells; waste water returned to shallow glacial deposits by way of cesspools. System locally out of balance, causing local salt water intrusion.

D—Phase 4 of development. Pumpage almost entirely from deep public-supply wells; waste water discharged to the sea by way of sewers. System out of balance; salty water actively moving landward.

E—Phase 4 of development. Pumpage almost entirely from deep public-supply wells; waste water discharged to the sea by way of sewers. System out of balance; may be subject to salt water intrusion in the future.

F—Very little groundwater development. Water supply derived from New York City municipal-supply system; waste water discharged to the sea by way of sewers. System in balance.

G—Very little groundwater development. Water supply derived from New York City municipal-supply system; waste water discharged to the sea by way of sewers. Large areas contain salty ground water owing to former intensive groundwater development and related saltwater intrusion.

maining alternative is the collection of rainfall in cisterns.

In 1609, a group of seven ships under the aegis of Sir George Somers left England for Virginia. Colonists on two of these ships would never see Virginia. A violent storm struck the little convoy in the mid-Atlantic, sinking one ship and scattering the others. The *Sea Venture*, battered and leaking, foundered on coral reefs surrounding the Bermuda Islands. The would-be Virginia colonists found themselves on a small group of islands roughly 600 miles east of the Carolinas (Fig. 11–27).

The violent storm which brought them to the islands was not typical of the weather in the region. During much of the year a persistent high pressure ridge and

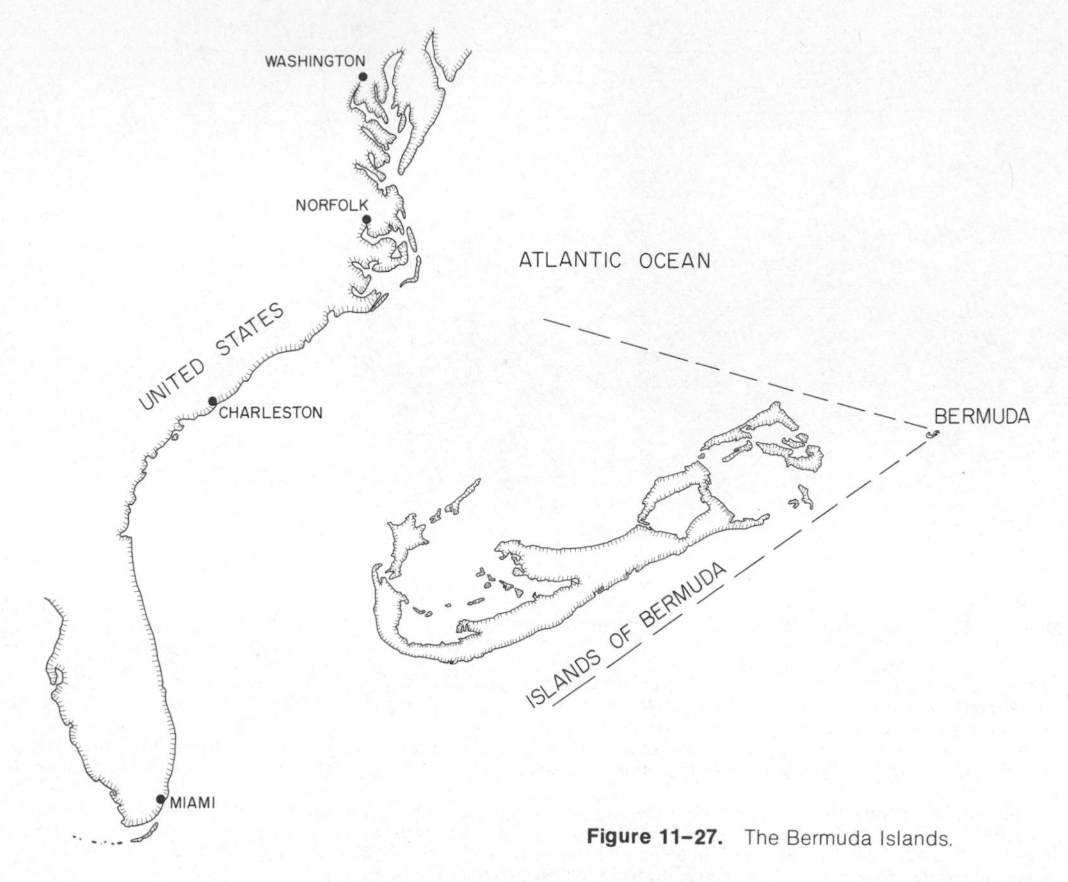

Figure 11–27. The Bermuda Islands.

warm Atlantic waters combine to produce one of the most delightful climates on earth. Winds are warm-without the moist stickiness of most tropical islands. Billowing white clouds float lazily in a bright blue sky.

Figure 11–28. Rolling hills near Hamilton, Bermuda, are tops of fossil sand dunes formed during Pleistocene glacial epochs, when sea level was much lower than at present. Only the tops of the highest dunes are now above sea level.

Figure 11–29. Cross-bedded sandstone in Black Watch Cut, Bermuda. The famous British regiment made the road cut through a prominent fossil dune while stationed in Bermuda many years ago. Sandstone of the fossil dune is weakly cemented. Workmen cut the sandstone into blocks with handsaws. Upon exposure to the air, the blocks harden somewhat. Most of the residential construction on the island consists of sandstone blocks.

The high-pressure ridge responsible for the clear skies is so persistent that it is called the *Bermuda High*.

However, not all effects of the Bermuda High are beneficent. Too many beautiful days means too little rainfall. The geology of Bermuda is such that neither surface water nor ground water contributes significantly to the water supply. The combination of unfavorable geology and erratic rainfall creates periodic water shortages.

Surface water is an unsatisfactory source for two reasons. First, because Bermuda is a small island, there is not enough area for the development of large watersheds. Secondly, rocks outcropping in the Bermuda Islands are fossil sand dunes formed during a lowering of sea level during Pleistocene glaciation (Fig. 11–28). The dunes are weakly cemented and very porous (Fig. 11–29). The sandstone in the dunes is largely composed of limestone fragments. An extensive system of solution cracks and caverns has developed inside the dunes (Fig. 11–30), which causes the tides to rise and fall in some of the marsh lands in central Bermuda. The cracks and caverns also contribute significantly to downward seepage of rainfall.

Because ground water and surface waters are unusable, water is collected in cisterns. This is accomplished in two ways. The largest collection systems consist of paved hillsides (Fig. 11–31), which cause

Figure 11–30. Partially flooded cavern in Bermuda. The cavern was formed as a result of solution of limestone during a time when sea level was lower than at present. The rising sea level has partially flooded this cavern. Sea water covers stalagmites and lower tips of some stalactites. Although they are not visible in this picture, stalagmites can be seen in the clear waters of the cavern.

Figure 11-31. Paved hillside in Bermuda prevents rainwater from seeping into the ground. The water is collected in a cistern at the base of the hill.

water to funnel into underground cisterns at the base of the hill.

The ridged and whitewashed roofs of Bermudian buildings add greatly to the beauty and charm of the island (Fig. 11-32). The ridges and whitewash, however, have an important practical basis. Roofs are important collectors of rainwater in Bermuda. Each house or building has one or more cisterns into which rainwater is funneled during the infrequent rains. Rooftop ridges keep rainwater from flowing off the roof and funnel the water into a downspout, which carries it to a cistern.

Owing to the dearth of wood on the islands, rooftops are shingled with thin slabs of sandstone. In its natural state, the porous sandstone would not protect the rooms beneath from a heavy dew. Several coats of whitewash, however, seal the shingles and at the same time provide a clean surface for collection of water.

Bermuda's rooftops come very close to the ideal type of solution to most environmental problems: a solution that is both effective and esthetically pleasing (Fig. 11-33).

6 SURFACE WATER

We have discussed that part of the hydrologic cycle which deals with water falling on the earth's surface

Figure 11–32. Bermudian ridged rooftops. Insets show details of ridges and shoulders, which funnel water into a downspout leading to a cistern where the water is stored. Roofs are shingled with thin blocks of sawn Bermuda sandstone. Several coats of whitewash seal the porous sandstones shingles, insure clean water and add considerably to the beauty and charm of the buildings.

Figure 11–33. St. Georges, Bermuda. Ridged, whitewashed roofs add much to the quaint charm of this town, first settled in the seventeenth century.

and infiltrating the soil mantle. We have mentioned another part of the cycle, in which water is evaporated or is absorbed by plants and transpired. A small amount is also detained and stored in ponds and lakes. The final residual of precipitation runs off the surface into rivers. Surface run-off, then, is considered to be rainfall excess.

Water that flows in a channel is called a river or stream. There are different kinds of rivers, depending on the relationship of the water table to the bed of the river channel. If the water table is below the bed of the river, water will flow down the channel only during and immediately after rainstorms (Fig. 11–34*A*). Such *ephemeral* rivers are common in arid and semi-arid regions. If the water table is above the bed of the river, ground water will flow out into the channel all year round (Fig. 11–34*B*), and the river is termed *perennial.* The low flow of a river which is maintained year round by seepage from the water table is called the *base flow.* If the water table fluctuates, water will occupy the river channel when the water table is higher than the bed, and the stream will be dry when the water table is below the level of the bed (Fig. 11–34*C*). Such *intermittent* rivers are common in regions of very porous rock or where rainfall is seasonal. They are becoming more common in humid regions where groundwater supplies are being depleted and the water table rises during spring rains and falls with use during drier parts of the year. (See Fig. 11–4.)

Because ground water supplies the base flow of a river all year round, the fact that the flow of rivers is the result of rainfall was not understood until 1674, when Perrault actually measured the rainfall over the

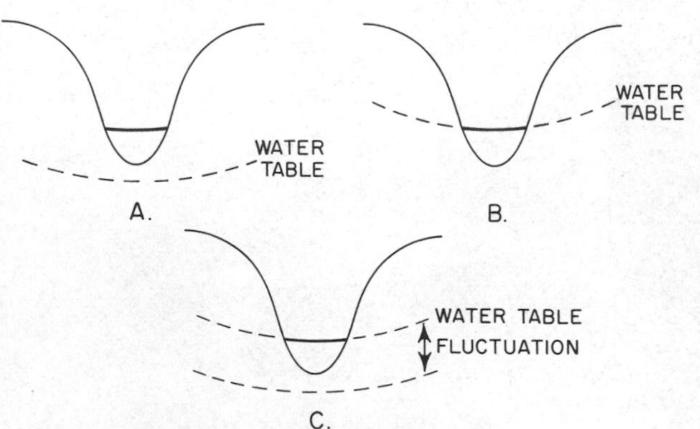

Figure 11–34. *A, B, C* shows relation of the water table to stream beds that results in ephemeral, intermittent and perennial rivers.

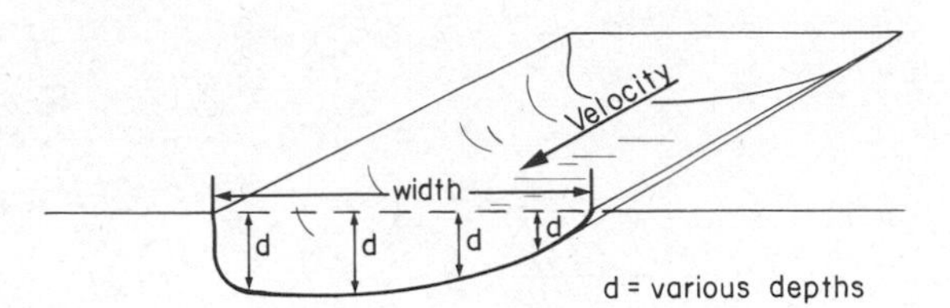

Figure 11–35. Cross-section of a river channel, and factors used in measuring discharge.

Seine River basin in France and compared it to the amount of water flowing in the river. He found that there was plenty of rainfall to supply the discharge of the river. Many other measurements since then have confirmed Perrault's demonstration that the main discharge of rivers of the world is from rainfall.

Discharge is the rate of flow of water through a given channel section (Fig. 11–35) and is measured in cubic feet per second (cfs) or in cubic meters per second (cms). The U.S. Geological Survey has set up more than 9000 gauging stations on rivers of the United States where discharge is measured. Their records are used to determine characteristics of discharge, which will aid us in predicting low flow for water use, peak flow for flood prediction and can be used for many other practical purposes.

Discharge and precipitation records allow us to work out a *water budget* for a drainage basin. The *drainage basin* is the area enclosed by a boundary along the highest points of a region such that the direction of movement of water downslope is toward the river of the given drainage basin (Fig. 11–36). Because precipitation records are in inches of rainfall (or snowfall), the

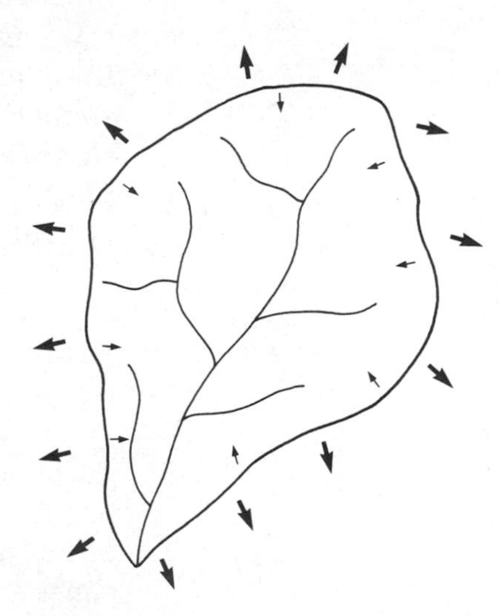

Figure 11–36. A drainage basin.

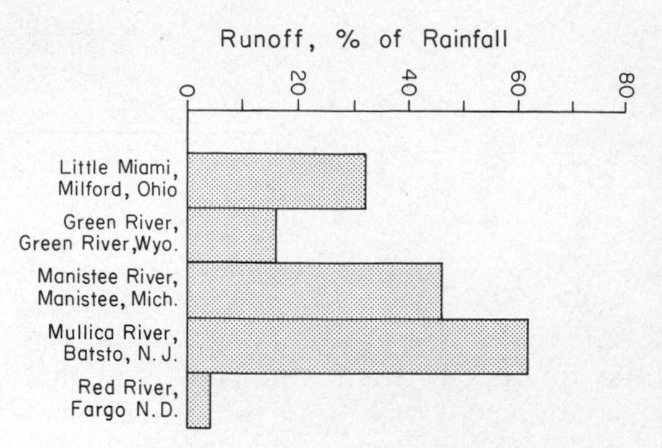

Figure 11–37. Water budget for some rivers of the United States, showing run-off as a per cent of rainfall.

discharge in cfs is converted to the same units. An inch of rainfall or run-off is defined as the water needed to cover the whole area of the basin to a depth of one inch. Another measure often used to calculate irrigation needs and reservoir storage is acre-feet. One acre-foot is the amount of water needed to cover one acre of land to a depth of one foot.

$$1 \text{ inch/sq mile} = 53.3 \text{ acre-feet} = 17.25 \times 10^6 \text{ gal.}$$

Having measured precipitation and discharge, we can then determine the *evapo-transpiration* in the basin by using the hydrologic equation, where evapotranspiration (ET) is equal to the amount of rainfall (RF) minus the run-off (RO) minus groundwater storage (GW):

$$ET = RF - RO - GW$$

This equation represents the water budget of the drainage area. If we assume that storage is in a steady state and is replenished as used, we can ignore it in the equation. Figure 11–37 gives the water budget of some rivers in the United States. Note the difference in run-off in various climatic zones. Over the United States as a whole approximately one third of the rainfall runs off in rivers.

Factors in Run-off

The total amount of run-off in a river depends on the geologic character, topography and orientation of the basin and the climate and rainfall characteristics. Type of bedrock and soil are important geologic factors because these influence the porosity and permeability of the surface. Topography is important because the height of the basin may influence the amount of rainfall. The slope determines to some extent how fast the water flows off and thus whether it will flow off slowly enough for some of it to sink into the ground. Vegetation not only intercepts rainfall but also retards flow on the surface, allowing time for infiltration. Orientation determines the amount of *insolation* (exposure to sunlight) and also the relation of the basin to the path of storms. Characteristics of rainfall such as duration, intensity and total amount are all very important in determining run-off, which we defined earlier as excess rainfall. Evaporation is an important part of the water loss in a watershed. Factors which control the amount of evaporation are climatic—air and water temperatures, wind, humidity and atmospheric pressure. Another important factor is soil moisture conditions previous to a storm. If a recent rain has already saturated the ground, another storm will quickly result in rainfall excess.

An example of the importance of previous rainfall is the floods of 1969 in Los Angeles County, California. These floods occurred despite preventive structural measures including five dams, levees and channel modifications on rivers of the region. After a winter drought, heavy rains began on January 18, 1969, in Los Angeles and the surrounding area. When the rain stopped on January 22, it looked as if the danger was over and the rivers would not overflow their banks. However, rains began again on January 23 and continued for three more days. As much as 47 inches of rain fell on the storm center. The resulting floods caused thousands of people to be evacuated and damages amounting to more than $28 million.

Urbanization

Man is an important factor in determining the amount of run-off. By changing the land surface, devegetating, bulldozing and building upon the land, man greatly affects the amount of water that infiltrates or runs off the surface. For example, clearing a forest reduces infiltration and increases surface run-off into streams. Cementing over the surface of an area results in more rainfall excess because the rain has no chance of sinking into the ground.

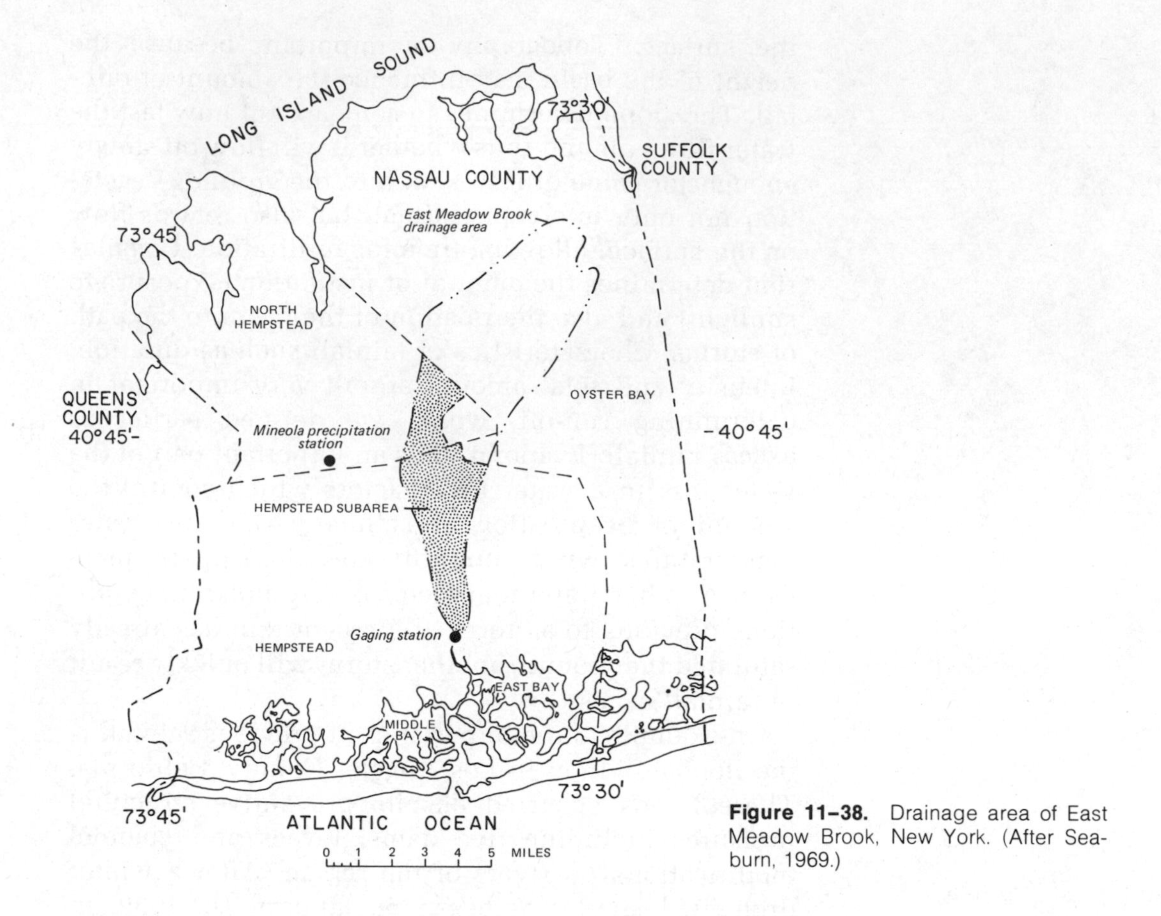

Figure 11–38. Drainage area of East Meadow Brook, New York. (After Seaburn, 1969.)

Various studies have shown that urbanization of a flood plain and areas adjacent to rivers causes an increase in river discharge and total run-off. A good example is a study in Nassau County, Long Island, New York, where rapid urban growth was taking place. The area of study, East Meadow Brook, is shown in Figure 11–38. This is a region of highly permeable sand and gravel. Under predevelopment conditions only 10 to 15 per cent of the discharge in East Meadow Brook was direct run-off from the surface. As much as 85 to 90 per cent of the stream flow was from groundwater discharge. The upper part of the basin, composing about 65 per cent of the drainage area, remains in the form of forested estates that allow natural infiltration. However, the lower third of the basin experienced rapid urbanization around the town of Hempstead. Table 11–2 gives the mean yearly run-off in acre-feet, the increase in rainfall over the period of study and the increase in run-off over the same period. The table in-

TABLE 11-2. CHANGES IN RAINFALL AND RUN-OFF, EAST MEADOW BROOK*

Mean Annual Run-off (acre-feet)	Year	Rainfall % increase 1937–43	Run-off % increase 1937–43
920	1937–43		
1170	1943–51	0.6	27
2200	1952–58	5.7	140
3400	1959–62	7.0	270

*Data from Seaburn, G. E.: Effects of urban development on direct run-off to East Meadow Brook, Nassau County, Long Island, New York. U.S. Geological Survey Professional Paper 627-B. Washington, D.C., U.S. Geological Survey, 1969.

dicates three periods of differing run-off patterns: the pre-1943 to 1951 period; 1951 to 1959; and post-1959. Although each period is marked by an increase in rainfall, the percentage increase in run-off of each period is much greater than that of rainfall.

An analysis of peak discharges (Table 11–3) shows that the peak flow in 1962 was almost 2.5 times the peaks in 1939. The hydrograph curves shown in Figure 11–39 give the run-off pattern for East Meadow Brook for two storms. Figure 11–39*A* shows that before

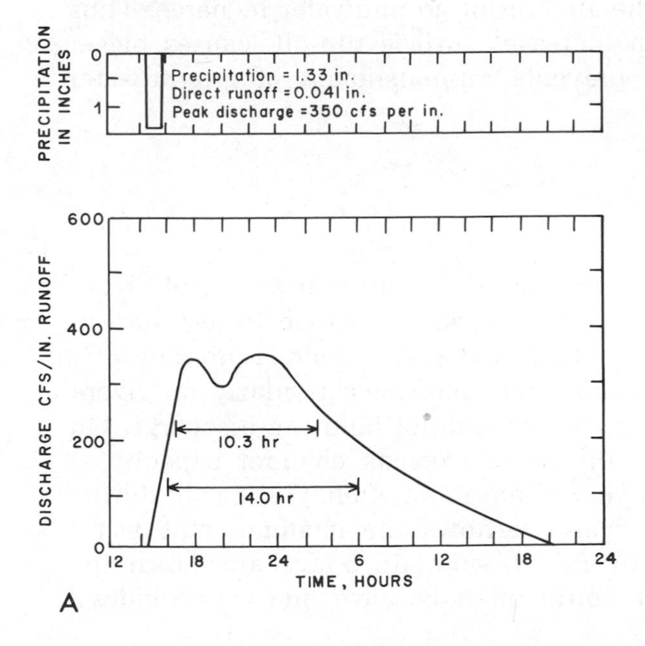

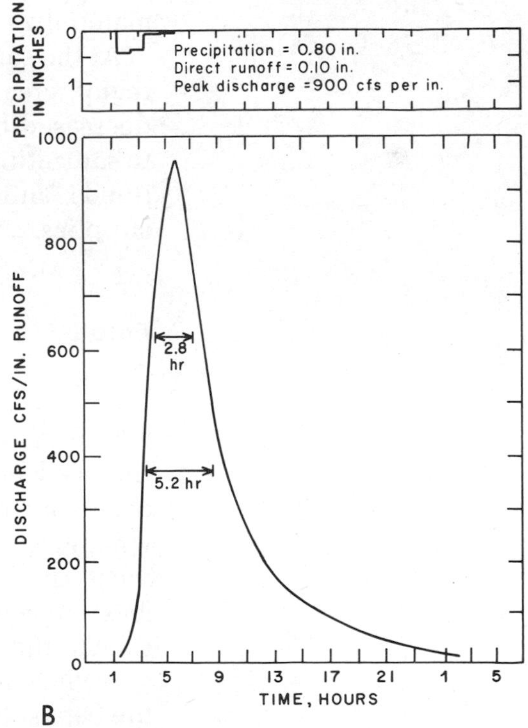

Figure 11–39. Hydrograph curves for two storms, East Meadow Brook, New York. *A*, Before urbanization. *B*, After urbanization.

TABLE 11-3. PEAK DISCHARGES AND UNIT HYDROGRAPH WIDTHS, EAST MEADOW BROOK*

Year	Peak Discharge (cfs/sq.mi.)	Width of Unit Hydrograph (hours)
1939	35.2	16.9
1947	48.3	13.4
1954	77.6	8.4
1962	80.8	6.4

*Data from Seaburn, G. E.: Effects of urban development on direct run-off to East Meadow Brook, Nassau County, Long Island, New York. U.S. Geological Survey Professional Paper 627-B. Washington, D.C., U.S. Geological Survey, 1969.

urbanization with a rainfall of 1.33 inches over the watershed, there was a total direct run-off of 0.041 inch. The peak flow was 350 cfs. After urbanization a rainfall of 0.81 inch gave a run-off of 0.10 inch and a peak flow of more than 900 cfs. Width of the storm hydrograph curve gives the time from the beginning of run-off to the end of the peak flow. The decrease in width of the storm curves (hydrographs, Fig. 11–39) indicates an increase in the speed at which water runs off the land. Urbanization results not only in more total run-off but also in larger floods that materialize more quickly.

At the same time that urbanization of the Hempstead region greatly increased the amount of run-off, it also decreased the amount of groundwater recharge. Thus urbanization increases surface run-off, causes bigger floods and prevents replenishment of groundwater supplies.

Floods

Let us examine more closely one aspect of rivers which man has influenced and which in turn has influenced man—that of flooding. Floods are naturally occurring events that take place regularly on rivers whenever the channel cannot hold the discharge. On most rivers, discharge exceeds channel capacity at least once a year, if not more often. Ordinarily during floods the river overflows the channel, and water spreads out over the flood plain, a level area bordering the banks. But quite often, because the river provides a

Figure 11–40. Levee along Choconut Creek, New York. The levee lies between the river and the floodplain on which the homes are built.

transportation route and a close supply of water and is easy to build on, man constructs cities along the banks. Everything is fine and the city grows, until there is a great storm in the basin. Then the river becomes swollen with water, overflows its banks and destroys the city and some of the people. A great cry rises up for protection from another such disaster. Engineers are contracted to contain the river by building dikes or levees along the banks (Fig. 11–40). They also construct a dam upstream, designed to hold the rampaging waters from the towns below. But they have forgotten the natural processes by which a river does its work. The dam above does collect water, but it also collects the sediment being carried by the river as it flows into the reservoir behind the dam. Because the water that flows out the dam spillway is clear (without a sediment load), as the stream moves down the channel, it scours and picks up a new load. Because the amount of discharge is small (most of the water is held in the reservoir), the river will drop this material when the velocity decreases. Thus it will build up its channel farther downstream, clogging it and raising the level of the bed. During high water, then, after a period of time, even minor floods may overtop the levee (see Fig. 5–31).

After the many disastrous floods on the Mississippi River, the Army Corps of Engineers was given the responsibility of protecting the towns and cities along the banks of the Mississippi from future floods. An early program involved building levees to contain the high discharges and prevent the river from overflowing onto its flood plain. In addition, over the years numerous upstream and tributary dams have been built. The containment of stream flow by the levees has led to deposition in the channel. For example, at St. Louis, the bed has built up as much as four feet in elevation. Thus, where previously (1927) the river channel at St. Louis could hold a discharge of 0.634×10^6 cfs, in 1973 a flow of 0.48×10^6 cfs would cause a flood.

Moreover, those who settle in the flood plain behind the dikes or levees find that the levees prevent the water from leaving the land and returning to the river channel after the peak flows have abated. The floods of 1973–74 on the Mississippi remained on the flood plain for months, preventing farmers from planting crops.

Of course, dams do help relieve flooding by containing some of the water that might have flowed down the river. However, it is impossible to build enough dams to prevent all floods. The June, 1972, disaster at Rapid City, South Dakota, occurred despite the fact that there was a big dam on Rapid Creek. Although the Pactola Reservoir behind the dam controls 76 per cent of the watershed, when a high intensity storm with a rainfall of almost 4 inches in 30 minutes occurred in the 60 square miles of drainage area between the reservoir and Rapid City, the resulting flood caused $150 million in damage and took 231 lives.

The point is that structural measures, though they may help, are not enough. Even though we have built dams, constructed levees and channelized rivers, flood damage in the United States increases each year as a result of a cycle of cause and effect. People build on a flood plain; a flood occurs with damage and loss of life. An outcry arises for remedial measures and protection, so dams, levees and channelways are constructed. These measures provide a false sense of security, so the flood plain is further developed, and the next flood causes even greater damage. How can we end this cycle? The only way is by trying to understand the natural laws that rivers follow and then harmonizing our management with nature (see Chap. 5).

Part of this harmonious management is based on the

realization that floods are a natural event and that flood plains are part of the river. Hence, in so far as possible, man must avoid overdevelopment of areas most susceptible to frequent high discharges. Wise flood plain management practices include the following:

1. Structural measures
   A. Dams, levees, dikes, channelization

      A dam will retard run-off and allow it to move downstream in a regulated manner, thus preventing high discharges. Levees and dikes will restrain the stream flow in a predetermined channelway. Channelization measures are an attempt to speed up the flow of water in the channel and move it out quickly so that it cannot be retarded and overflow the banks.
2. Nonstructural measures
   A. Flood plain delineation and zoning

      Careful delineation of actual high-velocity floodway zones is needed. Once the zones have been determined, these areas should be regulated in terms of development and urbanization.

   B. Flood insurance program

      A program of flood insurance has been set up by an act of the federal government. This act must be enforced so that the burden of flood damages is removed from the national government; those dwelling on the flood plain must assume responsibility for being in a hazardous area.

   C. Floodproofing

      This refers to construction methods that deter damage during floods. Adequate laws must be formulated and enforced to require construction in the flood plain that is adequate to resist flood damage.

   D. Flood warning system

      A flood warning system must be drawn up and implemented for each flood-prone area. This will enable those on the flood plain to be evacuated and save their belongings.

The main emphasis in such a program of management should be on nonstructural measures. Structural measures are useful in certain situations, but they have proved inadequate in others, and they give a false sense of security. It is only by a careful combination of the measures listed above that we can hope to alleviate future flood disasters and make the most productive use of our flood plains and the rivers of which they are a part. The overflow onto the flood plains renews the land and also the waters beneath the surface.

We could not do better than sum up the chapter on water with a quotation from Loren Eiseley:

> If there is magic on this planet, it is contained in water . . . Its substance reaches everywhere; it touches the past and prepares the future; it moves under the poles and wanders thinly in the heights of air. It can assume forms of exquisite perfection in a snowflake, or strip the living to a single shining bone cast up by the sea.*

*From Eiseley, Loren: The Immense Journey. New York, Random House, 1957, pp. 15–16.

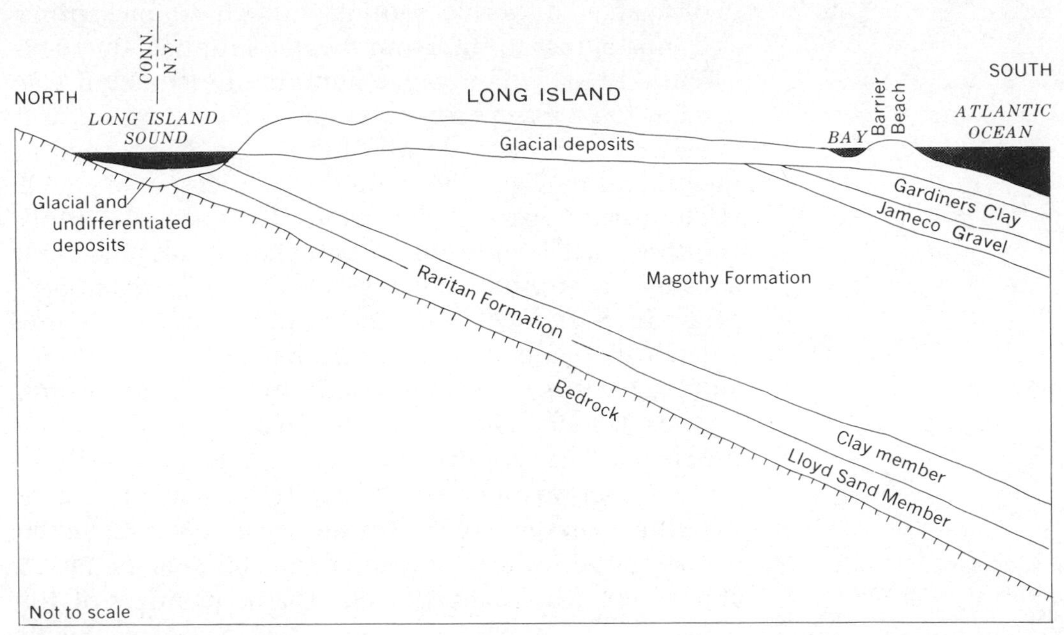

Figure 11–22. Diagrammatic section showing general relationships of the major rock units of the groundwater reservoir in Nassau County, Long Island, New York. (From Heath, R. C., et al., 1966.)

units are approximately 300 feet thick. Overlying these units is a complex unit, the Magothy formation, approximately 1000 feet thick, which consists of an assortment of sands and clays. The permeability of this unit varies widely depending on the local presence of sand or clay. The next water-bearing unit is a gravel and sand unit, the Jameco gravel. This unit is permeable and, because of the overlying clay unit, is an artesian aquifer.

The surficial deposits on Long Island are sands and gravels of Pleistocene glacial origin. These highly permeable deposits allow rapid infiltration of precipitation and are responsible for the strong recharge characteristics of the system (Figs. 11–23 and 11–24).

The groundwater reservoir units in this area contain mostly fresh water, except in the western part of the island in Brooklyn and Queens. On the contour map, Brooklyn and Queens show cones of depression that go below sea level. Consequently, the ground water in that part of the island is frequently salty from saltwater encroachment caused by overexploitation of groundwater supplies, and these reserves are practically destroyed. In addition to overexploitation, a lack of recharge exists in these areas simply because as high population density areas they are largely paved over so that most precipitation does not infiltrate but

Unit III

POLICY AND PLANNING

12

WHO IS RESPONSIBLE?

Waves break on a shore composed of old limestone reefs in Guadeloupe, French Antilles. Eroded remnants stand gauntly offshore.

1 PROLOGUE

An old man rots in jail in Pike County, Kentucky. He defended his land with a shotgun. He blew off the arm of the mine foreman who was threatening to bulldoze the old man's land to get to the coal seam below.

Who is responsible?

2 INTRODUCTION

The technical aspects of environmental problems are frequently complex, and in many instances solutions are not available. However, the question of responsibility—in terms of cause, cure and regulation—for environmental problems is an even more difficult topic. The issues are interwoven in the fabric of social and political structure. Environmental issues frequently arouse political passions of amazing variety.

Conservatives in Congress, with the support of the administration, "killed" a federal land planning bill because they feared it would encourage federal interference with "free enterprise." The bill called for federal money to support state land planning units. On the other hand, critics of the oil industry, during the 1973–1974 energy crisis, accused the industry of profit gouging and creating the crisis. In particular, the high level of oil company profits and the low level of investments in refineries and exploration were cited. However, a capitalist economic structure requires high levels of profit for high levels of investment capital.

Another example of political reactions to environmental issues occurred at the 1972 Stockholm Conference on the Human Environment (see also p. 434), a United Nations sponsored international meeting on environmental problems. Many underdeveloped nations viewed the movement to formulate international environmental regulations as an attempt of the developed nations to impede and frustrate underdeveloped nations' legitimate aspirations.

A full discussion of such problems is beyond the scope of this work. However, it must be understood that regardless of how practical, from a geologic, engineering or scientific viewpoint, a solution to a particular environmental problem may be, that solution must be politically reasonable also. Examples of political-geologic problems appear throughout this work, especially in units II and III. The interrelationship of politics and mineral resources has been much studied. However, in recent years the complexity of this relationship has increased tremendously. Prior to the Industrial Revolution, problems were essentially those of procurement of natural resources. Now the issues include not only resource procurement but also the consequences of resource use—waste disposal problems, air and water pollution and even such economic problems as the effects on the world's economic system of the vast accumulation of oil money in a few Mideast nations.

Responsibility for these problems must be shared. How is this to be accomplished? Is the human race capable of the type of cooperation necessary? At present, the situation is best described as chaotic, in part hopeless and in part hopeful.

3 ROLES OF GOVERNMENTAL UNITS

Types of national regulations vary, depending on the political structure of a given nation. In the United States there are federal, state and local laws.*

Radiation pollution is the only area in which initial federal laws took primary responsibility for regulation and control. This was the natural outgrowth of federal development of nuclear power. In all other areas—water, air and noise pollution—the federal government initially delegated authority to the states and only later accepted primary responsibility for regulation and control.

The federal government did, however, take an early interest in certain aspects of water pollution. In the late nineteenth century, several laws were passed relative to navigable waters. In 1886 a law which forbade dumping of refuse in New York Harbor was enacted, and in 1899, a broader statute, the Rivers and Harbors Act, was passed. The objective of both of these was to protect shipping against refuse in navigable waters. In fact, the 1899 Act exempts liquid waste that flows from streets and sewers. Although the original intent was not water-pollution control as such, the Rivers and Harbors Act has become one of the main tools used to deal with recent problems of water-pollution control. Seventy years passed before the Act was used against industrial water pollution. One of the unique provisions of the law is a bounty for whoever gathers evidence against the polluter. The fine is shared equally by the government and the informant. Some attempts have been made to apply the act to streams or rivers that are not navigable. In one case, in West Virginia, seepage from a solid waste landfill was polluting a small stream that ultimately fed into the Kanawha River, which is navigable and therefore covered by the 1899 Act. Proponents attempted to prove that the seepage from the landfill in the tributary ended up in the Kanawha and thus was polluting a navigable river. Ironically, the little stream is pristine

---

*For a review of the role of various government units, see Grad, Rathjen and Rosenthal (1971) and Environmental and Natural Resources Affairs of the 92nd Congress (1973). Much of this chapter is adapted from these works.

compared to the chemical sewer known as the Kanawha River.

By the early twentieth century the relationship between polluted water and certain diseases was recognized. The Public Health Services Act of 1912 led to joint federal-state action on national standards for treatment of drinking water. Enforcement was delegated to state and local authorities; the standards were federal. Chapter 14 describes a case in rural West Virginia where these national standards were never enforced even as late as 1970 because a provision of this law, and of later laws, exempted small (fewer than 200 connections) water systems. In these small systems only minimal regulations, including chlorination and the lowest permissible level for a few elements, are required. In many cases, even these regulations are not observed.

In 1948 the first federal Water Pollution Control Act was passed. This Act clearly asserts that the states are primarily responsible for water quality. The role of the federal government was confined to bringing suit against a polluter of interstate waters only after the state concerned was given adequate time to resolve the problem, and even then the state had to give permission for proceedings to begin. The original Act was scheduled to last for five years and was extended to 1956, at which time a broader water-pollution control law was enacted by Congress. Like the one before it, this act asserted the primacy of the states in water-pollution control. Among the features of this bill were intensified research and training efforts; grants to states and cities for water treatment plants; simplified procedure for federal action against interstate polluters; and a program of pollution control for federal installations. In 1971, this act was amended to include "navigable waters," a broader term than "interstate waters."

The Water Quality Act of 1965 was the next major water-pollution legislation. A national policy for water-pollution control was pronounced, marking the first time that the federal government took primary responsibility for water-pollution control. The responsibility for enforcement of this legislation was first invested in the Department of Health, Education and Welfare, then transferred to the Department of the Interior and finally by 1970 resided in the relatively new Environmental Protection Agency. The EPA is now the federal agency with overall responsibility for environmental regulations.

The Clean Water Restoration Act of 1966 provided for increased research and training programs and construction grants for water treatment plants and extended federal responsibility to include international waters. In addition, the Oil Pollution Act of 1924 was amended to include inland water as well as the original intercoastal waters.

In 1967, the Water Quality Act issued a deadline of June 30, 1967, for states to set water quality standards for interstate waters within their jurisdiction. Failure to do so or setting of unacceptably low standards would allow the federal government to set the standards. All states complied with the deadline.

The Water Quality Act of 1970 added controls and regulations to previous acts in areas such as oil pollution, hazardous substances and mine waters. Demonstration clean-up projects for the Great Lakes were also introduced.

The 92nd Congress undertook revision of water-pollution and water-quality legislation based on the previous 25 years' experience. The legislation has the following basic provisions:

1. The national goal is to eliminate all discharge of pollutants into receiving waters by 1985. An interim goal . . . provides for protection of aquatic life and wildlife and provides for recreation in and on the water to be reached by 1983.
2. Up to $18 billion is authorized for fiscal 1973, 1974, and 1975 for obligation in an accelerated program of construction of publicly owned waste treatment facilities incorporating the best practicable technology. Facilities which integrate treatment of municipal and industrial waste are to be encouraged. All point and nonpoint sources of pollution are to be identified and a plan developed to bring them all under control.
3. Areas with serious water quality problems are to be identified and attacked first.
4. All point source effluents are to be limited to levels achievable through use of best practicable technology by 1977 and best available technology by 1983.
5. To insure that point source effluents are limited, each source must obtain a permit which specifies the limita-

tion to be achieved and the measures that must be taken to demonstrate that it is being achieved.[1]

President Nixon vetoed this legislation because he considered it to be "budget-wrecking," but the veto was overridden by both houses of Congress.

All of these measures are indicative of increasing governmental activity in the area of pollution control. Figure 12–1 shows the Table of Contents of the 1138-page report summarizing the activities of the 92nd Congress in the area of environmental legislation. Analogous legislative activity is occurring in other areas of environmental regulation and control.

The first federal air-pollution legislation was passed in 1955. As in the case of water-pollution control legislation, the intent was clearly that the role of the federal government should be limited.

> The Committee recognized that it is primarily the responsibility of state and local governments to prevent air pollution. The bill does not propose any exercise of police powers by the federal government and no provision in it invades the sovereignty of states, countries or cities. There is no attempt to impose standards of purity.[2]

The Clean Air Act of 1963 initiated a movement toward federal primacy by increasing the emphasis on regional planning and on dealing directly with state governments. (In the 1955 Act the federal government dealt directly with local governments.) By 1965, amendments to the Act increased federal responsibilities, including setting of air quality standards for new automobiles and the right to bring suit in cases of violation. The 1967 Act continued the old provision of grants and so forth described earlier and empowered the Secretary of Health, Education and Welfare to establish air quality standards.

The Clean Air Act Amendments of 1970 provided further control of air quality standards for the nation. Responsibility for these programs was delegated to the Environmental Protection Agency (EPA). Unfortu-

---

[1]Environmental and Natural Resources Affairs of the 92nd Congress prepared by the Environmental Policy Division, Congressional Research Service Library of Congress. Washington, D.C., Government Printing Office, 1973, p. 569.

[2]Grad, F. P., et al.: Environmental Control. Priorities, Policies and the Law. New York, Columbia University Press, 1971, pp. 51–52. (From a Senate committee report during the 84th Congress, 1955.)

CONTENTS

Figure 12-1. Table of contents, Environmental and Natural Resources Affairs of the 92nd Congress. (Prepared by the Environmental Policy Division, Congressional Research Service Library of Congress, Washington, D.C., Government Printing Office, 1973.)

nately, the administration intervened to undercut the policies of the Congress and EPA by removing the actual decision-making power to the Office of Management and Budget. Clearly contrary to the legislative intent, such factors as "cost effectiveness" and not solely air quality became criteria. In addition, funds were frozen so that although authorized, they were not spent. In water-pollution control legislation, which was vetoed and the veto overridden, the administration froze the funds, in effect revetoing the bill. In 1974 these vetoes by impoundment of funds are being tested in the courts, because serious questions of constitutional law are at issue.

In the past, solid waste disposal was basically a local problem. However, the problem has reached regional and national proportions owing to several factors such as a vast increase in solid waste due to population growth, use of nonreturnable containers and decreased availability of land in urban areas.

The first solid waste disposal legislation was passed as a part of the Clean Air Act of 1965. Current federal legislation, the Resource Recovery Act of 1970, emphasizes the need to recycle solid waste. Another goal of the 1970 Act was that the EPA assist in closing many of the 16,000 open dumps in the country, thus encouraging alternate methods of solid waste disposal. The Act also authorizes studies to help develop these alternate methods. One such study was authorized to determine the following:

1. Means of recovering materials and energy from solid waste, the market potential of the recovered materials, and the impact of recycling on existing markets.
2. Changes in products, production, and packaging characteristics which would reduce solid waste.
3. Methods of collection which would facilitate volume reduction, reuse, or disposal of solid waste.
4. Recommended incentives and disincentives to accelerate recycling.
5. Effect of current public policies on recycling and the effects of eliminating these policies.
6. Necessity of and methods for imposing disposal fees on various goods.[3]

As stated previously, this problem is still being handled by local governments even though in many cases the issue has outgrown the locality. For example, Chicago is using downstate Illinois areas for disposal of sludge left from sewage treatment. The sludge is to be used as fertilizer. New York City dumps its sewage sludge at sea, a practice which is beginning to cause real concern as sludge volume increases. The "dead

[3]Environmental and Natural Resources Affairs of the 92nd Congress, 1973, p. 653.

sea" area is growing, and there is a possibility that the sludge is moving toward shore. The state of Connecticut has announced plans to construct plants at locations throughout the state for recycling and burning of all solid waste material. The energy produced will provide electric power generation. The plan is scheduled for completion by 1984.

The Atomic Energy Act of 1946 asserted the primacy of federal responsibility for radiation control. Prior to 1946, controls of radiation, if any, resided in the states. The atomic energy legislation of 1954 reasserted federal responsibility, but in 1969, federal control was challenged by the Minnesota Pollution Control Agency. This agency attempted to place more stringent controls on radioactive waste discharge from a nuclear power plant in Minnesota than were imposed by the AEC. The power company sued the state agency, and in 1972 the Supreme Court upheld the primacy of the AEC.

Since that time, the AEC has undertaken cooperative contracts with the states to share responsibility for radiation control and regulation. The first contract of this type was with the commonwealth of Pennsylvania, and by 1972 seven states had signed similar agreements. These contracts provide for state participation in monitoring radioactive discharge from nuclear power plants and require that states supply the following data to the AEC:

1. . . . the effects of discharges on normal background levels of radiation which exist in the particular state;
2. . . . the adequacy of controls being exercised by licensees over radioactive effluents at operating facilities; and
3. . . . the potential radiation exposure to the public attributable to facility operations.[4]

There is a great diversity in state and local regulations, ranging from virtually no legislation to elaborate controls. In New York, the state and New York City are responsible for all radiation not regulated by the AEC. This encompasses medical x-rays and medical applications of radioactive isotopes by state and city health authorities and control of industrial x-ray and isotopes application by the state Labor Department.

---

[4]Environmental and Natural Resources Affairs of the 92nd Congress, 1973, p. 746.

The federal government regulates nuclear reactors and radioactive waste.

By 1970 federal control of radiation, except radioactive waste disposal, was consolidated with other environmental matters in the Environmental Protection Agency. Radioactive waste disposal is still an AEC responsibility. (See Chapter 8.)

The United States does not at present have an overall energy and natural resources program. The 91st Congress (1969–1971) set into motion various studies of energy and natural resources. A bill to establish a Commission on Fuels and Energy failed to pass Congress. This was to have been a legislative-executive-public partnership to study the total energy situation.

Early in the 92nd Congress, a Senate resolution was passed that provided for a study by the Senate Committee on Interior and Insular Affairs covering the same areas as the aborted bill from the previous session. This committee, along with ranking members of the Senate Commerce, Senate Public Works and Joint Atomic Energy committees, would "make a full and complete investigation and study . . . of the current and prospective fuel and energy resources and requirements of the United States and the present and probable future alternative procedures and methods for meeting anticipated requirements, consistent with achieving other national goals."[5]

The study group was also authorized to investigate the following areas:

1. The proved and predicted availabilities of our national fuel and energy resources . . . as well as worldwide trends in consumption and supply;
2. Projected national requirements . . . to meet short range needs and to provide for future demand for the years 2000 to 2020;
3. The interest of the consuming public, including the availability in all regions of the country of an adequate supply of energy and fuel at reasonable prices and . . . the maintenance of a sound competitive structure in the supply and distribution of energy and fuel to both industry and the public;
4. Technological development affecting energy and fuel production, distribution, transportation, and/or transmission;

[5] Environmental and Natural Resources Affairs of the 92nd Congress, 1973, p. 6.

5. The effect that energy producing, transportation, upgrading and utilization has upon conservation, environmental and ecological factors;
6. The effect upon the public and private sectors of the economy of any recommendations made under this study, and of existing governmental programs and policies now in effect;
7. The effect of any recommendations made pursuant to this study on economic concentrations in industry particularly as these recommendations may affect small business enterprises;
8. Governmental programs and policies now in operation; and
9. The need, if any, for legislation designed to effectuate recommendations in accordance with the above and other relevant considerations.[6]

The resolution illustrates the close relationship of environmental, political and economic affairs. It had broad support from government, industry and consumer groups, and the Senate passed it without debate.

During the 92nd Congress, 32 hearings were held and more than 20 documents produced. The work of the committee will be continued in the 93rd Congress, which should lead to policy statements regarding various energy and natural resources issues.

The National Environmental Policy Act

In 1969, as a result of great pressure by those concerned with the deterioration of the American environment, Congress passed the National Environmental Policy Act (NEPA), which was scheduled to take effect on January 1, 1970. In this landmark bill Congress recognized that it is the responsibility of the government to "encourage productive and enjoyable harmony between man and his environment; to promote efforts which will prevent or eliminate damage to the environment and biosphere and stimulate the health and welfare of man" (Sec. 2) and to "assure for all Americans safe, healthful, productive, and esthetically and culturally pleasing surroundings" (Sec. 101 (b) 2).

The Act also requires (Sec. 102 (c)) that all federal agencies file a statement regarding the environmental impact of any contemplated action.

[6]Environmental and Natural Resources Affairs of the 92nd Congress, 1973, p. 7.

All agencies of the Federal government shall—

(C) include in every recommendation or report on proposals for legislation and other major Federal actions significantly affecting the quality of the human environment, a detailed statement by the responsible official on—

(i) the environmental impact of the proposed action,
(ii) any adverse environmental effects which cannot be avoided should the proposal be implemented,
(iii) alternatives to the proposed action,
(iv) the relationship between local short-term uses of man's environment and the maintenance and enhancement of long-term productivity, and
(v) any irreversible and irretrievable commitments of resources which would be involved in the proposed action should it be implemented.

Although Section 102 clearly requires compliance of all agencies, there was much ambiguity in other sections of NEPA. For example, it created the Council on Environmental Quality (CEQ) to set up guidelines for compliance with the Act and left it to CEQ, the agencies and the courts to define the scope of the Act more clearly. Environmental groups have used NEPA in their fight to protect the environment from projects they deem damaging. From 1970 to 1973, 149 litigations were brought before the courts against agencies for noncompliance with NEPA. The court decisions have clarified much of the vague language of NEPA.

These decisions have strongly upheld and enforced procedural standards and strict compliance with NEPA. They have established that all agencies, even the Environmental Protection Agency (EPA), must comply with the Act and file an impact statement of their major actions. The Act also applies to actions which are only partly federal, such as cooperative federal-state, federal-county or federal-urban projects. It applies to federal grants, contracts or loans to companies or private individuals; to federal permits or licenses; and to private actions on any federally controlled land. Thus NEPA applies to all Department of Transportation highway projects, to HUD (Housing and Urban Development) projects or urban renewal proposals and to all public service organizations that need federal licenses.

Two agencies have won recognition of noncompliance for purposes of national security: the AEC for Nevada nuclear tests and the Department of Defense for storage of chemical/biologic warfare weapons at

Rocky Mountain Arsenal, Colorado (see Chap. 2). Owing to the energy crisis of 1973, the Alaskan pipeline project was exempted from compliance with NEPA by a special act of Congress, nullifying its own previous action.

The courts also determined that NEPA applies to any part of a project remaining after January 1, 1970, that still needed funding authorization, even though the project had been approved before that time. Thus many projects previously proposed by the Army Corps of Engineers but not funded now had to supply impact statements before funding could be sought. NEPA was also considered applicable to continuing programs already in existence, such as the herbicide and pesticide projects of the Department of Agriculture and continuous dredging programs of the Army Corps of Engineers. It thus provides a useful device for continuous evaluation of actions previously deemed necessary.

Environmental impact statements should be written as early as possible in the planning of federal projects. They must provide a detailed description of the proposed project and a full disclosure of all known environmental consequences. Statements must present alternatives (including the possibility of no action at all) to the proposed project and the impact of the alternatives. They must discuss the "trade-offs" between short-term local uses and the long-term productivity. They must state any irreversible effects of the action, indicating in what ways the action limits future options. For example, the immediate leasing and development of offshore oil may result in short-term depletion of a valuable resource. Another example is the irretrievable loss of agricultural land (see Chap. 10) in return for a four-lane high-speed expressway.

Drafts of environmental impact statements are distributed to interested federal agencies, to local and state governments and to concerned individuals and groups for comment over a 90-day period. This commenting period is extremely important, and the results must be included in the final environmental impact statement. It provides time for examination of the proposal by the public and gives private citizens the opportunity to express their concern. It thus allows public challenge of the decisions of a government agency.

Failure to comply with NEPA, as many agencies

have discovered, can lead to court proceedings and an injunction to stop action until an adequate environmental impact statement has been prepared. Actually, the Act has no power to force an agency to abandon an action that is found to have dire environmental consequences. The Act requires only that all possibilities be examined, in the hope that a decision will be made in good faith.

The real significance of NEPA is its attempts to change the decision-making mechanism of government agencies. It provides an opportunity for agencies to examine the environmental implications of a proposed action, weigh the benefits and disadvantages and make a choice on whether to proceed with the proposal. Furthermore, in order to insure proper consideration of environmental effects, the Act provides for full disclosure to the public and offers individuals the right to actively participate in making the decision. Thus the Act makes public officials accountable for their actions.

4 SOME INTERNATIONAL ASPECTS

As environmental problems grow more complex, they go from local, to regional, to national, and finally to international in scope. In the United States, environmental standards and regulations have shifted from local and state responsibility to national responsibility as environmental problems have grown in scope and complexity. The logical next step would seem to be establishment of international standards and regulations, but this is not possible at present because there is no world government as such. Current international regulations consist of agreements between individual nations through treaties or other agreements. This approach, while helpful, is limited. For example, the treaty between the United States and the U.S.S.R. banning atmospheric testing of nuclear devices could not prevent France, China or any other nation that did not join the treaty from such action.

In spite of the absence of a world government, there is a tremendous ongoing effort to formulate global agreements through the United Nations. The most comprehensive United Nations effort to date was the 1972 Stockholm Conference on the Human Environment. At this meeting, basic global environmental problems were outlined and discussed, including the extent of environmental degradation and the basic programs necessary for the most immediate problems. The complexity and frustrations involved in global environmental problems are well summarized by a report of two senators who attended the conference.

> . . . We have become more aware of the growing division between the rich and the poor nations on the subject of environmental quality. It is extremely unfortunate that the desire for development has transformed the unifying potential of environmental concern into a politically divisive issue. We hope that this . . . Conference will be a major step in bridging the gap between the developed and the developing countries, and will act as a catalyst in formulating action for future international environmental agreements.[7]

While the comprehensive approach of the Stockholm conference is useful in focusing world attention on global problems, smaller conferences have the advantage of concentrating on specific areas. These conferences cover a wide variety of environmental issues such as fishery, whaling and ocean pollution. One interesting area is that of sea law (see Chap. 9). Traditionally, sea law was concerned only with questions of navigation and fishing. But since World War II, technologic advances and the discovery of oil and mineral deposits under the ocean have added a new dimension to sea law.

Figure 12–2 illustrates the broad division of water into inland, marginal and high seas. The inland waters, which include rivers, lakes and coastal bays, present few legal problems. The marginal seas extend three miles or more from the low-water mark. (For a description of the long and interesting history regarding the origin of the three-mile limit, see Shalowitz, 1962.) Although this limit attained widespread international acceptance by the end of the eighteenth century—primarily through the persistence of Great Britain and later the United States—it has never been formally recognized by international treaty. Currently a hotly debated subject, some coastal nations are claiming up to 200 miles as marginal seas. (The concept of a three-

---

[7]Environmental and Natural Resources Affairs of the 92nd Congress, 1973, p. 802.

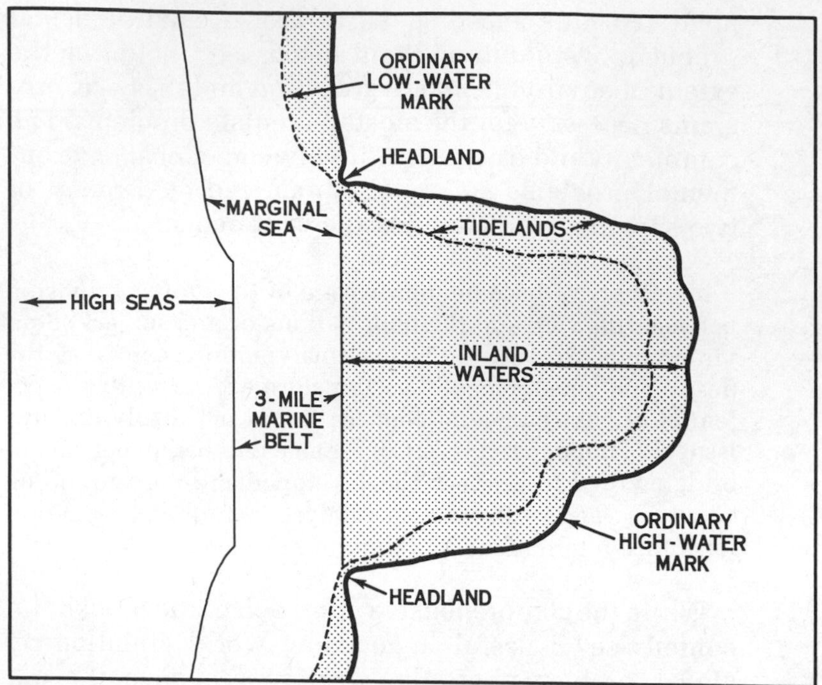

Figure 12–2. The 3-mile marginal belt and its relation to inland waters and the high seas. (From Shalowitz, A. L.: Shore and Sea Boundaries, Vol. I. Washington, D.C., U.S. Department of Commerce, Coast & Geodetic Survey, 1962.)

mile minimum is universally accepted; the argument is over the maximum distance.)

The region of the high seas, beyond the 3-mile limit, is characterized by *freedom;* that is, the right of all nations to use them. Traditionally, freedom of the high seas included navigation and fishing. In 1958, the first United Nations Conference on the Law of the Sea was held in Geneva. The four freedoms of the high seas were defined as follows:

1) freedom of navigation;
2) freedom of fishing;
3) freedom to lay submarine cables and pipelines; and
4) freedom to fly over the high seas.[8]

This conference also dealt with the question of the continental shelf and the development of natural resources by a coastal state.

In the late 1940's, federal and state governments in the United States battled over control of the sea bottom

[8]Shalowitz, A. L.: Shore and Sea Boundaries, Vol. I. Washington, D.C., U.S. Department of Commerce, Coast and Geodetic Survey, 1962, p. 256.

beyond the three-mile limit. The courts ruled that the federal government had dominion over these areas, including the oil contained in them. The next question is, how far offshore does United States (or any other coastal nation) control of the sea bottom extend?

The 1958 Geneva Convention defined the limit in terms of the continental shelf, which in turn was defined according to water depth. Specifically, the continental shelf is defined as:

> The sea-bed and subsoil of the submarine areas adjacent to the coast but outside the area of territorial seas, to a depth of 200 metres or, beyond that limit, to where the depth of the superjacent waters admits of the exploitation of the natural resources of the said areas.[9]

Note that a definite depth, 200 meters, is modified by an "exploitation" limit. In effect, as recently as 1958, 200 meters seemed a reasonable depth to use as a definition because it was based on contemporary oil drilling technology (see Fig. 8–17).

Subsequent technologic developments (among other factors) have increased the complexity of the United Nations Law of the Sea conferences, particularly in regards to the issues of imminent deep-sea mining of manganese nodules (Fig. 12–3) (see Chap. 9) and intensified deep-sea oil exploration. Drilling ships now

[9]Shalowitz, p. 246.

Figure 12–3. Assault barge —Howard Hughes is reported to be preparing this huge barge for submersion to the ocean floor, where it will dig for copper, nickel, and manganese. (From Anderson, A.: The rape of the seabed. Saturday Review/World, Nov. 6, 1973, p. 16. Photo by UPI.)

have the capability of drilling in water up to 1000 meters in depth.

Again, much of the debate splits along the lines of developed versus developing nations. The developing nations generally support proposals to control deep-sea mineral exploration through a United Nations agency that would do more than merely lease sea-bottom rights to large corporations. They envision a major United Nations role, which in time would generate funds for developing nations.

The developing nations generally take a position similar to the United States' statement regarding the seabed.

> . . . There are difficult problems because there is a sharper difference in the developing and developed country points of view in the deep seabed.
>
> We and the other developed countries take the position that we do not want an international organization to have a monopoly on the exploitation of this area. We want private enterprise and individual countries to be able to participate under international regulation and with reasonable contribution to the international community.[10]

As of 1974, there is no regulation or control of deep-sea mineral exploitation. It is essentially a "lawless" area, and the world faces the distinct threat of a "winner take all" struggle that could be environmentally and politically disastrous.

5 THE ROLE OF THE INDIVIDUAL

In today's world, individual effort is generally effective only insofar as it mobilizes groups, governments or other organizations. In most cases, individual attempts to block the actions of a corporation or governmental unit are doomed to failure.

The story of the old man in Kentucky recounted in the prologue is an extreme example, but it illustrates the general futility of individual action. The coal was removed from his land, and thus far nothing has roused the people of the Appalachian coal fields or the people of the United States to stop uncontrolled strip mining. Even the eloquent writings, movies and speeches of

[10]Environmental and Natural Resources Affairs of the 92nd Congress, 1973, p. 796.

Figure 12–4. Strip-mined area of Kentucky. (From Caudill, H. M.: My Land is Dying. New York, E. P. Dutton and Co., 1971.)

Harry M. Caudill, the lawyer from Whitesburg, Kentucky, have been to little avail (Fig. 12–4). Eventually such efforts may produce some collective action, but until then they are voices in a wilderness.

Although we are dealing with the geologic and technical aspects of environmental problems, any implementation of solutions to these problems requires a complex political process involving responsibility at all levels of society.

13

URBAN CASE STUDY— SAN FRANCISCO BAY AREA

This beautiful lake provides San Francisco (just below the horizon, right center) with water. It also covers the trace of the best known and heretofore most destructive fault in North America, the San Andreas.

1 INTRODUCTION

The San Francisco Bay area was chosen for an urban case study for three reasons. First, it is a large metropolitan area. About 5 million people live in an area of more than 7000 square miles. More people live in the Bay area than in Israel or Connecticut. The Bay area is larger than Connecticut and almost as large as Israel. The people of this area form an economic unit tied together by geographic and geologic commonalities.

Second, the geology of the Bay area is extremely varied and, as a consequence, presents a variety of problems. Major earthquakes are a well known and frequent hazard. On a day-to-day basis, landslides, land subsidence, mineral extraction, water supply, stream pollution, flooding, silt accumulation in the Bay and chemical pollution of the Bay are at least as important as the more dramatic earthquake hazard.

Third, the people of the San Francisco Bay area are doing something about their geologic hazards. They have formed the Association of Bay Area Governments

Figure 13–1. High-altitude photo of the San Francisco Bay region. The Bay is the dark area extending across the picture about one third of the way down from the top. Clouds cover the Pacific Ocean; a tongue extends through the Golden Gate. San Francisco lies on the point of land immediately south (left) of the Golden Gate. (USGS photo.)

and the San Francisco Bay Conservation and Development Commission. These agencies, in collaboration with the Department of Housing and Urban Development and Department of Interior have begun a pilot project, the San Francisco Bay Region Environment and Resources Planning Study, to find out exactly what the geologic problems are, the relative importance of each geologic problem and what can be done about them.

The study is experimental in nature. Earth scientists and government officials are joining together to improve regional urban development, planning and decision-making and to help society gain ultimate benefit from the physical environment, the landscape and the resources of their area.

The importance of the study extends far beyond the San Francisco Bay area. Methods and procedures developed here are expected to serve as models in other communities.

Clearly, this is an exciting program. For the remainder of this chapter we will concern ourselves with this area, its geology, its problems and, very importantly, with the attempts being made to optimize the livability of this beautiful region. Initially, we will review the geologic history of the area, discovering how the rocks of the Bay area were formed and how they came to be located there, some of them moving hundreds of miles to reach their present positions. We will then focus on a number of individual aspects of the Bay area geologic environment now undergoing intensive study.

This chapter will necessarily remain unfinished because the detailed study of the Bay area will continue well beyond the publication date of the book. Nevertheless, it is hoped that this chapter will provide a springboard for discussion and further study of the problem of the urban geologic environment.

2 GEOLOGIC HISTORY

Geologically speaking, the configuration of the San Francisco Bay area rocks is a relatively recent development. The San Andreas fault, the best known of the active California faults, first came to life in northern Mexico about 30 million years ago. The San Andreas fault grew northward, reaching the Bay area only a few million years ago. The northward extension of the fault into the Bay region coincided with the extinction of Coast Range volcanoes, which had belched lava and ash onto the surrounding land and sea for tens of millions of years.

Figure 2–6 in Chapter 2 shows present-day lithospheric plate boundaries. Mentally move North America eastward toward Europe, uncovering the East Pacific Rise until it extends all the way to Alaska. Extend the Mid-American trench northward along the west coast of Mexico, California, Oregon, Washington and Canada, and you will have a mental picture of what geologists think existed 40 million years ago along the west coast of North America.

Rocks bordering the California coast at that time consisted largely of volcanic rocks, sandstone, shales, cherts and some limestone. Many of these rocks had

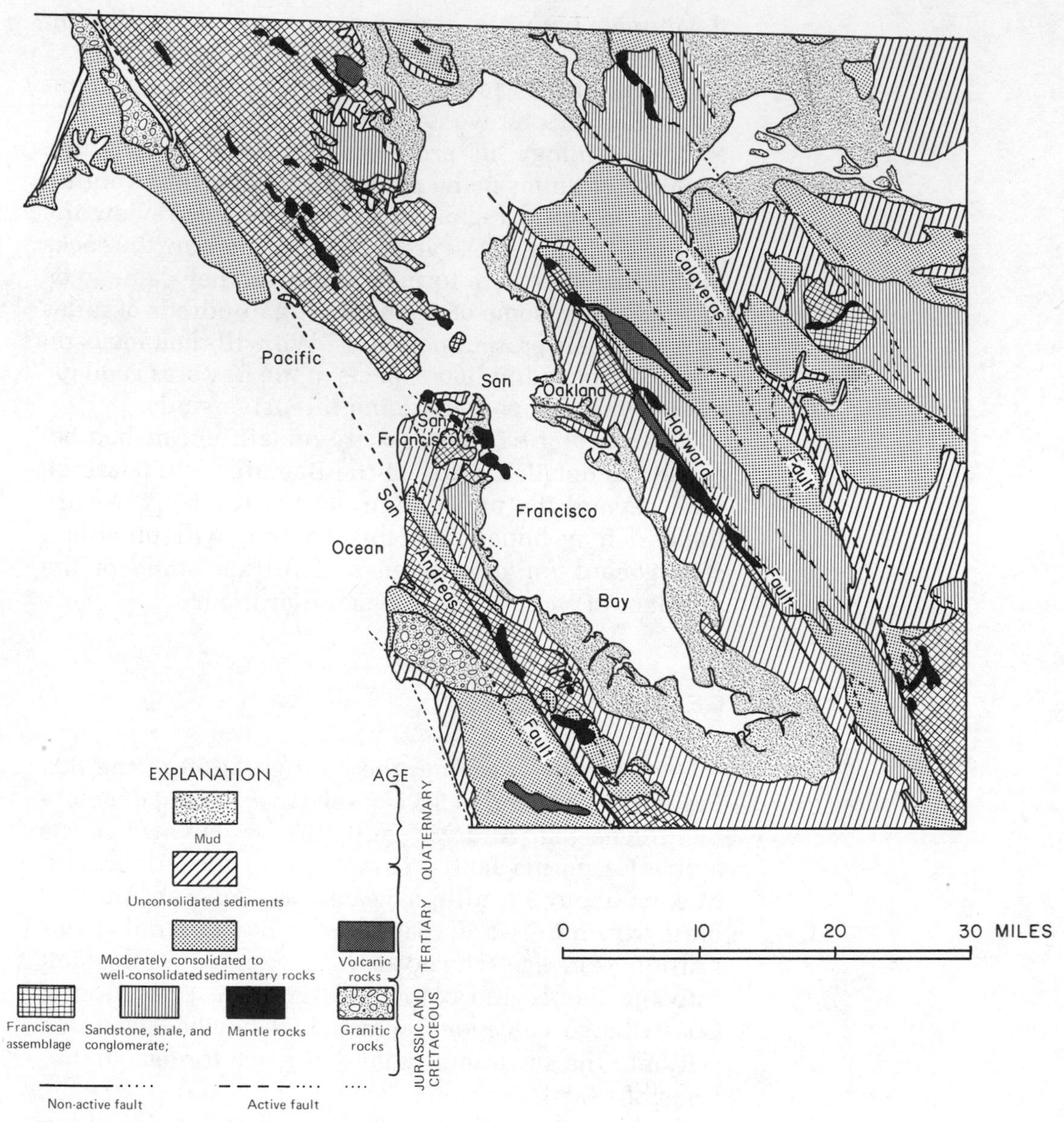

Figure 13–2. Geologic map of the San Francisco Bay area.

been erupted or deposited on Pacific plate sea floor 70 to 180 million years ago.

Pacific plate crust was slowly disappearing into subduction zones plunging beneath the westward-moving North American plate, but a thin veneer of rock was being skimmed off and plastered against the western margin of North America. These rocks, known today as the Franciscan assemblage (Fig. 13–2), comprise most of the bedrock of the Bay region. Here and there, slices of mantle rocks were squeezed up with the sea floor rocks and pushed up toward the surface. Erosion subsequently bared some of the mantle rock. Outcrop locations are shown in Figure 13–2.

Approximately 30 million years ago, rocks in the Bay area reached an important geologic milestone, which would change the geologic regime from compression, subduction and volcanism to shearing, transcurrent faulting and no volcanism. This transformation is continuing today. At that time the westward drift of North America, coupled with the insatiable appetite of subduction zones fringing its margin, brought the easternmost part of the East Pacific Rise (a spreading center) into the subduction zone. Changing stress patterns due to disappearance of the East Pacific Rise into the maw of the subduction zone gave birth to the San Andreas fault system.

At first, the San Andreas fault was probably a short transcurrent fault system joining broken ends of the East Pacific Rise where the rise was disappearing into the subduction zone along the coast of northwestern Mexico. As more and more of the East Pacific Rise disappeared beneath North America, the San Andreas system lengthened until today it extends from northern California into the Gulf of California. From there it extends to a point south of Baja California by means of a complex set of transcurrent faults linked by very short spreading segments.

The San Andreas still connects the broken ends of the East Pacific Rise. A short segment of the once lengthy Rise remains weakly active off the coasts of Northern California, Oregon, Washington and British Columbia. Weak subduction in this region is probably responsible for volcanism in that area (see Chap. 3).

The San Andreas, other active transcurrent faults in the Bay region and most of the inactive transcurrent faults (Fig. 13–2) were formed and set in motion by relative north-south slip of North American and Pacific plates. For our purposes, we include all Bay area transcurrent faults when we talk about the San Andreas fault system.

Volcanism continued along the California coast after the formation of the Franciscan assemblage and during the formation and extension of the San Andreas system. Many of the sedimentary rocks deposited along with the volcanic rocks during this period have not yet consolidated into really hard rock but are still somewhat *friable* or "crumbly." Present-day distribution of these rocks is shown in Figure 13–2.

Northerly motion of the Pacific plate relative to the North American plate has cut loose a slice of crust containing 80-million-year-old granites, Franciscan-age rocks and younger sedimentary and volcanic rocks from its home far to the south. This slice is located west of the San Andreas fault (Fig. 13–2) and is still moving slowly northward, setting off a big earthquake every 50 to 100 years as it jerks along.

The final episode in the geologic evolution of the Bay region occurred during the past few million years as San Francisco Bay sagged slightly, possibly because of local tensional forces perpendicular to the San Andreas fault system. Sediments eroded from surrounding highlands were carried by streams toward the Bay, where they were deposited in and around it. Much of the sediment being deposited today is very fine grained and forms extensive mud flats rimming the Bay. Areas underlain by mud and unconsolidated sediments are, of course, the most hazardous locations during earthquakes.

What is the long-term (millions of years) future of the San Francisco Bay area rocks? No one knows for sure, but one possibility is that the westward-moving American plate (north and south parts) will eventually overrun the remainder of the East Pacific Rise and become again a simple subducting margin as it was prior to subduction of parts of the rise.

3 ENVIRONMENTAL GEOLOGY

Because intensive coordinated study of the environmental geology of the Bay area is just beginning, it is impossible at this time to present more than a few tentative conclusions regarding solutions to Bay area problems. What we can do, however, is review the environmental geology of the Bay area and show what is being done to identify and find solutions to its problems.

In the Bay area, there are four major groups of geologically related environmental problems: earthquakes and earthquake-related problems, problems arising from erosional and depositional processes, problems related to obtaining an adequate supply of good quality water for household and industrial use and problems related to extraction of minerals from the area.

Let us begin with a brief discussion of earthquakes and earthquake-related problems.

Earthquakes

Earthquakes and earthquake damage were discussed at length in Chapter 2. Many of the examples used in that chapter were drawn from the San Francisco Bay area, so a detailed discussion of Bay area hazards is unnecessary in this chapter. This discussion will be limited to three aspects of earthquake and related problems: a look at the active faults in the region, a brief summary of the 1906 earthquake and a summary of present investigations of earthquakes and related phenomena in the area.

Active Faults

There are six recently active faults in close proximity to San Francisco Bay (Fig. 13–3): the San Andreas, Hayward, Calaveras, Pleasanton, Green Valley and Rogers Creek. All of these faults belong to the San Andreas system. The Pleasanton and Calaveras faults are probably closely related, and the Green Valley fault may be an extension of one or both of these. The Rogers Creek fault is probably an extension of the Hayward fault. Thus the six faults are distributed along three major crustal breaks, the San Andreas, Hayward–Rogers Creek and Pleasanton–Calaveras–Green Valley zones. The San Andreas and Hayward faults are the most hazardous. The Pleasanton–Calaveras–Green Valley zone has been much less active but is potentially dangerous.

The Hayward fault, a large active branch of the San Andreas system, could very well be the site of the next major earthquake in the Bay area. The Hayward marks the eastern margin of the lowlands surrounding the Bay. In Oakland, the Warren freeway and Lake Temescal lie within the fault zone. North of Oakland, the Berkeley Hills are part of an ancient eroded scarp formed by movement along the fault. The fault passes

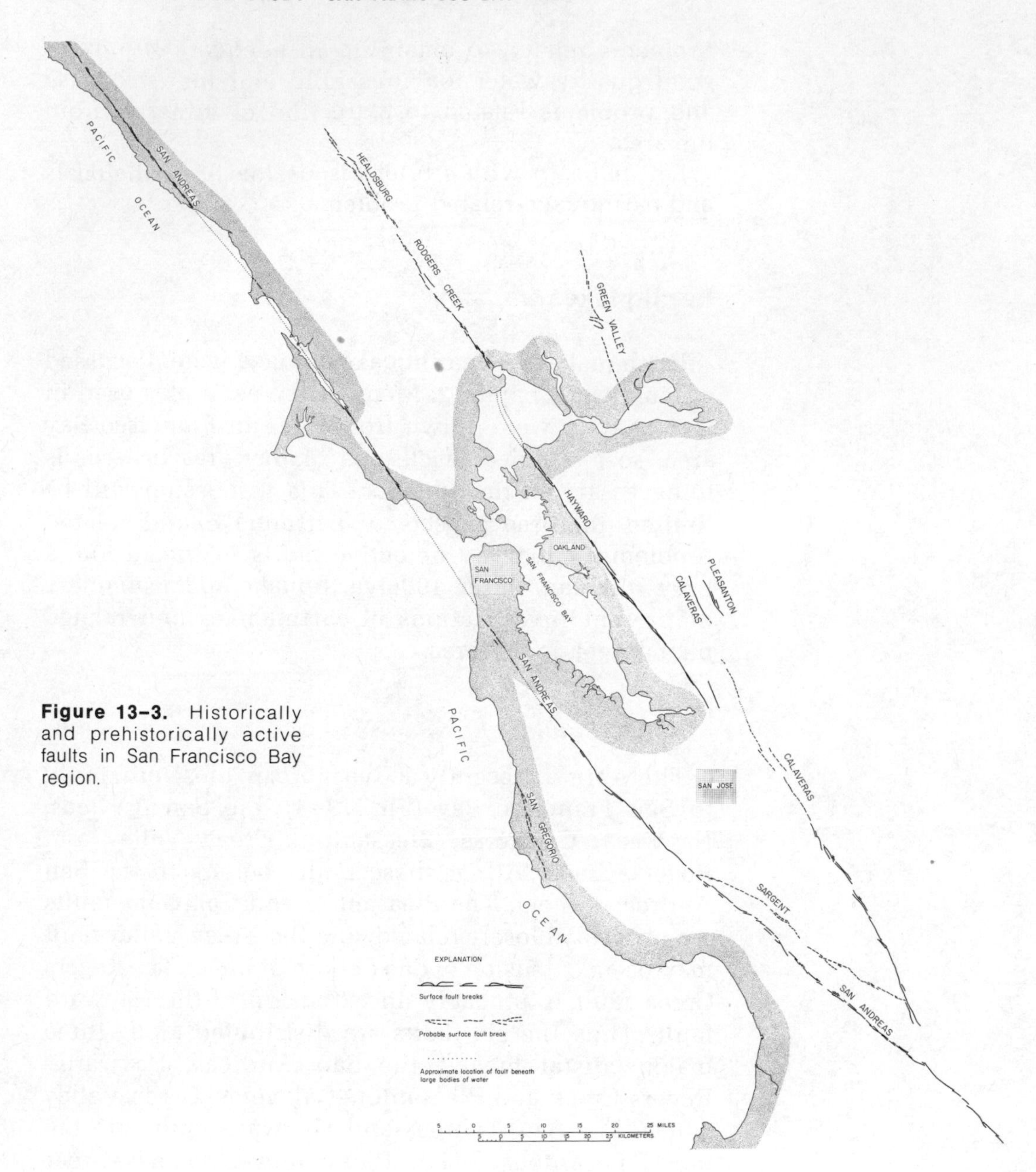

Figure 13–3. Historically and prehistorically active faults in San Francisco Bay region.

beneath the western part of the football stadium of the University of California campus in Berkeley.

Two great earthquakes have occurred along the Hayward fault. Details of the 1836 earthquake are little known, but it is recorded that large fissures appeared, and aftershock activity lasted for a month.

Before the 1906 earthquake, the 1868 earthquake resulting from movement along the Hayward fault was known as "the great earthquake." In the 1868 earthquake the East Bay town of Hayward suffered most. Almost every house was thrown off its foundations, and several buildings were entirely destroyed. Damage in San Francisco was mainly limited to structures on reclaimed or filled land, a fact which underscores the need today for careful compaction of all reclaimed land prior to starting construction. Strong tremors raced under the Bay and through San Francisco, but buildings on firm ground and rock suffered little damage.

1906 San Francisco Earthquake

California has had three monster earthquakes during recorded history: the 1857 quake in Southern California, the 1872 quake in Owens Valley near the Nevada line and the 1906 San Francisco earthquake. The 1872 Owens Valley, the 1811 New Madrid, Missouri, and the 1964 Alaskan earthquakes probably had greater magnitudes than the 1906 San Francisco earthquake, yet the 1906 earthquake is the most significant in United States history for several reasons:

1) It was the first great catastrophic earthquake to strike a major United States population center. More than 600 people died in the earthquake and subsequent fire, and the property loss was $400 million, a staggering sum in uninflated 1906 dollars. In addition to San Francisco, many other towns and cities along the San Andreas fault were damaged. The worst hit was Santa Rosa, north of San Francisco, where the entire downtown section was demolished.

2) Its contributions to seismology and geology. Observations of the great extent of surface rupture, horizontal movement of rocks along faults and elastic behavior of rocks prior to and during faulting were major "firsts" deriving from studies of this earthquake. Detailed contemporary descriptions of the earthquake and its effects prepared by University of California geologist Andrew Lawson and his colleagues of the State Earthquake Investigation Commission are still a gold mine of information about faulting and earthquakes.

3) Hazard recognition. The hazards of building on alluvium, either natural or man-made, were strikingly evident after this earthquake.

4) Prediction. Establishment of new triangulation

networks capable of measuring the slow deformation of rocks prior to an earthquake gained new impetus after this quake. Data collected from these triangulation networks form the basis of much of our knowledge regarding probable locations and dates of future large earthquakes.

5) Research. Before 1906, seismologic research was virtually nonexistent in the United States. The 1906 earthquake and its smaller successors attracted outstanding scientists to California institutions such as California Institute of Technology, University of California and Stanford. Pioneering research at these and other centers provided us with much of what we know today about earthquakes.

Present Investigation

Investigations in the Bay area parallel those discussed in Chapter 2. These include location of active and recently active faults, earthquake and microearthquake recording, drilling for rock samples, slope stability, rock and soil response to earthquake waves and distribution of earthquake activity (*seismicity*). The reader is referred to the appropriate sections of Chapter 2 for details of these studies.

Erosion and Deposition

Whereas earthquakes are dramatic, causing sudden damage, loss of life and disruption of normal activities, erosion and deposition slowly and insidiously transfer rock fragments from place to place. Over an extended period of time, erosion and deposition have the potential to cause greater damage and sometimes greater loss of life than earthquakes.

In the Bay area, numerous investigations related to erosion and deposition are underway. For convenience, these investigations have been grouped into four categories: floods, landslides, bay infilling and land pollution. Because water is the main erosive agent in the Bay area, it is not surprising that most of these investigations are also pertinent to another major problem area, water, which is discussed in a later section in this chapter. As you study this and following sections, bear in mind the close relationship of these two groups of investigation.

Floods

Most erosion and deposition takes place during floods. The slowly moving waters of ordinary stream flow are incapable of moving any but the smallest rock fragments. Thus it is important to establish the high-water mark of a "50 year flood"; that is, a great flood which will recur once every 50 years on the average. From these data, appropriate plans can be developed to control flood waters, and the land can be zoned to prohibit construction in areas likely to be flooded. In addition to high-water data, the quantity of stream-borne sediment will be determined as a function of flood size, the effect of urbanization on stream sediment will be determined, areas of active erosion and deposition will be located and the effects of erosion and deposition estimated. These data enhance the effectiveness of erosion control measures.

Landslides

Landsliding is a mechanism of mass movement of rock and soil, just as erosion is a mechanism of movement of rock and soil (see Chap. 5). Landslides are a serious problem; they caused more than $25 million in damage in the Bay area during the winter of 1968–69. At this rate, total damage due to landslides could exceed total damage due to more spectacular but less frequent earthquakes.

Landslides result from many factors. Given a metastable blanket of rock and soil covering a hillside, any of the following can trigger a landslide: an earthquake, a heavy rain, a poorly conceived excavation at the base of the slide or removal of vegetation whose roots bind soil and rock together and soak up some of the moisture. In the Bay area, zones of active landsliding are being mapped (Fig. 13–4). These field data, in conjunction with slope stability data and rock and soil properties data, will be used to evaluate landslide hazards in different areas.

Bay Infilling

Rock and soil eroded from hillsides surrounding the Bay eventually come to rest somewhere within the Bay. In 1835, San Francisco Bay extended over an estimated area of 680 square miles. Since then 280 square miles or slightly more than 40 per cent of the

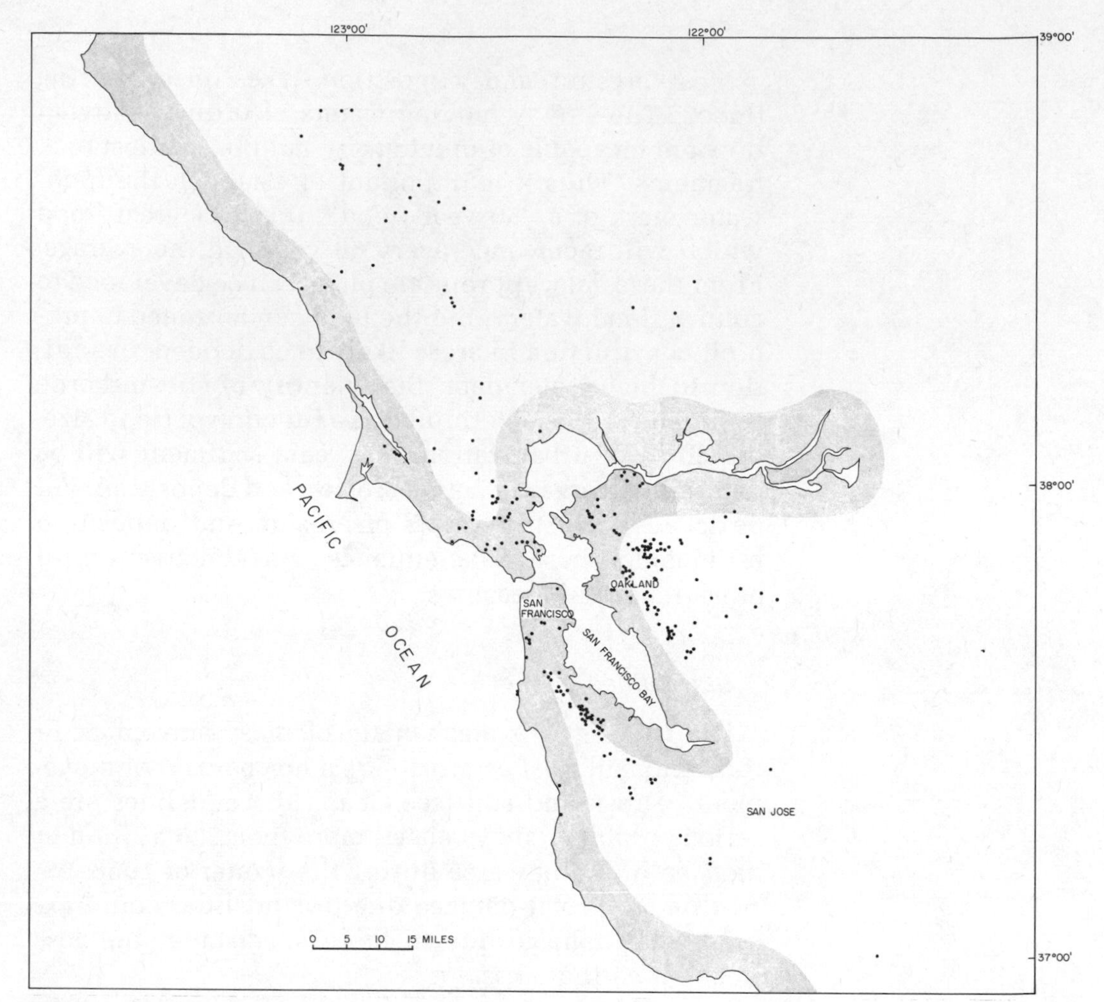

Figure 13–4. Damaging landslides in the San Francisco Bay region during the winter of 1968–69 caused more than $25 million in damage. Dots indicate landslide locations.

original area has been filled in. The average depth of the Bay has decreased by an unknown amount. The filling, although mainly deliberate, was enhanced by increased rates of erosion caused by man's removal of trees and other erosion-retarding vegetation. The filling of the Bay and subsequent construction on the fill have not only increased the earthquake damage potential in the Bay area but have also robbed residents of the Bay area of scenic marshland, once the home of abundant wildlife.

A plenitude of investigations cannot reverse the infilling, but perhaps a realization of the extent of infilling (Fig. 13–5) will help slow the destruction of the region's greatest resource, San Francisco Bay itself.

Land Pollution

Man is a geologic agent, too. He erodes the hillsides and transports sediments to other locations where he deposits them. But man is an erosional and depositional agent with a difference; he mixes impurities with his sediments, impurities that remain in the soil for years and which sometimes have an adverse effect on the rock, soil, water and life in the vicinity of his deposits. This is land pollution.

The potential for land pollution exists around every solid waste disposal site, landfill and septic tank. The San Francisco Bay region contains 77 landfill sites, ranging in area from about one acre to 500 acres. Collectively, landfill sites occupy about 5000 acres, or a little less than 8 square miles. Studies are planned to determine the drainage path of any chemical substances leached from the landfill materials and to sample waters containing the leached chemicals. Other objectives of the study include location of all septic tanks and identification of areas of high land pollution risk. The study will "red flag" areas susceptible to land pollution and areas already polluted so that local authorities can take corrective action.

Water

As we have seen in Chapter 11, water is the lifeblood of civilization. We need it for drinking and for food production, we need it to carry off our wastes and we need it for transportation and industry. The intricacy of water usage patterns makes it very difficult to study this most important of man's resources. The Bay region is no exception in this respect. The interaction of surface waters in streams, lake and Bay with subsurface waters filling pores of fringing sediments; the presence of industrial, domestic and agricultural waste in abundance; and the extensive demands put on the system by a large population create an immense ecologic problem. One of the major difficulties of the Bay region study is simply identification of all of the component elements and determination of how these components interact among themselves and with other factors discussed in preceding sections of this chapter. The following sections summarize some of the more important aspects of the regional study.

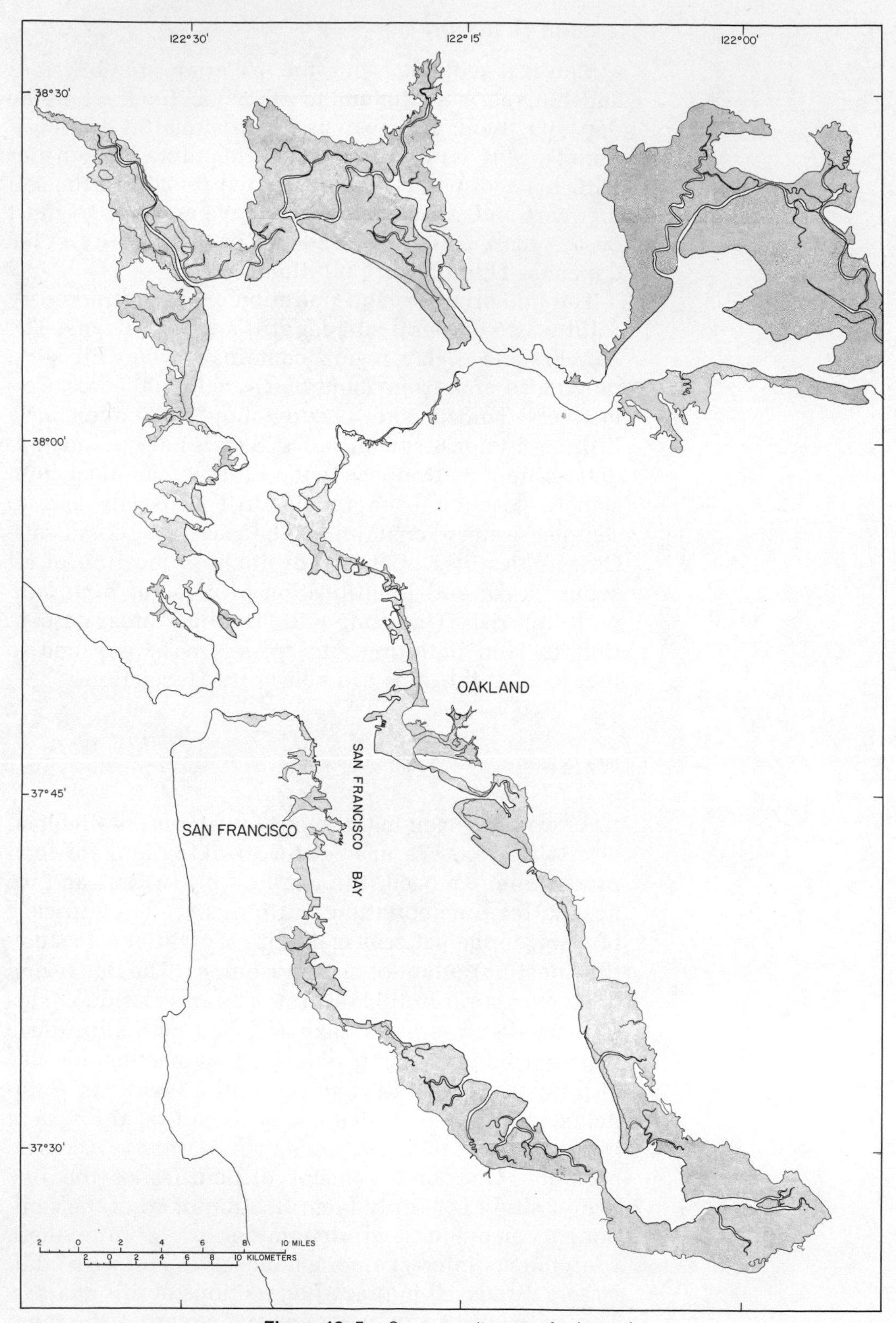

Figure 13–5. *See opposite page for legend.*

Supply

Water is available in abundant quantities in the Bay region, but not all of it is suitable for human and industrial consumption. The problem is to identify available sources of water and to determine the capacities of water supply systems and the impact of such factors as increasing population, industry, sedimentation and pollution on existing systems. Studies now underway will provide data regarding source of water, population served, water consumption, estimated future population and estimated future water needs.

Gound Water

Water for agricultural, industrial and domestic consumption comes from either surface waters such as streams or lakes or ground water, which is defined as water occurring in pores and cracks beneath the surface of the ground. The Mediterranean climate of the Bay region—winter rains and dry summers—causes streams in the area to dry up in the summer. This lack of an adequate summer supply of surface water forced Bay region inhabitants to tap not only local groundwater reservoirs but also to import water from distant groundwater reservoirs.

As shown in Chapter 11, ground water and groundwater geology are complex subjects. Management of a groundwater reservoir requires definition of the boundaries of the reservoir, usually by drilling, determination of storage and transmission characteristics of aquifer rock and determination of rates of withdrawal and recharge of the reservoir.

With these basic factors determined, we can then measure or define natural discharge (e.g., springs) and recharge, chemical and biologic properties of the water and susceptibility to pollution from septic tanks, industrial waste and seawater invasion. With all these data at hand, an effective management program can be planned.

Studies underway in the Bay region will locate areas underlain by aquifers, determine depths and potential yields of the aquifers and show changes in water table elevation during historic time—an important factor in evaluating recharge of an aquifer. These data will be integrated into a plan for optimal utilization of the groundwater resources of the region and strategies for combating pollution.

Figure 13–5. Sketch showing extent of original marshlands surrounding San Francisco Bay. These are largely filled now, and in some areas the coastline extends into the Bay well beyond the original outer limits of the marshes.

Land Subsidence

It may seem strange to include land subsidence under the heading "Water," but the principal cause of land subsidence is withdrawal of fluids from beneath the ground. Withdrawal of ground water for domestic, agricultural and industrial use has resulted in widespread subsidence and even faulting where subsidence has been uneven.

Parts of the Bay region containing extensive, thick deposits of unconsolidated and semiconsolidated sediments are extremely susceptible to subsidence due to groundwater withdrawal. At the southern end of the Bay, where groundwater withdrawal has already caused subsidence (Fig. 13–6), it has been necessary to construct levees in order to prevent flooding of subsided areas by Bay waters.

Contemporary investigations will determine extent

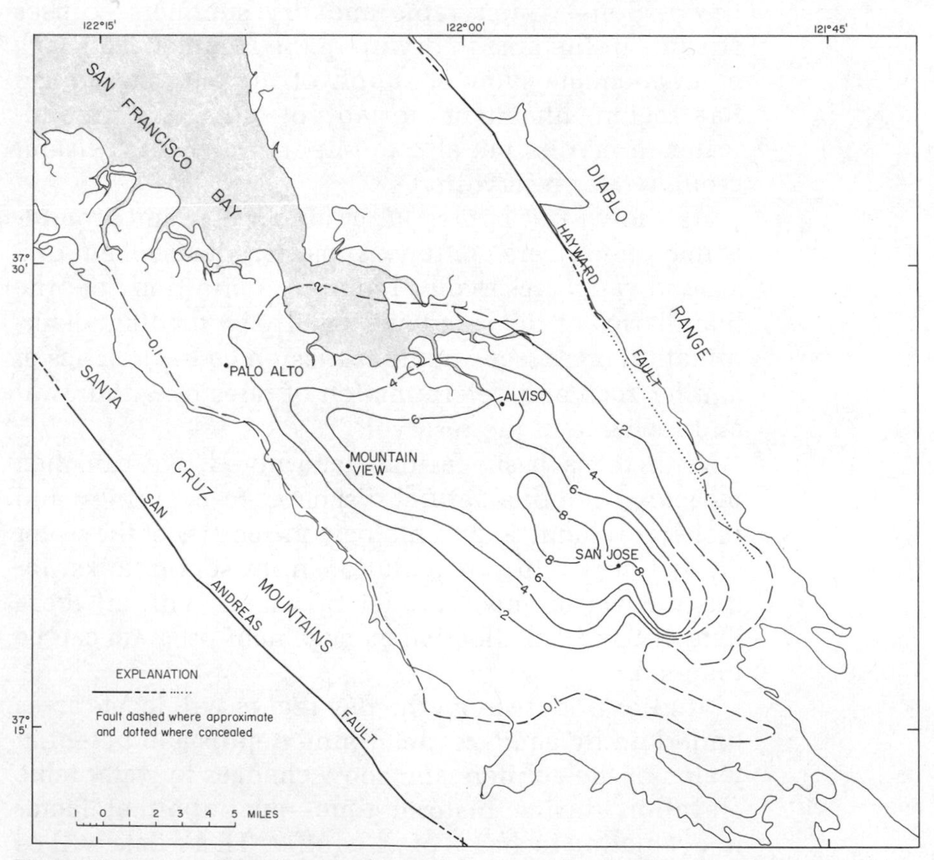

Figure 13–6. Land subsidence in the southern part of the San Francisco Bay region. It has been necessary to build levees to keep Bay waters from flooding sunken areas.

of existing subsidence and identify areas of potential subsidence.

Aquifer Recharge

In the Bay region, groundwater reservoirs have been pumped for years with little drop in the water table so that groundwater is regarded as a renewable resource. All resources, however, are subject to overuse, and water is no exception. More people, greater demands, loss of some aquifers because of pollution and reduced recharge resulting from paving large areas of permeable soil through which water once seeped, and even flood control levees which constrain flood waters to a narrow channel and do not let them spread over floodplain sediments—all of these can restrict the availability of ground water. If groundwater aquifers are to continue to provide adequate water, care must be taken to insure that inflow of water into the aquifer is adequate and not polluted.

San Francisco Bay region studies will identify areas of natural and artificial recharge and areas of potential recharge, either natural or artificial. Pollution is the subject of a separate group of investigations.

Water Pollution

Everywhere one hears "pollution"! But what is pollution? If normal sedimentation is considered pollution, the Mississippi River pollutes the Gulf of Mexico with more sediment daily than man contributes to the world's oceans in a lifetime. Any definition of pollution depends on the use one plans to make of the "polluted" water (or air or land). Some "pollutants" are acceptable in certain circumstances but not in others. So the first problem in a study of water pollution is to determine the purposes for which the water is (or might be) used and what unacceptable levels of pollutants are.

In the Bay region, this means setting up criteria for recreational areas, wildlife refuges, fish habitats and hunting and fishing grounds as well as criteria for ground water. Bay region environmental geologists plan to measure pollutional loadings in stream, lake and Bay waters and to survey waste disposal procedures. From these data, they propose to develop a suite of alternative plans or strategies for development and use of the water resources of the region.

Mineral Commodities

The San Francisco Bay region produces a variety of mineral commodities. The mineral industries draw energy, water and manpower from the region. They return finished products, money from sales outside the region and, of course, water, air and soil pollution. The intimate relationship of the mineral industries with other aspects of the region's geologic, sociologic and economic environments necessitates consideration of the mineral industries in any "grand plan" for the region.

Evaporation of sea water—actually Bay water is almost as salty as sea water—yields salt, magnesium, synthetic gypsum and bromine. Evaporation ponds rim large segments of the Bay.

The region imports most of its fossil fuel. Small seams of coal have been mined in the region, and a few wells have produced relatively small amounts of natural gas and petroleum.

Stone and stone products compose a large segment of the region's mineral industries. Several types of stone are quarried for construction purposes. Limestone is quarried for cement production, and clay is mined for use in ceramics. Diatomite, a rock composed of siliceous skeletons of microscopic marine plants, is mined in the region and is used to make filters, fireproof insulation, fire brick and for a variety of other purposes. Some talc and asbestos are found in altered mantle rocks (see Fig. 13–2) outcropping in the area.

A variety of metal ores are also mined in the region. Some mantle rocks contain economic deposits of chromite, the principal ore of chromium. Altered marine rock plastered against the continental margin during a subduction phase contains manganese ores. Other rocks contain veins of *magnesite,* an ore of magnesium. Diminishing reserves and competition from magnesium extracted from sea water have greatly reduced the importance of these deposits during the last few decades. *Cinnabar,* a mercury ore, was used by Indians as a paint pigment. Since commercial production of mercury began in 1845, rich mines in the southern Bay region have produced large amounts of mercury emplaced during the volcanic phase preceding the development of the San Andreas fault.

Bay area environmental geologists are inventorying the mineral resources of the region. Upon completion of the inventory they plan to investigate the social and economic impact of the mineral industry and other environmental factors affecting the Bay region environment as a result of mineral extraction. In the third phase of their investigation, they will develop a strategy or strategies for mineral extraction compatible with maintaining a quality environment.

4 CONCLUSION

The necessary scientific and technologic tools are available for the collection of geologic data pertinent to environmental studies. But will these tools be used? Will enough money be made available to support the investigators? And will results of the investigations be applied by executive and legislative bodies of the Bay area governments?

The evidence is not everywhere encouraging. A prominent elected Bay area official appeared on nationwide television and announced that earthquake hazards in San Francisco were far less than scientists had suggested. Smog has come to the Bay area. Only a few years ago, one could look east from the Stanford University campus and on most days see the mountains across the Bay. Now it is possible to see the mountains on significantly fewer days. The population of the Bay area continues to grow, straining water resources, encouraging construction in marginal regions and polluting land, water and air. Vested interests fight for their rights, sometimes with flagrant disregard for the rights of others. Money plays a big role. How far can the Bay area go in tearing down and replacing old unsound structures or in renovating and reinforcing them to resist earthquake stresses? Relocation of a freeway in a landslide-prone zone is a multimillion dollar undertaking. The freeway was probably placed where it is because more desirable locations were in heavily populated regions.

In the final analysis, it is teamwork that counts in the solution of environmental problems. The San

Francisco Bay area is no exception. The team consists of geologists, sociologists, biologists, lawyers, economists, legislators, government administrators and others. The Bay area contains an outstanding cadre of geoscientists. Although the nongeoscience members of the team are not personally known to the authors, they are probably equally capable. The end result of the Bay area study will depend on how well these people work together. And frequently, getting people to work together is the toughest step of all.

14

RURAL CASE STUDY— BUFFALO CREEK, LOGAN COUNTY, WEST VIRGINIA

One of the Carolina Bays in southeastern North Carolina. Pieces of a comet striking the earth probably excavated this and several hundred similarly shaped lakes in the region.

1 INTRODUCTION

Smoldering slag heaps
Rise high.
Swirling death fumes
Swirl high.
Proud dirtied men
Stand high.
Against red-dog ridges,
Beneath blackened skies
The strong and straight
Stand weeping;
The crushed and crumpled
Lie still.

R. Michael Leonard (1972)

On Saturday morning, February 26, 1972, a dam, constructed of mine waste material, in Buffalo Creek hollow, Logan County, West Virginia, gave way, and the ensuing flood caused extensive loss of life (125 known dead) and property. The verse above is from a poem written by a student who spent his spring vaca-

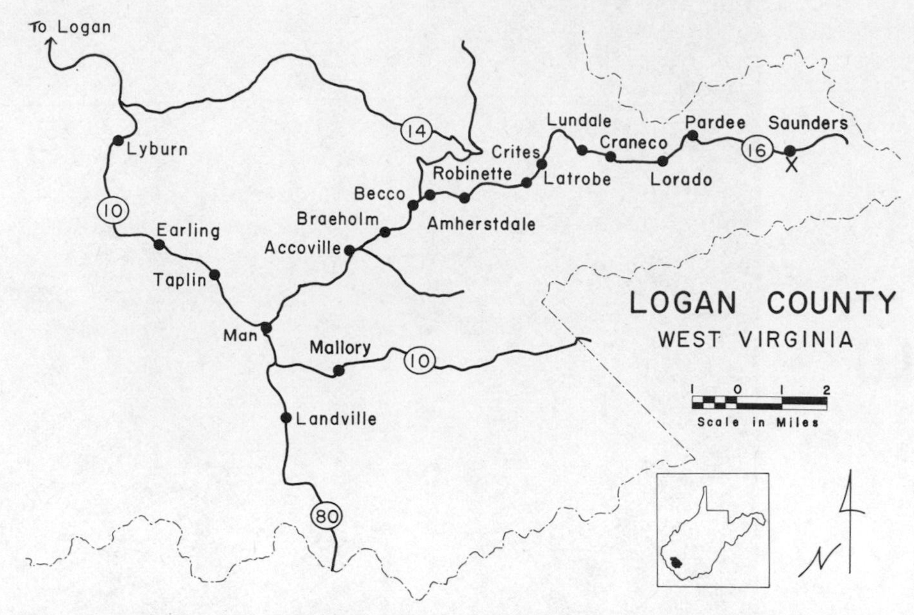

Figure 14–1. Map showing Buffalo Creek Hollow, towns and an index map.

tion in Buffalo Creek hollow as a Salvation Army volunteer. Although the verse is primarily an anguished reaction to the flood disaster, it also says much about the general conditions in the area. Frequently a disaster is only an indicator of less obvious problems. Buffalo Creek hollow is an eloquent example of this.

2 GEOGRAPHIC AND GEOLOGIC SETTING

Buffalo Creek hollow, Logan County, is located in southern West Virginia in the heart of the central Appalachian coal fields (Fig. 14–1), a dissected plateau region of sharp relief. The elevation ranges from 600 feet above sea level in the north to 2750 feet in the south. While this is not great relief, the elevation changes are rapid, and adjoining valleys (hollows) are effectively separated from each other. The hollows are narrow, steep walled and sinuous.

Buffalo Creek is a tributary of the Guyandotte River, which flows northwest to the Ohio River at Huntington, West Virginia. There are 5000 people living in fourteen communities along Buffalo Creek in a 15-mile stretch from Saunders to Man, where Buffalo

Figure 14–2. An undisturbed hollow adjacent to Buffalo Creek Hollow.

Creek flows into the Guyandotte. These communities were mostly built as company towns, and originally all the houses and utilities were owned by the coal companies. Locally, these communities are still called "camps." During the 1950's, the coal companies sold the houses to the tenants and the utilities to various private companies. There never were any sewage facilities, and Buffalo Creek is used as an open sewer. Almost all the houses, roads, railroads, and so forth are located on the valley floor, filling the valley completely (Figs. 14–2 and 14–3).

The rocks in this area are flat-lying sedimentary rocks, mostly sandstones and shales with interbedded coal seams. (See Chapter 8 for a description of the origin and occurrence of coal.)

3 THE CASE

Mining

Buffalo Creek is rich in coal, with approximately fifteen operating mines. In the upper reaches of the hollow, near Saunders, the Buffalo Creek Mining Company, one of many mining companies owned by

Figure 14–3. The mouth of Buffalo Creek.

Figure 14–4. Aerial view of Middle Fork Hollow. (From Report of the Citizens Commission Investigation, 1972.)

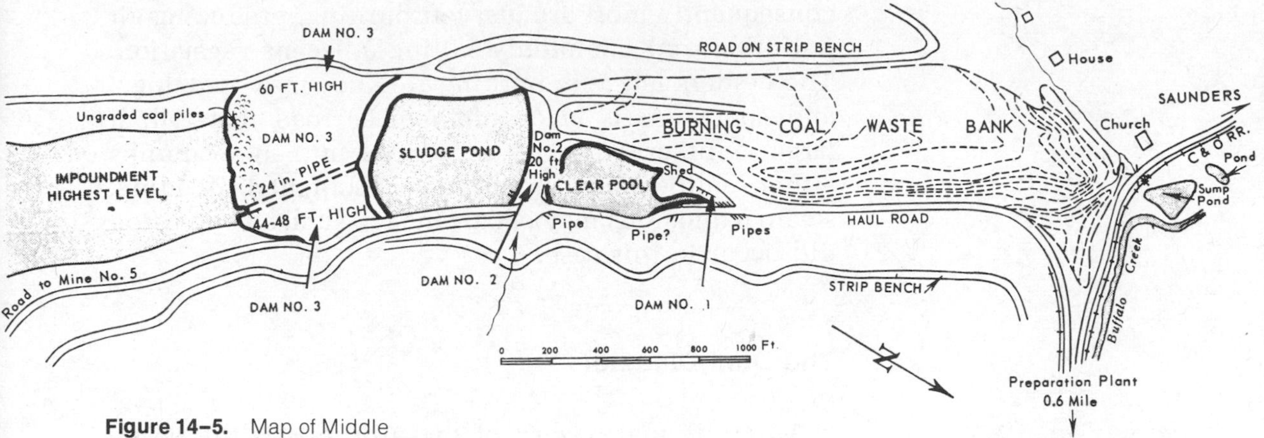

Figure 14–5. Map of Middle Fork Hollow, showing location of dams. (Modified after Davis et al., 1972.)

the Pittston Company of New York, has eight mines—five deep mines and three surface mines. Pittston is the largest independent coal producer in the United States. The operations in Buffalo Creek hollow were started by the Lorado Mining Company in 1945. In 1964, Lorado was bought out by the Buffalo Creek Mining Company, which was bought by Pittston in 1970.

By 1972, the Pittston operations were generating approximately 1000 tons of waste material per day from 5000 tons of unwashed coal. In addition to the solid waste, a mixture of water and sludge is produced by the coal washing operation. This preparation plant used approximately one half million gallons of water per day.

The solid waste material is composed primarily of coal, shale and some sandstone and is dumped at some site convenient to the preparation plant. These dumps of mine waste material are locally called "slate" or "slag" dumps. In this particular case, the waste material was dumped in the mouth of Middle Fork hollow at Saunders (see Figs. 14–4 and 14–5). The Lorado Mining Company had started the dumping in about 1946. Over the years this waste pile grew to fill the hollow, approximately 400 feet across, 200 feet high and more than 2000 feet up the hollow. After a few years, the dump caught fire and has continued to burn since that time. The Appalachian countryside is dotted with hundreds of these burning, smoking mine waste dumps, which result when waste materials undergo spontaneous combustion. It is very difficult to extinguish these fires except by spreading out the material;

consequently most are just left burning, producing air pollution and commonly killing adjacent vegetation. The cinders left after burning are known as "red dog," and some use is made of them as road material. If large volumes of water are forced into these dumps either deliberately or through accidental flooding, steam explosions may occur. In fact, such an explosion did occur in this case.

The Dam Disaster

An additional hazard of these dumps is the possibility of impounding water behind them. There are 35 such impoundments in this area of southern West Virginia and adjoining counties in eastern Kentucky. In many cases they are the result of blocking the natural drainage and are not deliberately constructed impoundments. However, sometimes the mining companies use the mine waste material to build settling ponds for the water and sludge from the preparation plants. Pressure from the state began in the early 1950's, in the form of water pollution regulation, to at least clean up the sludge in the water before returning it to the stream. In the late 1950's, the first dam (Dam #1) was built in Middle Fork hollow above the burning dump (see Fig. 14–5). In 1966–67 Dam #2 was built. In a couple of years, the settling pond behind Dam #2 filled, and a third dam (Dam #3) was started several hundred feet up the hollow from #2. Once again mine waste material was dumped and bulldozed to form the dam. Dam #3 was constructed without any foundation work or engineering specifications. The waste material was dumped and bulldozed over the sludge behind Dam #2. Trees, trash and anything else in the path of the dam was simply buried as the dam grew (Fig. 14–6).

By 1970, Dam #3 was closed and water and sludge accumulated in the pond behind it. Generally these settling ponds do not hold large volumes of water for extended periods. However, this dam did hold a large volume of water, which had reached a minimum of 31 feet or a maximum of 48 feet behind the dam before it gave way. The minimum figure was given by the newly appointed mine superintendent (the previous superintendent disappeared temporarily after the disaster); the

Figure 14–6. Remains of Dam #3, showing buried trees.

latter figure comes from a United States Geological Survey report. These depths correspond to 91 million and 132 million gallons of water respectively. By 1972, Dam #3 had grown to 60 feet high, 465 feet across the hollow and 480 feet thick on the right side and 360 feet on the left side, looking upstream (see Fig. 14–5).

In this structure there was no provision for overflow. According to eyewitnesses, there were no overflow pipes in the structure until about a day before the dam collapsed. One 24-inch pipe was installed two feet below the top of the dam. The other two dams did have overflow pipes in them. In essence, these "dams" were built without any engineering. During investigations of the disaster by the Citizens Commission to Investigate the Buffalo Creek Disaster, testimony from the bulldozer operator who built Dams #1 and #2 showed this clearly. His complete instructions were to bulldoze the mine waste material so that the top of the structure was 14 feet wide—wide enough for the bulldozer.

During the three days preceding the dam break, rainfall measured 3.7 inches. This is not an unusual amount of rain; in this area, 3.7 or more inches of rain can be expected to fall during a three-day period on the average of every two years. In the 17-year period, 1955–1972, there have been 8 rainfalls equal to or greater than 3.7 inches over a 3-day period. In addition, there was some run-off from melting snow during this period. Although the total run-off was heavy prior to the dam collapse, it was not unusual. In fact, Buffalo Creek was not near flood stage and was receding prior to the disaster.

Dam #3 had a history of breaking and slumping. Eyewitness and government reports show that it slumped and broke on one side (the same side that gave way in February, 1972) in the spring of 1971. There was not much water behind it at that time, and only minor flooding occurred in Saunders. Because of this and other occurrences of slumping and breaking of other mine-waste dumps and dams, the people of the Buffalo Creek area and of Saunders in particular were aware of the dangers. They were aware of the danger that morning, and some of the men were keeping an eye on the dam. Thanks to that vigilance, most of the people from Saunders were evacuated to the school at Lorado, three miles down the hollow, prior to the dam failure. There was never a warning issued by the officials of the coal company, although they too were aware of the danger; in fact, every effort was made to publicly appear as "business as usual." The deep mines were worked that night, and the pumping of water and sludge into the pond behind Dam #3 continued at least until the evening before the disaster.

Figure 14–7. View from top of dump looking toward remains of Saunders.

Figure 14–8. View of remains of dump at Saunders. In the foreground, there had been a railroad and a road.

The deep-mine crews were going to work that Saturday morning, although the strip-mine crews were not, because of water in the surface mines—not because of the dangerous condition of the dam. That morning all the local company officials were at the office in Lorado. Crews were being readied to install another overflow pipe in the top of the dam. (The overflow pipe shown

Figure 14–9. View of remains of Dam #1. The telephone pole is located next to the shed on map, Figure 14–5.

Figure 14–10. Thirty-ton dump truck washed 1.8 miles from Dam #1, Figure 14–9.

in Figure 14–5 had been installed only a day or so prior to the disaster.) The crews assembled at Lorado never made it to the dam—it gave way before they left the office.

At 8 AM on Saturday, February 26, 1972, the dam slumped, and 130 million gallons of water and sludge, and approximately 6 million (according to the government report) cubic feet of mine waste material poured into Buffalo Creek hollow. Dams #1 and #2 were immediately overtopped. The water and sludge pooled momentarily behind the burning dump at the mouth of the hollow until it exploded and washed out. The village of Saunders, which consisted of a church and two dozen houses, was completely destroyed (Figs. 14–7 and 14–8). In fact, all the structures in the first 2.5 miles below the dam in Buffalo Creek hollow proper were destroyed. In this stretch of the hollow, most of the scouring and deposition of the coarser waste material occurred. Not even building foundations were left, and the railroad and road were almost totally removed.

Figure 14–11. View several miles below dams.

Lorado, 2½ to 3 miles below the dam site, was half destroyed. Lundale and Crites were almost completely destroyed. It took three hours for the flood to reach Man, 15 miles away. In those three hours, 125 people

Figure 14–12, *A* and *B*. Destruction in Lorado, approximately three miles below dams.

died, hundreds were injured, 500 houses and trailers destroyed; a total of $50 million in physical damage and an immeasurable amount of misery. Most of the deaths were women, children and older people. This time a mine disaster had struck directly at the miner's family instead of the miner.

Was this simply an isolated incident or accident? The result of carelessness and irresponsibility of one company? The answer is clearly—no. One need only

Figure 14–13. Aerial view of Lundale.

look at the history of the area to see a pattern of neglect and irresponsibility.

Mine Waste Materials

As discussed earlier, mining operations produce large quantities of waste material. The haphazard dis-

Figure 14–14. View at Crites.

posal of this material over the years has created many problems. The dumping of the solid waste has left thousands of unsightly dumps throughout Appalachia. Many of these dumps burn, creating additional problems of air and water pollution. The seepage associated with the dumps, whether burning or not, is frequently very acid. For example, at Earling, the local grade school is built on an old waste dump on the flood plain of the Guyandotte River. The seepage pools at the base of the dump, adjacent to the river, have a $pH = 2.5$ ($pH = -\log [H^+]$). The rocks in the pools are covered with a yellowish limonitic (limonite is an amorphous hydrated iron oxide) coating. Iron and other elements are leached out of the dump by the acid water and eventually end up in the stream. In other cases it can be shown how ground water (see Chap. 11) is contaminated by this process. In general, this leads to water supplies high in iron and sulphate. In the next section we will see that approximately 20 per cent of the municipal water supplies of Logan County are high in iron and sulphate.

The problem of mine waste dumps is not new, not unrecognized. The latest flurry of activity began after the disaster in Aberfan, Wales, in 1966, when a mine waste bank flowed through the village, killing 122 people, mostly children in a grade school. The United States Geological Survey and Bureau of Mines began a survey of mine waste banks and impoundments in southern West Virginia, eastern Kentucky and southwestern Virginia. Parts of a letter dated May 6, 1967, from the Secretary of the Interior to the West Virginia senators, congressional delegation and state and local officials reads as follows:

> Following the recent disaster at Aberfan, Wales, caused by instability of a mine waste dump, the Geological Survey and Bureau of Mines have examined a number of waste piles in West Virginia. In the belief that the subject is of interest to you, I am outlining the conclusions reached by the inspection team as reported to me by the Directors of the Geological Survey and Bureau of Mines.
>
> The Geological Survey and Bureau of Mines have completed an inspection of 38 coal waste banks in southern West Virginia. The inspection was made of only a small fraction of the total number of mine dumps, selection being made on the basis of previous observations.
>
> The inspection revealed that 30 of the banks showed some

signs of instability ranging in severity from bent trees and slumps and bulges on the face of the bank to actual mud flows and slides that have damaged structures; four of these banks are sufficiently unstable to warrant specific attention. Corrective measures such as reduction of the angle of bank slopes, provision of adequate spillways, and prevention of excess water percolation through the bank could increase the stability of all but a few of the 30 banks. Five conditions that accounted for most of the instances of instability are:

1. Lack of adequate, protected spillways on 12 banks which blocked small valleys or on banks which contained settling ponds for coal wash water was the most prevalent problem. The lack of protective spillways could lead to breaching and erosion of the bank by heavy rainfall with potentially adverse local effects downstream.

2. Seven banks showed distinct signs of slope failure from overloading. Most of these cases involved oversteepening of slopes resulting from aerial tramway dumping, and a few involved overloading resulting from dumping of filter cake excavated from settling basins or from emplacement of new tramway dumps on top of old banks of questionable strength.

3. Six burning waste banks represent cases of potential instability because heavy rainfall conditions may trigger internal explosions that can initiate large slides. Adverse effects resulting from burning banks have been the cause of considerable property damage and loss of life in the bituminous region in years past.

4. Five banks emplaced on very steep or unstable foundations have experienced slope failures, some of which have caused minor damage to structures.

5. Several banks on steep slopes are over 700 feet high, and although they appear to be stable, their height alone warrants additional monitoring.

When there is incipient danger to housing, the relocation of the houses may be the economic solution.

Four specific cases are of sufficient immediate importance to merit timely corrective measures. In each case the operators have been apprised of the situation.

With the expansion of the bituminous coal industry to near record production from large mines with associated large waste piles, and with increased coal content in the waste material, problems of instability may increase. However, by engineering improvements in bank development and by zoning regulations to exclude man-made structures from unstable areas, the problems involved can be greatly reduced or eliminated.

The Bureau of Mines will continue to observe mine dumps for possible critical conditions, and will receive counsel from the Geological Survey where unstable conditions that might endanger life are recognized.

Similar letters were sent to officials in Virginia and Kentucky. The letter clearly describes the types of problems and indicates official awareness of these problems.

Municipal Water Supplies

A study of the municipal water supplies of Logan County, conducted during the spring of 1971, revealed some startling data about the water quality of many areas of the county. This study was conducted by a group of professors and students from area colleges and was initiated following newspaper reports of "worms" in the drinking water of Mallory (see Fig. 14–1). The "worms" in fact are midge fly larvae, which have infested the water supply of this town for years. In addition, there have been cases of a salamander coming out of a water tap. The attitude of the local health official and the owner of the water system was one of a lack of concern based on the fact that people had lived with this for years, so why worry about it now.

Analyses of the water supply for Mallory showed that the drinking water was not chlorinated on a continual basis and contained less than the legally required 0.2 ppm (part per million = 0.0001 per cent) residual chlorine, which is considered to be the minimal concentration necessary to control bacteria. In Buffalo Creek hollow, every community where water was analyzed gave the same result—a lack of regular if any chlorination of the water. In the overall results for Logan County, 74 per cent of the municipal water systems (there are 35 separate water systems in the county, but most are owned by a few water companies) contained less than the legal minimum of 0.2 ppm residual chlorine. These figures do not appear in the state health department analyses because samples for the state health laboratories, which are collected by the local health official, are always found to be chlorinated because the water company is notified in advance as to when the official sample collection will be taken.

The immediate conclusion to be drawn from the preceding observations is obvious; many of these water supplies are bacterially contaminated. Although total bacteria count is not specifically controlled by law, it is an important measure of water quality. A count of 100 bacteria per milliliter or less is deemed acceptable for good water quality. Fifty-one per cent of the analyzed municipal water systems contained bacterial concentrations of greater than 100/ml. In addition to the total bacteria counts, specific tests for *E. coli (Escherichia coli)* were carried out. *E. coli* is the best known indicator of domestic sewage pollution, and 11 per cent of the analyzed samples contained *E. coli* bacteria. It is clear from the above data that sewage is contaminating part of the water supplies.

The maximum concentrations of iron and sulphate allowed by state and federal regulations are 0.3 ppm and 250 ppm respectively. These limits were exceeded in 28 per cent of the analyzed samples in the case of iron and 18 per cent in the case of sulphate. This is probably owing to contamination derived from acid mine drainage.

4 THE CURE?

Let us first look at the regulatory legislation and see whether the present laws are adequate. The Citizens Commission to Investigate the Buffalo Creek Disaster pointed out in their report that both state and federal regulations exist to cover this type of mine-related structure. The Bureau of Mines had responsibility under Section 77.200—"All mine structures, enclosures, or other facilities (including custom coal preparation) shall be maintained in good repair to prevent accidents and injuries to employees." And Section 77.216 states, "If failure of a water or silt retaining dam will create a hazard, it shall be of substantial construction and shall be inspected at least once each week." Dam #3 at Saunders was known to be hazardous, having failed once in 1971, and in addition, as pointed out earlier, many other hazardous impoundments were known in the area.

The state has responsibilities in several areas. The Water Resources Division under the state Water Pollution Control Act (Chapter 20, Article 5A of the West Virginia Code) has jurisdiction to prevent the release of pollutants, such as silt from coal washing operations, into streams. Because the dam structure had failed before, releasing pollutants to the streams, the Water Resources Division had ample grounds to impose penalties and insist on proper and safe structures. This was never done.

The Public Service Commission has the responsibility to issue permits and inspect any dam holding 15 feet of water or more (Chapter 61, Article 3, Section 47 of the West Virginia Code). The company never applied for a permit for any of the dams, although Dam #3 was clearly intended to impound more than 15 feet of water.

Therefore it is clear that at least minimal regulations existed that would justify both state and federal action at Dam #3, but none was taken. Additional regulation may be helpful, but enforcement even of existing regulations is lax.

This raises the basic question of responsibility and ownership of resources such as coal. Can a dominant industry owned by large absentee corporations responsible only to their stockholders ever be effectively regulated on a local level without the active participation of the local workers and people? This question must be considered along with the other alternatives of regulation and control of companies dealing with natural resources of a region. The active participation of the local workers and residents means, in essence, democratic control of the coal industry by these people. In conclusion, a quote from the Citizens Commission report (p. 30) summarizes this nicely:

> There is a basic question raised anew by Buffalo Creek, the latest assault by the coal operators in their long slaughterhouse in death, injury and disease: Whether the people of Appalachia and West Virginia can any longer afford this senseless destruction of their lives, their land, and their democratic institutions; or whether the ownership and operation of the coal mines should be brought under democratic control to benefit all the people. All too clearly the tragedy of Buffalo Creek has torn away the mask, revealing the ugly truth that powerful coal interests dominate the government, the environment, and the West Virginia way of life to the detriment of all its citizens. Discussion and action are needed now to transform King Coal, the tyrant, into Citizen Coal, the servant of all—before and not after another Buffalo Creek disaster.

15

THE GEOLOGIST IN PLANNING AND MANAGEMENT

Uneven land subsidence caused this church in Mexico City to tilt. Note the base of the church.

1 SOME ENVIRONMENTAL PROBLEMS

A severe famine struck the Deccan area of India in the winter of 1972–73. Many farmers left their parched acres and moved to the cities to seek relief. Others stayed with their land and tried to survive. Thousands died of starvation in the cities and towns; additional thousands died in rural areas.

Ironically, there was enough food in Indian storehouses to feed the starving people. The Indian government, anticipating periodic times of low agricultural production, had wisely stockpiled during recent years of high agricultural production. Now a bad year had come and people were starving in spite of the stockpiles. Why?

The answer was rather simple. Although food stocks were adequate to feed the people, transportation facilities were inadequate to carry sufficient amounts of food from the storehouses to cities and farms where people were. The Indian government had considered only a part of the problem – the quantity of food necessary to be stockpiled – and had ignored another equally important part of the problem—getting the food to the starving.

Early enthusiasm for nuclear reactors has waned as unforeseen and potential dangers appeared during development. It is now clear that large quantities of heat given off by reactors can seriously alter biologic habitats in rivers and lakes whose waters are used to cool the reactors. There is some question as to the safety of the reactors, and the amount of used but still radioactive fuel that can be disposed of without pollution is undetermined. Also, low-cost uranium fuel is not as plentiful as was once thought.

Waters backing up behind the Aswan Dam in Egypt will irrigate thousands of dry Saharan acres and provide much needed food for Egypt. But the Dam also prevents the annual flooding of the lower Nile, an event which yearly replenishes the fertility of the soil along its banks. The annual replenishment of the soil by the flood waters has been responsible for the persistence of Egyptian culture. Other ancient Middle Eastern cultures collapsed as nutrients were used up or salts polluted once rich farmlands. It seems a safe bet that accumulation of undesirable salts and leaching of unreplenished nutrients from soils below the Dam will drastically reduce fertility of the soils.

Increased evaporation of lake waters behind the Dam will reduce the downstream flow and allow salt water to invade parts of the delta. The salt water may destroy a thriving fishing industry. Is it worth it? Only time will tell.

Factors in Planning

The preceding examples resemble one another in that each problem contained elements which were not identified or were not seriously considered during planning and early phases of implementation. The problems were multidisciplinary in nature. The agricultural officials who helped grow the Indian rice and the bureaucracy who stored it were seemingly unaware of the limitations of Indian transportation, probably because transportation was outside their areas of expertise. Engineers and physicists responsible for the design and construction of nuclear reactors were similarly ignorant of the effects of heat released by reactors into nearby bodies of water, just as geologists were ignorant of the extent of low-cost ores. Biologists, if they were consulted, would have been unfamiliar with the "jargon" of the engineers; physicists and geologists would have been unable to translate the physical scientists' parameters into meaningful biologic data. Knowledgeable agricultural experts, soils scientists and biologists apparently were not consulted in the case of the Aswan Dam. Nationalistic and political considerations concerning the Dam probably overrode any suggestion that side effects of the Dam should be studied, just as political considerations have often been responsible for smokescreens surrounding certain technologic developments in the United States.

Teamwork

Solutions to complex environmental problems lie in teamwork, just as teamwork between engineers, geologists, doctors and people from many other disciplines was necessary to solve the complex problem of transporting men to the moon and getting the greatest scientific return out of the few hours there and the few hundred pounds of rocks which they brought back to earth.

In the case of the Deccan famine, a team might include not only agriculture and transportation experts but also groundwater geologists to investigate the availability of groundwater for irrigation, engineers and geologists to study the possibility of reservoirs, meteorologists to investigate possibilities of climate modification and population experts to determine the optimal population of the Deccan.

An Aswan Dam team might include engineers, agricultural experts, economists to predict economic losses as well as gains, geologists to study effects of the Dam on deposition, erosion, loss of water from the reservoir due to evaporation and seepage, and public health experts to investigate the impact of Dam waters on disease-producing organisms, sewage disposal and so on.

The nuclear reactor team should include economic geologists to investigate nuclear fuel resources, hydrogeologists to investigate the impact of reactors on nearby ground and surface waters, doctors, engineers, physicists, zoologists, botanists, sociologists, economists and others.

The preceding examples show that the *geologist serves as a member of the environmental team*. This is the role of the geologist in planning and managing our culture and in helping us to live in harmony with our environment.

The three examples are all related in some manner to the solid earth. Indeed, *most environmental problems are related to the solid earth*. Some deal with soils, others with surface waters, ground water, rocks, mineral resources, tectonics, erosion and deposition or some combination of these. Thus, the geologist is a ubiquitous member of the environmental team.

In many instances, flora and fauna quickly reestablish themselves after an environmental "goof." This is because living organisms have a short recycling period, usually of the order of a few years. In geologic terms, even man has a short recycling period. If the human race were eliminated today, an intelligent equivalent of man would probably evolve to fill his ecologic niche within a few million years—a relatively short time in comparison with other geologic phenomena.

Man's ignorance and destructive instincts, however, are capable of creating environmental perturbations that will require very long periods of time to restabilize. The ocean could require thousands of years to cleanse itself of pollution acquired within a few decades. Man will dissipate in less than a century the natural gas which slowly accumulated inside rocks over hundreds of millions of years. Most deposits of iron ore formed during the early part of the earth's history. Abundant plant and animal life during the past billion years have changed the earth's atmosphere, increasing the amount of free oxygen and decreasing the amount of carbon dioxide. As a result of these changes, formation of large iron ore deposits may be a thing of the past. In a few centuries, a dot on the geologic time line, man will probably rob his children and his evolutionary successors of the earth's readily available iron.

If we ignore fundamental aspects of public health in our environmental plans and policies, we risk at worst a severe epidemic. Epidemics can be terrible. Social order is disrupted. Personal suffering is immense. But the earth's population reestablishes itself within a few decades. If, on the other hand, we ignore key aspects of geology in our environmental plan, *resulting deterioration of the basic part of our environment, the earth, can be permanent.*

2 GEOLOGICALLY ORIENTED PROBLEMS

In previous sections, it has been pointed out that the solution to environmental problems usually requires a team of scientists and administrators. In some problem areas, the geologist plays a prominent role; in others, his participation is minimal. Although problems discussed in this section are those in which geologists play prominent roles, it should not be forgotten that geologists provide basic data inputs into many other problem areas as well.

Texas Coast

In the Texas coastal zone, a spectrum of active geologic, physical, biologic and chemical processes has created 135 recognized environments. Some of these are relatively stable; others are very delicately balanced.

Erosion and deposition constantly alter the shoreline. Active faults are abundant. Mineral production, largely oil and gas, produces an income of about $1 billion per year.

Hurricanes striking the Texas coast frequently disrupt human activities with the force of their winds and by the extensive flooding caused by the accompanying rainfall. There is a wide range of climate. Commercial fishermen operating offshore and in the sounds and bays catch fish (including shellfish) valued at over $200 million per year. Low-lying flat fertile soils of the zone produce $500 million worth of agricultural products each year.

The largest petrochemical complex in the world is situated along a ship channel connecting Galveston Bay and the city of Houston. About one third of Texas population and industry are located in the coastal zone, much of it in the Galveston–Houston area. The active attack on the diverse problems of this complicated region has earned University of Texas environmental geoscientists an eminent position among the world's environmental scientists. Results of some of their investigations in the Galveston–Houston area are summarized in succeeding paragraphs.

Erosion and Deposition

Give a man an ax and he will start cutting trees. Primitive forest-dwelling aborigines do it, the Romans deforested Europe with their iron axes, and the pioneers in America cleared vast areas in their search for cheap farmland. One person with an ax, however, is ineffective in comparison with modern man and a bulldozer when it comes to clearing land. He clears not only trees but in many cases clears the land of all vegetative cover.

In the Texas coastal zone, construction of roads, brine disposal and other human activities have resulted in destruction of vegetation and excessive erosion. Devegetation of barrier island flats and dunes makes these areas extremely susceptible to erosion by wind and water. Their destruction by erosion removes

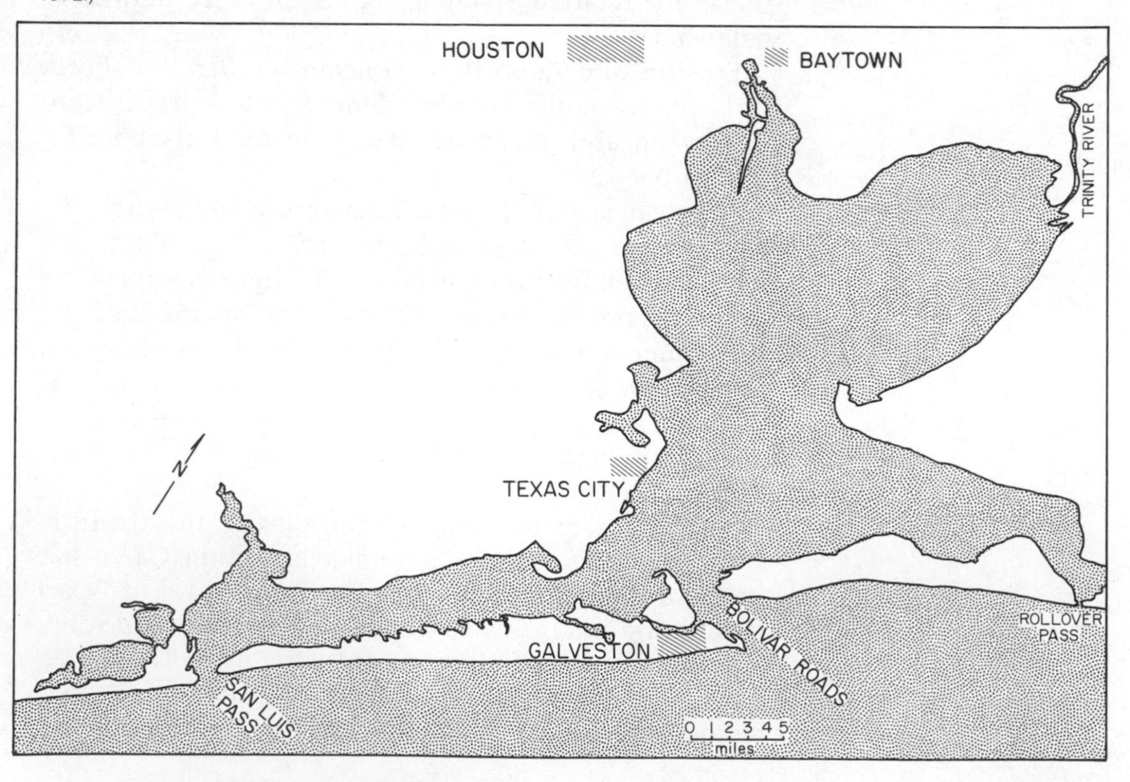

Figure 15–1. Galveston Bay and vicinity. Rapid deposition is taking place on either side of the passes and at the mouth of the Trinity River. (Adapted from Fisher, W. L., et al.: Environmental Geologic Atlas of the Texas Coastal Zone–Galveston-Houston Area. Austin, Texas, University of Texas Bureau of Economic Geology, 1972.)

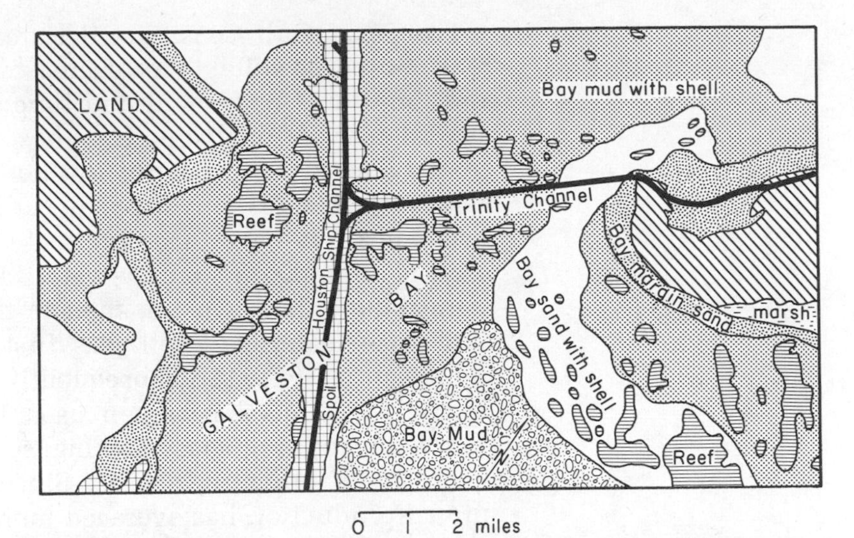

Figure 15–2. Distribution of shell reefs (mostly oyster) in northern Galveston Bay. Sedimentation deriving mainly from spoil banks dredged from channels is covering and killing the living reefs. (Adapted from Fisher, W. L., et al.: Environmental Geologic Atlas of the Texas Coastal Zone—Galveston-Houston Area. Austin, Texas, University of Texas Bureau of Economic Geology, 1972.)

a protective barrier between people inland and the furies of the frequent hurricanes.

Active erosion in one area usually causes excessive deposition in other areas. In the Galveston–Houston area, sediments are rapidly accumulating in small deltas on the offshore and inshore sides of the main passes or water outlets (Fig. 15–1). The Trinity River mouth is another area of rapid sediment accumulation.

Accumulation of sediment in some areas threatens important shellfish reefs. Built mostly of oyster shells, reefs in the Galveston–Houston area cover 23 square miles of bay-estuary-lagoon bottoms and constitute an important resource.

Reef-threatening sedimentation derives mainly from spoil banks made up of mud, sand and organic matter dredged from channels crossing the inland water system. Sediment eroded from the spoil banks is deposited on nearby bottoms. This mantle of sediment kills the living reefs in the vicinity (Fig. 15–2).

Shell Use

In addition to dangers arising from excessive sedimentation, oyster and clam reefs are extensively dredged for their shells.

The Texas Gulf Coast is a flat area devoid of "hard rock" outcrops suitable for aggregates for construction and limestone for cement and lime manufacture. The nearest conventional source of these rock types lies in central Texas, 150 miles to the northwest. Lacking a nearby source of conventional carbonate rock, construction firms have turned to reef shell. Shell occurs in reefs or banks in shallow waters of the area.

Inhabitants began using shell in the late 1850's for road base material. Shell was first used in cement manufacture in 1916. Subsequently, shell and shell-derived products have been used to make lime, in petroleum refining, glass-making, soap-making and in a variety of other industries. Since the late 1950's, annual production has averaged more than 10 million cubic yards. At the time of this writing, overall production is declining. Galveston Bay, a prolific source in the past, is no longer dredged for shell. Not only must area residents find substitute materials in the near future but oysters and clams, an important biologic resource, remain in danger in spite of recent conservation measures.

Channelization

When man attempts to modify his environment he usually fails to consider possible future detrimental effects of his actions. We have already seen how erosion and deposition of spoil bank material dredged from channels endangers living reefs in the area. This is but one of several side effects deriving from channelization. Spoil banks act as dams isolating the bay-estuary-lagoon system into compartments and alter water circulation, salinity and temperature patterns. Drainage patterns on land have also been modified, sometimes to the detriment of the environment.

Extension of the channels up natural waterways has usually required straightening of the natural meandering character of rivers, streams or bayous. The straightening process destroys one of nature's ways of reducing the intensity of flooding. Just as a skier slows himself by skiing a meandering course, water in a meandering river channel cannot attain the velocity and destructive power of an equal amount of water flowing down the same grade in a straight channel. The net result is more flooding, greater erosion, greater deposition and generally negative environmental impact.

Land Subsidence

Much of the Galveston–Houston area is subsiding (Fig. 15–3) owing mainly to groundwater withdrawal. Withdrawal of water from sand beds reduces the water pressure within the sand below that in adjacent clay horizons. Pressure differences cause water to be squeezed out of the compressible clay into the relatively incompressible sand; as the water is squeezed out, the clay compresses and the land subsides. As can be seen in Figure 15–3, subsidence has occurred over a large area. Especially large amounts of subsidence have occurred near Texas City, Baytown and in the Buffalo Bayou area, where heavy concentrations of industry create great demands for water.

On a local scale, petroleum withdrawal, solution

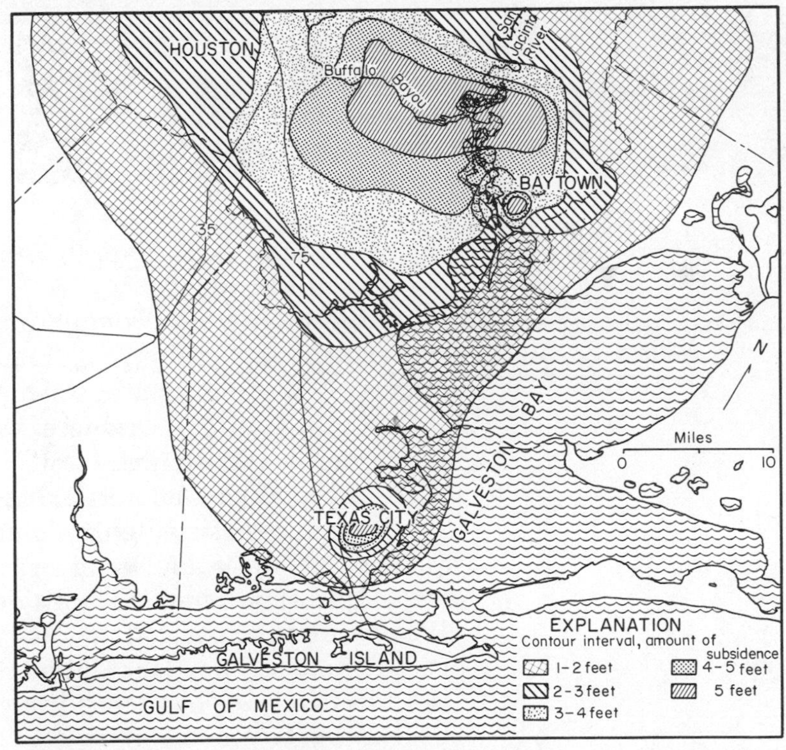

Figure 15–3. Land subsidence in Galveston-Houston area. (Adapted from Fisher, W. L., et al.: Environmental Geologic Atlas of the Texas Coastal Zone—Galveston-Houston Area. Austin, Texas, University of Texas Bureau of Economic Geology, 1972.)

Figure 15–4. Subsidence in Long Beach, California, caused by withdrawal of petroleum from underlying rocks. (Courtesy of D. R. Allen, Department of Oil Properties, City of Long Beach, California.)

mining of sulfur (see Chap. 9) and mining of salt have also caused subsidence.

Withdrawal of fluids from porous, unconsolidated or semiconsolidated soil or rock frequently results in subsidence of the ground surface. Another example is seen in Southern California, where withdrawal of petroleum from the Wilmington oil field created a vast, bowl-shaped depression in Long Beach. Subsidence up to 29 feet has caused great damage (Fig. 15–4). Subsidence due to groundwater withdrawal in the San Jose, California, area necessitated construction of dikes to prevent invasion of low-lying areas by waters of San Francisco Bay (see Chap. 13).

Land subsidence has caused sewer and drainage line misalignments, affected regional drainage patterns and increased susceptibility of low-lying areas to flooding, especially in coastal areas prone to storm surges (see

Chap. 4). The projected ultimate amount of land subsidence will double the original amount of land subject to hurricane flooding.

Active Faults

Up to 1160 linear miles of active or potentially active faults have been recognized in the Galveston–Houston area. Most currently active faults are located in areas of greatest surface subsidence. These have probably been activated by the process of subsidence. Other active faults are thought to be due to loading of the earth's crust by newly deposited sediments, gulfward creep of coastal zone rocks, upward movement of salt domes and tectonic forces.

Active faults have caused numerous breaks in pavement, buried pipelines, water mains and foundation failures. There appears to be no danger due to earthquake vibration, as the contribution of tectonic forces to faulting is very small.

The Salton Sea

The Salton Sea (Fig. 15–5) formed in 1905 as a result of a geologic accident. Prior to 1905, the Imperial and Coachella valleys resembled many other dry basins in the southwestern United States. These valleys lay in a tectonic depression formed by shifting of crustal blocks in the San Andreas fault system (Chaps. 2 and 13). The floors of the valleys were and are below sea level. It is the second lowest area in North America, Death Valley being the lowest.

Large-scale farming began in 1901 in Imperial Valley with construction of an irrigation canal bringing Colorado River water into the Valley. Additional outlets from the river were constructed in 1904. Some of the new outlets lacked permanent locks or gates to shut off excess flow of river waters into the canal.

In the spring of 1905, melting snows in Colorado, Utah and New Mexico sent an unusually large amount of water into far-off tributaries of the Colorado. Upon reaching Southern California, the flood waters broke through the poorly constructed headworks and surged into the new outlets, widening and deepening them. By the summer of 1905, flow in the river south of the new

Figure 15-5. The Salton Sea as seen from a satellite. The Coachella Valley is in the upper left-hand corner; the Imperial Valley covers much of the lower right-hand portion of the picture. Dark checkerboard areas are irrigated farmlands. (Photo courtesy of NASA.)

outlets dried to a trickle as most of the river water flowed through its newly cut channel and into the Imperial Valley. By the summer of 1906, the new channel was one half mile wide and in one place had cut a gorge 50 feet deep.

The flow was finally stopped after a railroad trestle was built across the new channel and carload after carload of rock and gravel were dumped into the river. More than 6000 carloads eventually went into the construction of a levee blocking the river from its new channel.

Initially, fresh water filled the new lake, but high evaporation and solution of salt residue left on the valley floor by evaporation of ancient lakes (see Chap. 4 regarding Pleistocene lakes in the western United States) rapidly raised the salinity of the lake waters. The lake, which today covers 360 square miles, was named the Salton Sea.

Game fish and their supportive food chains were introduced, and the area became a major recreational site, especially during the winter. This vacationland is now in danger. Salinity increases threaten to destroy critical elements of the food chain. Pesticides and fertilizers used in irrigation are polluting the water. Left unchecked, the game fish industry probably will not survive beyond 1987, and the United States will lose a unique asset.

Several mechanisms have been proposed to save the Salton Sea. A dike could be built separating a relatively fresh part of the sea from a very saline part. Diversion of all relatively fresh irrigation water into the fresh portion would maintain the low salinity level. At best, this plan might control salinity (but not fertilizer and other pollutants) for 50 years, and it would cost more than $100 million.

Another possible mechanism derives from hot brine pools beneath the Imperial Valley. The brines are much hotter than the boiling point of water at sea level. The pressure of overlying rocks prevents formation of steam, however, until the brines are pumped up to higher levels, where lower pressures prevail. The brines contain dissolved minerals, some of which are in demand.

It has been suggested that the hot brines be used to provide minerals, power and fresh water for the Salton Sea. The plan is as follows: First, the steam would pass through turbines and provide electric power. Then the condensed steam would be used for industrial purposes and irrigation. Finally, the water would flow into the Salton Sea, where it would help offset evaporative loss of fresh water and resulting increases in salinity. Minerals could be extracted from the residual brine left after formation of steam. Thus the brines could serve many purposes.

There are numerous problems, however. The brines are corrosive, which makes them difficult and expensive to utilize. A chemical processing plant and power plant would together cost an estimated $50 million. Disposal wells for the concentrated salts and other facilities would raise the cost above $90 million. The maximum fresh water output would not completely offset salinity increases due to solution and evaporation and would extend the recreational life of the Salton Sea only until about 2000 AD.

The future of the Salton Sea is far from clear.

Limestone Terrains

Limestone, dolomite, salt and gypsum are chemically derived rocks. Limestone and dolomite are soluble in acids; salt and gypsum are soluble in plain water. (Ground water is usually a weak acid by virtue of certain organically derived elements present.) The solubility of these rocks creates special problems in areas where they lie near the surface of the earth.

Salt and gypsum are geologically rare, but limestone (calcium carbonate) outcrops over large areas of the United States. Much of Florida is underlain by limestone, extensive limestone formations outcrop in valleys of the Appalachian Mountains from Alabama and Georgia to Maine and limestone outcrops extensively in areas of Kentucky and in many other localities in the United States.

Because of the solubility of limestone, dolomite and gypsum, a particular kind of landscape called *karst* is developed on these rocks. The creation of solution landforms is aided if the rocks are highly jointed and thin-bedded. In this way, water can enter through the joints and bedding, enlarging them by solution.

Characteristic surface forms of a karst region are *sinkholes,* disappearing streams and dry valleys. Sinkholes are open depressions developed either by solution of rock on the surface or solution of the rocks beneath the ground and collapse of the surface. Often infiltration and sinkhole collection of rainfall is so effective that there is little or no surface drainage. If there are rivers, they may disappear underground, leaving dry valleys. Alternatively, a river may suddenly rise as a spring from an underground river. Caves or caverns are also associated with solution features. Such underground passages often serve as conduits or pipelines for drainage.

Weathering of limestone terrains in temperate zones often results in a thick red claylike soil cover, which is well suited for agriculture. Many of the beautiful and productive farms in the Pennsylvania, Maryland and Virginia valleys (and elsewhere) owe much to underlying limestone. Chemical fertilizers, pesticides and other agricultural chemicals must be used with caution in limestone terrains. Clastic rocks such as muds, sands and gravel naturally filter water passing through their small pores, but in limestone terrains, the overlying soil is the only effective filter. It has been found that in areas with soils less than 20 feet thick, chemical concentrations can easily exceed the absorptive capacity of the soil and enter the underlying limestone. Once the chemicals are in the system of cracks, crevasses and caverns formed by solution in the limestone, no effective barriers exist to keep the chemicals out of water wells and springs in the vicinity.

Many urban areas located on limestone terrain simply divert excess storm water into available sinkholes. This water, carrying petroleum products, various chemicals, trash, bacteria, organic matter, solids and sediment, can do irreparable damage to groundwater reservoirs.

Differential settlement of structures built on carbonate rocks is widespread. The exposed top of the limestone strata dissolves irregularly, leaving jagged highs separated by deep open joints. Soil cover, however, makes it difficult to locate the highs and the joints. Segments of foundations over joints tend to settle owing to soil compaction, whereas nearby segments located over bedrock highs do not. The result is often a cracked foundation. Because soils compact slowly, the building may be completed before foundation failure occurs.

In extreme cases, cavern roof collapse can result in catastrophic failure of a structure. Cows, tractors and entire buildings have fallen into subterranean caverns following roof collapse. Parts of Route 202 near Philadelphia annually disappear. Rocks under the reservoir behind the Anchor Dam near Thermopolis, Wyoming, contained unrecognized "stringers" of dolomite ($Ca, MgCO_3$) within the Tensleep Sandstone formation. Some of the dolomite had dissolved, creating a network of open subterranean channelways. As the reservoir began to fill after completion of the dam, the weight of the water caused a cavern roof to collapse. According to an eyewitness, it was like "someone pulled the plug in a bathtub," as the water drained from the partially filled reservoir. Impermeable soil dumped into this and other joints and caverns finally plugged the holes in the leaky reservoir rocks, but the geologic investigation and engineering work delayed use of the dam for several years and significantly increased the overall costs.

Waste Disposal

Disposal of waste is a primary problem facing everyone, from the smallest level of domestic waste to the largest industrial waste. Waste has increased tremendously as a result of population growth, the Industrial Revolution and affluence. It is estimated that each United States citizen creates 6 to 18 pounds of solid waste daily, including personal, industrial and agricultural types. In addition to the space problem, there is also an economic one. Not counting private garbage collection, local governments spend about $3 billion annually on garbage collection and disposal.

In the past much sewage disposal has been by septic tanks and water systems, causing our rivers to become overtaxed and highly polluted. Solid waste has usually been incinerated or put into open dumps.

Newer methods of waste disposal include effluent spraying, sanitary landfills, disposal wells and recycling. Spray irrigation of liquid wastes has been used successfully on a modest scale at Seabrook Farms, New Jersey, and at Penn State University. The first major large-scale use will be tried in Michigan to dispose of waste of a whole county. Such a method accomplishes several beneficial purposes in addition to sewage disposal: it increases the groundwater supply, and it fertilizes the soil by restoring nutrients. Purification occurs as the sewage water infiltrates through the ground. Precautions must be taken to insure a satisfactory soil cover, adequate depth to groundwater table and no surface run-off. Type of soil and rock and permeability rate of each are important. A geologist should examine each area to be sprayed.

Sanitary landfills are a common modern method of solid waste disposal. Each day's waste is dumped into a hole or channel that has been dug, then it is covered, and the surface material is compacted. Landfill sites must be carefully chosen for their geologic-hydrologic characteristics. The porosity and permeability of underlying rock and soil must be investigated as well as topography and surface drainage. Depth to the water table and groundwater flow directions and rates should be determined. Great care must be taken to prevent contamination of surface and ground water from landfill. Because of the vast amount of waste, adequate landfill sites are becoming increasingly harder to find.

Disposal wells are used to inject large amounts of liquid waste beneath the earth's surface. This method has long been used by oil companies to dispose of brine. In the Texas oil fields alone there are an estimated 20,000 brine injection wells. However, use of disposal wells for industrial waste is more recent. At present there are about 124 such wells in the United States. It is considered one of the best ways to get rid of radioactive wastes. Again, geologic inspection of a site is necessary to determine the regional hydrodynamics, groundwater circulation pattern, geologic structure and types of rocks involved. The importance of the geologic aspects of the region is indicated by the Denver earthquakes caused by the disposal well for the Rocky Mountain Arsenal (see Chap. 2).

All in all, because of the numerous geologic-hydrologic complications and the possibility of pollution of either surface or ground water, waste disposal is a serious problem, one in which geologists have a vital role.

Foundations

Most foundation problems occur in buildings constructed on clay. In some areas, the correlation between foundation problems and underlying clay is so good that geologists map the location of clay horizons on the basis of cracks in foundations and walls of buildings constructed on the clay horizon.

Clays consist of very small flat particles. The crystal lattice of clay minerals is such that it readily accepts water. When the clays get wet, they swell; as they dry out, they shrink, often developing deep cracks. When the clays outcrop on a natural slope, alternate swelling and shrinkage causes the surface clay to move slowly downhill, or to *creep*, which imposes stresses on foundations and walls of buildings whose footings are not anchored on solid rock below the clay horizon. As a result, floors begin to slant, walls separate from each other and from floors, and foundations may shift and crack, all to the chagrin of the building's owner and occupants.

Many other rock and soil materials contribute to structural failures. Caverns in limestone, permafrost (permanently frozen ground occurring in polar regions), subsidence due to fluid withdrawal and other factors contribute to a multiplicity of foundation problems.

Foundation problems account for a greater economic loss worldwide than many other more spectacular and widely publicized geologic hazards. The importance of foundation problems remains unrecognized because of the relatively small cost of each event in comparison with the large cost of a more dramatic but far less frequent geologic catastrophe.

3 PLANNING AND MANAGEMENT ROLES

Preceding parts of this chapter have reviewed general environmental problems and possible interaction of geologists with other environmental scientists and have discussed specific scientific and engineering problems in which geologists play important roles.

So far, we have looked only at problems and seen how geologists can be of value. This is like studying a proposed building and deciding that a certain kind of brick is essential. We still must work out the details of using this brick in the construction of our building. Similarly, we must look at means of effectively using geologists within the structure of our society. It is of no value to have environmental geologists if they are not utilized to make timely investigations and if results of their investigations are not transmitted to and applied by contractors, governmental agencies and interested citizens' groups. In planning and management, the geologist may appear as a planner, an investigator, a "Paul Revere" sounding an immediate alert, a framer of laws and a leader of crusades against potential dangers.

Planning

It is important to plan timely and relevant investigations. It is also difficult in many cases to decide exactly which problems have highest priority. Only close harmony of administrators, engineers, geologists and environmental scientists can produce an optimal plan. Many geologists, however, are not accustomed to this role and it may require time to produce a cadre of experienced environmental geologic planners.

Investigations

The geologist is most comfortable in this, his traditional role. Investigation encompasses data collection, map-making and interpretation of rock characteristics in terms of real and potential geologic problems. The recently completed investigation of the Galveston–Houston area of the Texas coast is an example of an environmental geologic investigation at its best. This investigation encompassed a survey of impact of geology on the environment, physical properties of rock of the region, specific environments and biologic assemblages, current land use, mineral and energy resources, active geologic processes, man-made features and water systems, rainfall, discharge, salinity, topography and bathymetry.

The investigation now underway in the San Francisco Bay area (Chap. 13) is even larger in scope and involves a more populous area, in which geologic hazards are intensified by the presence of the San Andreas fault. Other smaller investigations are equally important. Although the Houston–Galveston and San Francisco Bay areas have spectacular problems that serve as outstanding examples, we must not limit investigations to such cases but must also investigate the less dramatic problems in areas of lower population density.

Early Warnings

Geologic investigations frequently turn up evidence of a hazard well in advance of the actual event. Five years before the 1964 Alaskan earthquake (Chap. 2), two U.S. Geological Survey geologists published a report which stated that the Bootlegger Clay was unstable and prone to sliding. During the earthquake, the Bootlegger Clay became "quick" and caused extensive landsliding. A later survey report of the disaster concluded that most of the severe damage in the Anchorage area resulted from failure and flow of the Bootlegger Clay.

Many years ago, geologists sounded the alarm regarding areas of high earthquake risk in the San Francisco Bay area, but construction continued. The San Fernando earthquake of 1971 (Chap. 2) revealed many inadequacies of construction codes, but progress in amending codes and reinforcing potentially dangerous construction proceeds at a snail's pace.

Codes and Ordinances

Geologic considerations are essential to the establishment of useful codes and ordinances. The characteristics of the various soils, the danger of landsliding due to changes in the slope of the land during construction, earthquake hazards, fluid disposal problems, availability of water, devegetation and many other aspects of geology should be considered in the preparation of codes and ordinances. The Environmental Policy Act passed by Congress in 1969 is a good beginning, but this law must be complemented and followed up on state and local levels in order to achieve maximum effect. Such implementation will be difficult because in many cases the necessary background information is lacking and because the law encompasses much unexplored territory.

Preparation of a *matrix* is one method of evaluating the impact of a proposal. A matrix is a "checkerboard" on which proposed actions are listed along one side and environmental factors that might be affected are listed down another side. The example shown in Figure 15–6 is for a phosphate mining lease. Activation of phosphate mining would require construction of buildings, highways, bridges and power lines; it would entail blasting and drilling, excavation, mineral processing, transportation, tailings disposal and possible spills and leaks of contaminated fluids. These factors appear at the top of the checkerboard. Environmental factors which would be affected by one or more of the mining factors are listed along the side of the matrix. The numbers in the checkerboard squares show *magnitude of impact* and *importance of impact* on a 10-point scale. For example, *mineral processing* will have a small impact (1) on *water quality,* and the importance of the problem is not great (1). On the other hand, *trucking* will have a moderate impact (5) on *rare and unique species,* and the problem is very important (10) in the area of the lease. Thus a matrix provides a comprehensive overview of the problems which will be encountered.

The phosphate lease matrix is relatively simple. A

	Industrial sites and buildings	Highways and bridges	Transmission lines	Blasting and drilling	Surface excavation	Mineral processing	Trucking	Emplacement of tailings	Spills and leaks
Water quality					2/2	1/1		2/2	1/4
Atmospheric quality						2/3			
Erosion		2/2			1/1			2/2	
Deposition, Sedimentation		2/2			2/2			2/2	
Shrubs					1/1				
Grasses					1/1				
Aquatic plants					2/2			2/3	1/4
Fish					2/2			2/2	1/4
Camping and hiking					2/4				
Scenic views and vistas	2/3	2/1	2/3		3/3		2/1	3/3	
Wilderness qualities	4/4	4/4	2/2	1/1	3/3	2/5	3/5	3/5	
Rare and unique species		2/5		5/10	2/4	5/10	5/10		
Health and Safety							3/3		

Figure 15–6. An environmental planning matrix. (Adapted from Leopold, L. B., et al.: A procedure for evaluating environmental impact. Washington, D.C., U.S. Geological Survey, Circular 645, 1971.)

matrix for a region such as the San Francisco Bay area might include nearly 100 environmental and an even greater numer of "impact" factors. The large number of factors reflects the complexity of the problem of dealing with our environment. Without utilization of a device or procedure such as a matrix, however, our codes and ordinances are certain to omit important considerations and thus fall short of our intent.

Direct Action

It is sometimes necessary for a geologist to appeal directly to the people or to "lead a crusade" himself. One of the best known, most effective examples of direct appeal was that of David Evans, a consulting geologist, who pointed out on television the association of earthquakes and fluid injection at the Rocky Mountain Arsenal.

Failure of geologists and engineers to step forward can be disastrous. In 1963, a mass of rock broke away from a hillside in Italy, slid into the waters of the Vaiont Reservoir and sent a wave crashing over the top of the dam and into the valley below, where the raging mixture of water, mud and debris caused 2117 deaths.

The disaster occurred because rocks adjacent to the reservoir consisted of clay strata dipping into the reservoir, some of which were known to have low shear resistance. Heavy rains caused the clays to swell and decreased resistance of the mass to sliding. A large mass began to creep in September and accelerated in early October. On October 8, the day before the slide, dam engineers realized the danger and attempted to lower the water level behind the dam. They were only partly successful because of heavy rains.

Five years later, the Italian government charged eight people with manslaughter and negligence for building a dam in the area despite knowledge of the instability of the mountainside. It seemed a pitifully small conclusion to such a great disaster.

BIBLIOGRAPHY

Chapter 2

Clark, W. B., and Hauge, C. J.: When the earth quakes . . . you can reduce the danger. Cal. Geol. *24*(11):203–216, 1971.

Cluff, J. S.: Peru earthquake of May 31, 1970; Engineering Geology Observations. Bull. Seismol. Soc. Amer. *61*:511–534, 1971.

Cox, A., Dalrymple, G. B., and Doell, R. R.: Reversals of the earth's magnetic field. Sci. Am. *216*:44–61, 1967.

A descriptive narrative of the earthquake of August 31, 1886 (Charleston, S.C.). Dawson Pamphlets *39*, no. 21, 1886.

Detwyler, T. R.: Man's Impact on Environment, New York, McGraw-Hill Book Co., 1971.

Eckel, E. B.: The Alaskan earthquake March 27, 1964: Lessons and Conclusions. Washington, D.C., U.S. Geological Survey Professional Paper 546, 1970.

Ericksen, G. E., and Plafker, G.: 1970, Preliminary report on the geologic events associated with the May 31, 1970 Peru Earthquake. Washington, D.C., U.S. Geological Survey Circular 639, 1970.

Grantz, A., et al.: The San Fernando, California earthquake of February 9, 1971. Washington, D.C., U.S. Geological Survey Professional Paper 733, 1971.

Hansen, W. R., et al.: The Alaskan earthquake March 27, 1964: field investigations and reconstruction effort. Washington, D.C., U.S. Geological Survey Professional Paper 541, 1966.

Hill, M. R.: Earth hazards—an editorial. California Division of Mines and Geology Mineral Information Service *18*(4):57–59, 1965.

Iacomi, R.: Earthquake Country. Menlo Park, California, Lane Books, 1964.

Officers of the Geological Survey of India: The Binar-Nepal earthquake of 1934: Mem. Geol. Survey, India *73*, 1939.

Oldham, R. D.: 1899, Report on the great earthquake of 12th June 1897. Memoir of the Geological Survey of India, *29*:1–379, 1899.

Pakiser, L. C., et al.: Earthquake prediction and control. Science *166*:1467–1474, 1969.

Phinney, R. A. (ed.): The History of the Earth's Crust. Princeton, N.J., Princeton University Press, 1968.

Plafker, G., Ericksen, G. E., and Fernandez Concha, J.: Geological aspects of the May 31, 1970, Peru earthquake. Bull. Seismol. Soc. Amer. *6*:543–578, 1971.

Richter, C. F.: Elementary Seismology. San Francisco, W. H. Freeman and Co., 1958.

Takeuchi, H., Uyeda, S., and Kanamori, H.: Debate about the earth. San Francisco, Freeman, Cooper and Co., 1967.

Whitcomb, J. H., Garmany, J. D., and Anderson, D. L.: Earthquake prediction: variation of seismic velocities before the San Fernando earthquake. Science *180*:632–635, May, 1973.

Wilson, J. T., et al.: Continents Adrift. San Francisco, W. H. Freeman and Co., 1972. (This book is a collection of 14 articles which originally appeared in Scientific American.)

Chapter 3

Anderson, D. L.: The plastic layer of the earth's mantle. *In* Wilson, J. T., et al.: Continents Adrift. San Francisco, W. H. Freeman and Co., 1970.
Bullard, F. M.: Volcanoes in History, in Theory, in Eruption. Austin, Texas, University of Texas Press, 1962.
Clague, D. A., and Jarrard, R. D.: Tertiary Pacific plate motion deduced from the Hawaiian-Emperor chain. Geol. Soc. Amer. Bull. *84*:1135–1154, 1973.
Crandell, D. R., and Waldron, H. H.: Volcanic hazards in the Cascade Range. *In* Olsen, R., and Wallace, M. (eds.): Geologic Hazards and Public Problems: Proc. Conf., 1969.
Day, A. L., and Allen, E. T.: The volcanic activity and hot springs of Lassen Park. Washington, D.C., Carnegie Inst., 1925.
Fenner, C. N.: The Katmai region, Alaska, and the great eruption of 1912. J. Geol. *28*:569–606, 1920.
Grant, M.: Cities of Vesuvius: New York, The Macmillan Company, 1971.
Green, D. H., and Ringwood, A. E.: The genesis of basaltic magmas. Contrib. Mineral. Petrol. *15*:103–190, 1967.
Hatherton, T., and Dickinson, W. R.: The relationship between andesitic volcanism and seismicity in Indonesia, the Lesser Antilles and other island arcs. J. Geophys. Res. *74*:5301–5310, 1969.
Holmes, A.: Principles of Physical Geology. New York, Ronald Press, 1965.
Huber, N. K., and Rinehart, D. D.: Cenozoic volcanic rocks of the Devils Postpile Quadrangle, Eastern Sierra Nevada, California. U.S. Geological Society Professional Paper 554-D, 1967.
Kinoshita, W. T., et al.: Kilauea volcano: the 1967–68 summit eruption. Science *166*(3904):459–468, 1969.
Loney, R. A.: Structure and composition of the southern coulee, Mono Craters, California. Geol. Soc. Amer. Mem. *116*:415–440, 1968.
Macdonald, G. A.: Volcanoes. Englewood Cliffs, N.J., Prentice-Hall, Inc., 1972.
Macdonald, G. A., and Hubbard, D. H.: Volcanoes of the national parks in Hawaii, 5th ed. Hawaii National Park, Hawaii Nat. Hist. Assn., 1970.
Marinatos, S.: Thera, key to the riddle of Minos. National Geographic *141*(5):702–726, 1972.
Moore, J. G., Nakamura, K., and Alcaraz, A.: The 1965 eruption of Taal volcano. Science *151*(3713): 955–960, 1966.
Ricter, D. H., et al.: Chronological narrative of the 1959–60 eruption of Kilauea volcano, Hawaii. Washington, D.C., U.S. Geological Survey Professional Paper 537-E, 1970.
Shoemaker, E. M., Roach, C. H., and Byers, F. M.: Diatremes and uranium deposits in the Hopi Buttes, Arizona. Geol. Soc. Amer., Buddington vol., 327–355, 1962.
Smith, R. L.: Ash flows. Geol. Soc. Amer. Bull. *71*:795–842, 1960.
Thomas, G., and Witts, M. M.: The Day the World Ended. New York, Ballantine Books, 1970.
Trofimov, V. S.: Origin of diamantiferous diatremes. *In* Symposium on volcanoes and their roots. I.A.V.C.E.I. Symposium Abst., 48–50, 1969.
U.S. Geological Survey: Volcanoes of the United States. Washington, D.C., U.S. Geological Survey, 1969.
Verhoogen, J., et al.: The Earth, an Introduction to Physical Geology. New York, Holt, Rinehart and Winston, 1970.

Chapter 4

Bryson, R. S.: "All other factors being constant . . ." Theories of global climatic change. *In* Detwyler, T. A. (ed.): Man's Impact on Environment. New York, McGraw-Hill Book Co., 1971.
Denton, G. H., and Porter, S. C.: Neoglaciation. Sci. Am. *222*(6):101–110, 1970.
Ericson, D. B., and Wollin, G.: Pleistocene climates and chronology in deep-sea sediments. Science *162*:1227, 1968.
Gilluly, J., Waters, A. C., and Woodford, A. O.: Principles of Geology, 3rd ed. San Francisco, W. H. Freeman and Co., 1968.

Glueckauf, E.: Compendium of Meteorology. Boston, American Meteorological Society, 1951.
Gross, M. G.: Oceanography. Columbus, Ohio, Merrill Publishing Co., 1971.
Ludlam, F. H., and Scorer, R. S.: Cloud Study: A Pictorial Guide. London, John Murray, 1957.
Revelle, R.: The ocean. Sci. Am. *221*(3):54–65, 1969.
Stewart, R. W.: The atmosphere and the ocean. Sci. Am. *221*(3):76–87, 1969.
Stommel, H.: 1957, A survey of ocean current theory. Deep Sea Research *4*:149–184, 1957.
Strahler, A. N.: The Earth Sciences. New York, Harper & Row, 1963.
The Times Atlas. World Climatology, London, The Times Publishing Co., 1958.
Williams, J.: Oceanography, An Introduction to the Marine Sciences. Boston, Little, Brown, 1962.

Chapter 5

Holeman, J.: The sediment yield of major rivers of the world. Water Resources Research *4*(4):737–747, 1968.
Judson, S.: Erosion of the land, or What's happening to our continents? Am. Sci. *56*(4):356–374, 1968.
Livingstone, D. A.: Chemical composition of rivers and lakes. U.S. Geological Survey Professional Paper 440-G, 1964.
Schumm, S. A.: The disparity between present rates of denudation and orogeny. U.S. Geological Survey Professional Paper 454-H, 1963.
Taylor, F. A., and Brabb, E. E.: Map showing distribution and cost by counties of structurally damaging landslides in the San Francisco Bay Region, Calif., winter of 1968–69. U.S. Department of Interior, Geol. Surv. MF–327, 1972.

Chapter 6

Bloom, A. L.: The Surface of the Earth. Foundations of Earth Science Series. Englewood Cliffs, N.J., Prentice-Hall, Inc., 1969.
Mann, K. H.: Case History—The River Thames. *In* Oglesby, R. T., Carlson, C. A., and McCann, J. A. (eds.): River Ecology and Man. New York and London, Academic Press, 1972.
McCormick, C. L.: Probable causes of shoreline recession and advance on the south shore of eastern Long Island. *In* Coates, D. (ed.): Coastal Geomorphology. Binghamton, N.Y., Publications in Geomorphology, 1973.
Thomann, R. V.: The Delaware River—A Study in Water Quality Management. *In* Oglesby, R. T., Carlson, C. A., and McCann, J. A. (eds.): River Ecology and Man. New York and London, Academic Press, 1972.

Chapter 7

Gerstenberg, R. C.: The profit system and America's growth. The New York Times, p. 29, March 4, 1974.
Gilluly, J., et al.: Principles of Geology, 3rd ed. San Francisco, W. H. Freeman and Co., 1968.
Grauband, S. R., ed.: The no-growth society. Daedalus Journal of the American Academy of Arts and Sciences *102*:4, 1973.
McGauhey, P. H.: Manmade contamination hazards. Groundwater *6*:3, 1968.
Meadows, D. H., et al.: The Limits to Growth. New York, The New American Library, 1972.
National Academy of Science Committee on Geological Sciences: The Earth and Human Affairs. San Francisco, Canfield Press, 1972.
Witkin, R.: U.S. pipes methane from coal mine. The New York Times, January 29, 1974.

Chapter 8

Abelson, P. H.: Scarcity of energy. Science *169*(3952): 1267, 1970.

Abelson, P. H.: Let the bastards freeze in the dark. Science *182*(4113):657, 1973.

Abelson, P. H.: Media coverage of substantial issues. Science 941, 1974.

Bateman, A. M.: Economic Mineral Deposits. New York, John Wiley & Sons, 1950.

Bolin, B.: The carbon cycle. Sci. Am. *223*:125–132, 1970.

Bullard, E., Everett, J. E., and Smith, G.: The fit of the continents around the Atlantic. Phil. Trans. Roy. Soc. Lond. A 258, 41–51, IV, 1965.

Cargo, D. N., and Malloy, B. F.: Man and His Geologic Environment. Reading, Mass., Addison-Wesley Publishing Co., 1974.

Carter, L. J.: Rio Blanco: stimulating gas and conflicts in Colorado. Science *180*:844–848, 1973.

Cheney, E. S.; U.S. energy resources: limits and future outlook. Am. Sci. *62*:14–22, 1974.

Cloud, P., and Gibor, A.: The oxygen cycle. Sci. Am. *223*(3): 111–123, 1970.

Cook, G. L.: Oil shale—an impending energy source. Journal of Petroleum Technology, 1325–1330, Nov. 24, 1972.

Coppi, B., and Rem, J.: The Tokamak approach to fusion research. Sci. Am. *229*(1):65–75, 1972.

Ewing, M., Worzel, J. L., Ericson, D. B., and Heezen, B. C.: Geophysical and geological investigations in the Gulf of Mexico, Part I. Geophysics *20* (1):1–18, 1955.

Farney, D.: Ominous problem: what to do with radioactive waste. Smithsonian *5*(1):20–27, 1974.

Gates, D. M.: The flow of energy in the biopshere. Sci. Am. *225*(224): no. 3, 88–103, 1971.

Gustavson, M. R.: Toward an energy ethic. Trans. Am. Geophysical Union *54*:676–681, 1973.

Hammond, A. L.: Breeder reactors: power for the future. Science *174*:807–810, Nov. 19, 1971.

Hammond, A. L.: Dry geothermal wells: promising experimental results. Science *182*:43–44, 1973.

Hammond, A. L.: Zero energy growth: Ford study says it's feasible. Science *184*:172, 1974.

Hammond, A. L.: Academy says energy self-sufficiency unlikely. Science *184*(4140):964, 1974.

Hammond, A. L., Metz, W. D., and Maugh, T. H. III: Energy and the Future. Washington, D.C., American Association for the Advancement of Science, 1973.

Holmes, A.: Principles of Physical Geology. New York, Ronald Press, 1965.

Hubbert, M. K.: Energy resources. *In* National Academy of Science: Resources and Man, Washington, D.C., 1969.

Hubbert, M. K.: The energy resources of the earth. Sci. Am. *225*(224): no. 3, 60–70, 1971.

International Petroleum Encyclopedia. Tulsa, Oklahoma, Petroleum Publishing Co., 1973, 1974.

Levorsen, A. I.: Geology of Petroleum. San Francisco, W. H. Freeman and Co., 1967.

McKenzie, G. D., and Utgard, R. O.: Man and His Physical Environment. Minneapolis, Burgess Publishing Co., 1972.

Morrow, W. F., Jr.: Solar energy: Its time is near. Technology Rev. *76*(2):30–43, 1973.

National Academy of Science Committee on Geological Sciences: The Earth and Human Affairs. San Francisco, Canfield Press, 1972.

Nephew, E. A.: The challenge and promise of coal. Technology Rev. *76*(2): 20–29, 1973.

Pimental, D., et al.: Food production and the energy crisis. Science *182*: 443–449, Nov. 2, 1973.

Rose, D. J.: Energy policy in the U.S. Sci. Am. *230*(1):20–29, 1974.

Schubert, G., and Anderson, O. L.: The earth's thermal gradient. Physics Today, 28–37, 1974.

Seaborg, G. T., and Bloom, J. L.: Fast breeder reactors. Sci. Am. *223*(5):13–21, 1970.
Singer, S. F.; Human energy production as a process in the biosphere. Sci. Am. *223*(3):175–190, 1970.
Starr, C.: Energy and power. Sci. Am. *225*(224): no. 3, 37–49, 1971.
U.S. Bureau of Mines Information Circular 8535: Washington, D.C., U.S. Department of Interior, 1972.
Vedder, J. G., et al.: Geology, petroleum development, and seismicity of the Santa Barbara Channel region, California. U.S. Geological Survey Professional Paper 679, p. 77, 1969.
Walsh, J.: Problems of expanding coal production. Science *184*:336–339, 1974.
Wilson, R. D., et al.: Natural marine oil seepage. Science *184*(4139), 857–865, 1974.

Chapter 9

Brown, H.: Human materials production as a process in the biosphere. Sci. Am. *223*:194–208, 1970.
Browne, C.: Tojo: The last Banzai. New York, Paperback Library, 1972.
Dudley, D. R.: The Romans: 850 B.C.–A.D. 337. New York, A. A. Knopf, 1970.
Flawn, P. T.: Mineral Resources. Chicago, Rand-McNally, 1966.
Flawn, P. T.: Environmental Geology. New York, Harper & Row, 1970.
Flawn, P. T.: Mineral resources and multiple land use. *In* Nichols, D. R., and Campbell, C. C. (eds.): Environmental planning and geology. Washington, D.C., U.S. Geological Survey and U.S. Department of Housing and Urban Development Publication, 1971.
LeMoreaux, P. E., and Simpson, T. A.: Birmingham's Red Mountain cut. Geotimes, 10–11, 1970.
MacIntyre, F.: Why the sea is salt. Sci. Am. *223*:104–115, 1970.
Pratt, C. J.: Sulfur. Sci. Am. *222*:62–77, 1970.
Sunset Editors: Gold Rush Country. Menlo Park, California, Lane Books, 1972.
United Nations Survey of World Iron Ore Resources. New York, United Nations Publications, 1970.
Voskuil, W. H.: Minerals in World Industry. New York, McGraw-Hill Book Co., 1955.
Wenk, E., Jr.: The physical resources of the ocean. Sci. Am. *222*:167–176, 1969.

Chapter 10

Allaway, W. H.: pH, soil acidity and plant growth in soil. *In* the 1957 Yearbook of Agriculture. Washington, D.C., Government Printing Office, 1957.
Broadbent, F. E.: Organic matter in soil. *In* the 1957 Yearbook of Agriculture. Washington, D.C., Government Printing Office, 1957.
Cruickshank, J. G.: Soil Geography. New York, John Wiley & Sons, Inc., 1972.
Jenny, H.: Factors of Soil Formation. New York, McGraw-Hill Book Co., 1941.
Russell, M. B.: Physical properties in soil. *In* the 1957 Yearbook of Agriculture. Washington, D.C., Government Printing Office, 1957.
Schwab, G. O., Frevert, R. K., Barnes, K. K., and Edminister, T. W.: Elementary Soil and Water Engineering. New York, John Wiley & Sons, Inc., 1971.
U.S. Department of Agriculture, Soil Conservation Service: Soil classification—a comprehensive system, 7th Approximation. Washington, D.C., Government Printing Office, 1960, 1967.

Chapter 11

Domenico, P. A.: Concepts and Models in Groundwater Hydrology. New York, McGraw-Hill Book Co., 1972.

Donn, W. L.: The Earth. New York, John Wiley & Sons, Inc., 1972.
Gilluly, J., et al.: Principles of Geology, 3rd ed. San Francisco, W. H. Freeman and Co., 1968.
Heath, R. C.: The changing pattern of groundwater development on Long Island, New York. Washington, D.C., U.S. Geological Survey Circular 524, 1966.
Holmes, A.: Principles of Physical Geology, 2nd ed. New York, Ronald Press, 1965.
Hydrogeology and effects of pumping from Castle Hayne aquifer system, Beaufort County, North Carolina. Raleigh, N.C., North Carolina Department of Water and Air Resources, September 1971.
Leopold, L. B., and Langbein, W. B.: A Primer on Water. Washington, D.C., U.S. Geological Survey, 1960.
McGuiness, C. L.: Groundwater research in the U.S.A. Earth Science Review 3:181–202, 1967.
Schneider, W. J., and Spieker, A. M.: Water for the cities – the outlook. U.S. Geological Survey Circular 601-A. Washington, D.C., U.S. Geological Survey, 1969.
Seaburn, G. E.: Effects of urban development on direct runoff to East Meadow Brook, Nassau County, Long Island, New York. U.S. Geological Survey Professional Paper 627-B. Washington, D.C., U.S. Geological Survey, 1969.
Strahler, A. N.: Introduction to Physical Geography, 3rd ed. New York, John Wiley & Sons, 1973.

Chapter 12

Anderson, A.: Chaos at sea. Saturday Review/World, November 6, 1973.
Anderson, A.: The rape of the sea bed. Saturday Review/World, November 6, 1973.
Blake, N. A.: Water for the Cities. New York, Syracuse University Press, 1965.
Caudill, H. M.: My Land is Dying. New York, E. P. Dutton and Co., 1971.
Curb on strip mining. The New York Times, July 18, 1974.
Doumani, G. A.: Ocean Wealth – Policy and Potential. Rochelle Park, N.J., Spartan Books, 1973.
Environmental and Natural Resources Affairs of the 92nd Congress. Prepared by the Environmental Policy Division, Congressional Research Service Library of Congress. Washington, D.C., Government Printing Office, 1973.
Foley, C.: Secret Hughes vessels will mine ocean ore. New York Post, November 1, 1973.
Goldman, C. R., et al. (eds.): Environmental Quality and Water Development. San Francisco, W. H. Freeman and Co., 1973.
Grad, F. P., et al.: Environmental Controls: Priorities, Policies and the Law. New York, Columbia University Press, 1971.
Hardin, G.: Exploring New Ethics for Survival: the Voyage of the Spaceship Beagle. New York, Viking Press, 1972.
Landau, N. J., and Rheingold, P. D.: The Environmental Law Handbook. New York, Ballantine Books, Inc., 1971.
McCullough, D. G.: The lonely war of a good angry man. American Heritage, December, 1969.
Murphy, E. F.: Man and His Environment: Law. New York, Harper & Row, 1971.
1899 pollution act splits fines 50–50. The Washington Post, May 30, 1971.
Rich and poor at seabed parley debate way to exploit minerals. The New York Times, July 2, 1974.
Sax, J. L.: Defending the Environment: a Strategy for Citizen Action. New York, Alfred A. Knopf, 1971.
Shalowitz, A. L.: Shore and Sea Boundaries, vol. I. Washington, D.C., U.S. Department of Commerce, Coast & Geodetic Survey, 1962.
Sloan, I. J.: Environment and the Law. Dobbs Ferry, N.Y., Oceana Publications, 1971.

Chapter 13

Atwater, T.: Implications of plate tectonics for the cenozoic tectonic evolution of western North America. Bull. Geol. Soc. Amer. *81*:3513–3536, 1970.

Brown, R. D., Jr.: Faults that are historically active or that show evidence of geologically young surface displacement, San Francisco Bay Region. A progress report, Oct., 1970. U.S. Geological Survey Misc. Field Studies Map MF–331, 1970.

Brown, R. D., Jr., and Lee, W. H. K.: Active faults and preliminary earthquake epicenters (1969–1970) in the southern part of the San Francisco Bay Region. U.S. Geological Survey Misc. Field Studies Map MF–307, 1971.

Nichols, D. R., and Wright, N. A.: Preliminary map of historic margins of marshland San Francisco Bay, California. Washington, D.C., U.S. Geological Survey Open File Map, 1971.

Poland, J. F.: Land subsidence in the Santa Clara Valley, Alameda, San Mateo and Santa Clara Counties, California. Washington, D.C., U.S. Geological Survey Map MF–336, 1971.

Schlocker, J.: Generalized geologic map of the San Francisco Bay Region, California. Washington, D.C., U.S. Geological Survey Open File map, 1970.

Taliaferro, N. L.: Geology of the San Francisco Bay Counties. California Department of Natural Resources, Bulletin 154, pp. 117–150, 1951.

Taylor, F. A., and Brabb, E. E.: Map showing distribution and cost by counties of structurally damaging landslides in the San Francisco Bay Region, California, winter of 1968–1969. Washington, D.C., U.S. Geological Survey Map MF–327, 1972.

Chapter 14

Davis, W. E.: Coal waste bank stability. Mining Congress Journal, July 1968.

Davis, W. E., et al.: West Virginia Buffalo Creek Flood: A Study of the Hydrology and Engineering Geology. Washington, D.C., U.S. Geological Survey Circular 667, 1972.

Disaster on Buffalo Creek 1972. Report of the Citizens Commission Investigation, 1972.

Letter of May 6, 1967, from the Secretary of the Interior to West Virginia Officials.

Report of Governor's Ad Hoc Commission to Investigate Buffalo Creek Disaster, 1972.

Chapter 15

Fisher, W. L., et al.: Environmental geologic atlas of the Texas Coastal Zone —Galveston-Houston area. Galveston, University of Texas Bureau of Economic Geology, 1972.

Henkel, D. J.: Geology and foundation problems in urban areas. *In* Nichols, D. R., and Campbell, C. C. (eds.): Environmental planning and geology. Washington, D.C., U.S. Geological Survey and U.S. Department of Housing and Urban Development, 1971.

Koenig, J. B.: Salton Sea: a new approach to environmental problems in a major recreation area. *In* Nichols, D. R., and Campbell, C. C. (eds.): Environmental planning and geology. Washington, D.C., U.S. Geological Survey and U.S. Department of Housing and Urban Development Publication, 1971.

Leopold, L. B., et al.: A procedure for evaluating environmental impact. Washington, D.C., U.S. Geological Survey Circular 645, 1971.

Miller, R. D., and Dobrovolny, E.: Surficial geology of Anchorage, Alaska and vicinity. Alaska, U.S. Geological Survey Bulletin 1093, 1959.

Parizek, R. R.: An environmental approach to land use in a folded and faulted carbonate terrane. *In* Nichols, D. R., and Campbell, C. C. (eds.): Environmental planning and geology. Washington, D.C., Geological Survey and U.S. Department of Housing and Urban Development publication, 1971.

Turner, Collie and Braden, Inc., Consulting Engineers: Comprehensive study of Houston's municipal water system for the City of Houston. Phase I, Basic Studies. 1966.

INDEX

Numbers in *italics* refer to illustrations; numbers followed by a (t) indicate tables.